Distribution System Planning

Distribution System Planning

Evolution of Methodologies and Digital Tools for Energy Transition

Marie-Cécile Alvarez-Hérault
Victor Gouin
Trinidad Chardin-Segui
Alain Malot
Jonathan Coignard
Bertrand Raison
Jérôme Coulet

iSTE

WILEY

First published 2023 in Great Britain and the United States by ISTE Ltd and John Wiley & Sons, Inc.

ISTE Ltd
27-37 St George's Road
London SW19 4EU
UK

www.iste.co.uk

John Wiley & Sons, Inc.
111 River Street
Hoboken, NJ 07030
USA

www.wiley.com

Library of Congress Control Number: 2022946208

British Library Cataloguing-in-Publication Data
A CIP record for this book is available from the British Library
ISBN 978-1-78630-791-0

Contents

Foreword

The energy transition is now a reality in France, Europe and the rest of the world. It responds to environmental protection objectives that require the reduction of greenhouse gas emissions and the preservation of natural resources, to the willingness of countries to secure their own supplies, and the aspirations of their citizens choosing to take control of their "energy destiny", which are notably oriented toward solutions that favor shorter circuits, with production located as close as possible to consumers. International agreements, in particular the Kyoto agreement of 1997, then the Paris Agreement following the COP 21 in 2015, as well as the decisions of the European Union, such as the Green Deal, specify the objectives and measures to be taken. All of these lead to profound changes in the energy landscape and major developments in electricity systems. Renewable energies (RE), mainly wind and photovoltaic, are developing rapidly. The same is true of low-carbon usage solutions, such as electric vehicles and heat pumps. Decentralized storage systems are emerging. Local logics are developing with self-consumption or energy communities. In addition, support for the hydrogen industry is opening up complementary perspectives.

The integration of renewable energies and the charging requirements of electric vehicles compel network operators to develop new solutions. Renewable energy plants have different characteristics than the generation plants used previously: they are decentralized and their production is intermittent. Moreover, charging electric vehicles differs from other consumption, as it does not always take place in the same place and potentially requires high power. As most of the renewable energy plants (more than 90% in France) and, of course, all the charging stations for electric vehicles are connected to the distribution networks, the management of these networks must be radically transformed.

More intelligent, more dynamic and more flexible management of distribution systems is needed. In particular, we need to forecast production and consumption, develop observability and congestion detection capabilities, be able to quickly adjust power produced, consumed or network schemes, and evolve voltage control devices, automation and protection systems.

As part of their public service mission, electricity distribution networks have a fundamental role to play in facilitating the achievement of objectives set for the energy transition, in the best possible conditions of economy, security and quality of supply. They will be able to play this role to the full, provided innovative solutions are designed, industrialized and deployed. The deployment of smart meters, digital transformation and advances in artificial intelligence open up major prospects for improvement that will be used in the service of these ambitions.

The challenges are great, because we must not forget that the electrical system is generally considered to be the most complex system ever built by man: it is vast, comprising millions of kilometers of lines and cables with ramifications that go from local to international, a superposition of several voltage levels, numerous possibilities of reconfiguration, and a complex supervision and protection system. It is an essential infrastructure, essential for the functioning of our modern societies, whose integrity must be guaranteed in all circumstances.

Distribution network planning is a key step in their management. The stakes are high: a recent study conducted amongst European associations of distribution network operators estimates that the investment requirements for the networks of European Union countries and the UK amount to approximately 400 billion euros over the period 2020–2030. Effective planning is therefore essential for optimal allocation of the considerable resources that need to be mobilized.

The paradigm shift implied by the energy transition requires the renewal of distribution network planning methods. Indeed, traditional models have reached their limits and must be adapted to take into account a growing number of uncertainties, integrate flexibility levers and in particular decentralized storage means, take advantage of the large volumes of collected data or simulate the contribution of increased and distributed intelligence. Moreover, specific models must be imagined with the emergence of the *microgrid* concept, the creation of energy communities and the development of multi-energy networks. We must also consider the possible appearance of direct current or hybrid networks.

This book therefore aims to address the challenges of planning distribution networks that must facilitate the decarbonization of energy systems and meet the new expectations of citizens and various users, creating value for all of them, while continuing to guarantee a quality and resilient electricity supply.

After a general introduction on electrical networks and their operations, the fundamental principles of distribution network planning are described. Next, the impact from the introduction of distributed energy resources on the planning models is explained along with a presentation of the different possible solutions, followed by some case studies to better understand the evolution of these models. The book continues with a presentation on the mathematical modeling of the problem and possible resolution methods, detailing the resolution algorithms on the cases studied. Finally, the last part of the book is devoted to new trends and emerging concepts, such as the coupling of planning and operation tools, data-centric software solutions, network architectures enabled by direct current, multi-objective and multi-sectored approaches, and the multiplication of players and their interactions.

It is a rich and topical work. It is intended for all players in the field, whether they are decision-makers, engineers, researchers, teachers or students. Happy reading!

Nouredine HADJSAID
G2Elab, Grenoble INP, Grenoble

Pierre MALLET
Enedis, Paris

September 2022

List of Notations

A notation can have several different definitions depending on the chapter and formula in which it is used and on its type (index or quantity).

Indices

Indices	Description
a	MV line
b	Type of conductor
c	Conductor (phases 1, 2, 3, neutral or earth)
d	MV distribution substation (in all chapters except in Chapter 2 where it means fault)
f	MV feeder
g	LV feeder
h	Component requiring maintenance
i	Customer
j	Line
j	Electrical branch connecting two electrical vertices
k	Vertex of network graph
l	Edge of network graph
n	Year
p	Study period
s	Primary substation
t	Time step
u	Usage (conventional consumption, generation, storage, electric vehicles, flexibility, etc.)
z	Zone bounded by NOS

Greek letters

Notation	Description	Chapter(s)
$\alpha_{temperature}$	Temperature coefficient of resistivity (°C^{-1})	1
α_{works}	Weighting coefficient for the cost of works on streets already used by the network (%)	4
$\alpha_{simulated\ annealing}$	Coefficient of progression of simulated annealing	6
$\alpha_{batteries}$	Weighting coefficient to ensure that batteries are not used more than necessary	6
ε^+ and ε^-	Deviation variables	6
φ	Phase shift between current and voltage	1, 2
γ	Angle of currents	5
$\eta_{batteries}$	Battery performance	6
Ω_{L_r}	All reinforced or added lines	4
Ω_S	All sources	5
Ω_{Cl}	All clients	5
Ω_{other}	Set of intermediate vertices	5
$\Omega_{\text{branch types}}$	Type of possible branches	5
$\Omega^{taps\ on}$	Range of variation of the on-load tap changer	5
$\Omega^{taps\ off}$	Range of variation of the off-load tap changer	5
Ω_{type}	All possible DER	6
$\Omega_{\tau_{int}}$	Set of functions $f_{\tau_{int}}$ allocating power and DER type to a vertex	6
$\rho_{resistivity}$	Resistivity of the material at 20°C (Ω·m)	1
ρ_n	Spearman's rho	6
θ	Temperature (°C)	1
θ_k	Voltage angle	5
τ_{DER}	DER penetration rate (%)	4
τ_{rebound}	Curtailment rebound rate (%)	4
τ_{deferral}	Deferral rate of curtailment (%)	4
τ_{fail}	Electrical conductor failure rate (number of incidents/year/km)	2, 5
τ^{avg}	Annual rate of change in average consumption (%)	2, 5
τ^{peak}	Annual rate of change in consumption at the peak (%)	5
ω	Electrical pulsation of the three-phase sinusoidal system	1

Notations

Notation	Description	Chapter(s)
$\underline{a}$	Rotation operator	1
C	Capacity (F)	1
c	Linear capacity (F/km)	1, 5
CU	Utilization coefficient	2
C_{CAPEX}	Capital cost (€)	2, 5
C_{OPEX}	Operational cost (€)	2, 5
C_{TOTEX}	Total cost (€)	2
C_{inv}	Investment (€)	2, 5, 6
C_c	Cost of line (€/km)	2, 5
C_t	Cost of trench (€/km)	2, 5
C_{RCS}	Cost of an RCS (€)	2, 5
C_{peak}	Cost of joule losses at peak (€/kW)	2, 5
C_{ENS}	Cost of energy not supplied (€/kWh)	2, 5
C_{PC}	Cost of interruptions due to failures (€/kW)	2, 5
C_m	Maintenance cost (k€)	2, 4
$C_{reinforcement}$	Investment in the cost of changing/adding a line (k€)	4
$C_{flexibility}$	Investment related to the implementation of flexibility (k€)	4
C_{losses}	Cost of power losses (k€/kWh)	4
C_s	Total discounted cost of the solution no. s (k€)	4
C_{ni}	Cost of unproduced energy by a producer (k€/kWh)	4
$C_{\lim}$	Maximum cost for storage remuneration (k€/kWh/an)	4
C_{FF}	Fixed remuneration cost of flexibility (k€)	5
C_{VF}	Variable cost of compensation of flexibility (k€/kWh)	5
$C_{annuity}$	Investment annuity (k€)	6
CU	Coefficient of use	2
ENS	Energy not supplied (kWh)	2, 4
E_{losses}	Energy lost through technical losses (kWh)	4
E_F	Total energy flexibility (kWh)	5
E_{un}	Total unproduced energy (kWh)	4
E_{S2}	Stored energy, both in charge and discharge in absolute value (in kWh)	4

Notation	Description	Chapter(s)
E_G	Set of graph arcs	5
e	Graph arc	5
Fp^{conso}	Power factor for consumption	5
Fp^{prod}	Power factor for generation	5
ΔV	Voltage variation (drop or rise) (in V)	1
δV	Voltage gap (in V)	1
f	Power system frequency (Hz)	1
f_{costs}	Objective function of cost	5
f_{CAPEX}	Objective function of capital expenditure	5
f_{OPEX}	Objective function of operating expenditure	5
f_{losses}	Objective function of power loss costs	5
f_{ENS}	Objective function of the cost of energy not supplied	5
f_{PO}	Objective function of the cost of a power outage	5
f_{PL}	Objective function of the balancing of PL products	5
F_a	Average boundary between curtailment and reinforcement	6
$\underline{G}$	Complex quantity that can represent either a voltage or a current	1
G	Non-oriented graph	5
$Gain_{OPEX}$	Difference between the operating expenditures of two solutions	2
$\mathrm{H}_{\max}$	Equivalent time of use at maximum power (hours)	2
HC	Hosting capacity (%)	4
i	Discount rate (%)	2, 4, 5
I	Current (A)	1
I_{th}	Maximum allowable thermal limit (A)	5
I_{max}	Maximum current (or "peak") (A)	5
J	Current in a branch (A)	5
L_{tot}	Total length of the system (km)	2, 4, 6
L_{zone}	Zone length (km)	2
L	Length (km)	2
m_{error}	Margin of error (central limit theorem)	6
n_y	Year the work is initiated	2
$\mathrm{N_S}$	Number of secondary windings	1
$\mathrm{N_P}$	Number of primary windings	1

Notation	Description	Chapter(s)
$N_{feeders}$	Number of network feeders	2
N_{cust}	Total number of customers	2
N_{CCO}	Number of customers cut off	2
$\mathrm{N_d}$	Number of defaults in the year	2
$\mathrm{N_M}$	Number of network components for which maintenance work is required	2
$N_{periods}$	Number of periods	2, 5
$\mathrm{N_z}$	Number of zones to have the same PL	2, 5
$\mathrm{N_{RCS}}$	Number of RCSs	2, 6
N_{arcs}	Number of arcs	5
$N_{vertices}$	Number of vertices	5
N_{years}	Total study period in years	2,5
N_{cond}	Number of conductor types	5
N_{PS}	Number of primary substations	5
N_{sf}	Number of secured feeders	5
N_{SS}	Number of secondary substations	5, 6
$N_{\mathrm{max\ branches}}$	Maximum number of branches on a single arc	5
$N_{branches}$	Number of branches	5
N_{lines}	Number of lines	5
$N^{transfo\ HV/MV}$	Number of HV/MV transformers	5
$N^{taps\ on}$	Number of taps of the on-load tap changer	5
$N^{MV\ feeders}$	Maximum number of MV feeders per transformer	5
$N^{taps\ off}$	Number of taps of the off-load tap changer	5
$N^{LV\ feeders}$	Maximum number of LV feeders per transformer	5
N_{CR}	Number of checkerboard rows (architecture with loops)	6
N_{CC}	Number of checkerboard columns (architecture with loops)	6
$\mathrm{n_{loops}}$	Number of loops (architecture)	6
N_{lps}	Number of loops (architecture)	2
$\mathrm{N_{trials}}$	Number of trials in the Monte Carlo simulation	6
$\mathrm{p}_{degradation}$	Probability of accepting a degradation of the objective function for the simulated annealing (%)	6
$\mathrm{p}_{default}$	Probability of default (%)	6
P	Active power withdrawn or injected (kW)	1, 2, 4, 5, 6

Notation	Description	Chapter(s)
P_{losses}	Technical losses by Joule effect (kW)	1, 6
PL	Product power × length (MVA.km)	2
P_{lmax}	Peak losses (kW)	2
P_{cut}	Power cut due to failures (kW)	2
$P_{allowed}$	Maximum power admissible (kW)	2, 6
P_{avg}^{feeder}	Average power of the feeder studied (in kW)	2
P_{avg}^{cut}	Average power cut	2
P_{avg}^{z}	Average power of zone z	2
P_{max}	Maximum power (kW)	2, 4
P_{avg}	Average power (kW)	2, 5
P_{min}	Minimum consumption (kW)	4
$P_{installed}$	Installed power (kW)	4
P_{margin}	Power margin (in W)	6
P_{cur}	Activated power curtailed (kW)	6
Q	Reactive power consumed or injected (kVAr)	1, 2, 4, 5, 6
$r_{\text{transformation}}$	Transformation ratio	1
r	Linear resistance (in Ω/km)	1, 2, 5
R	Resistance (in Ω)	1, 2
CBR_r	Cost–benefit ratio	2
S	Apparent power in complex (in VA)	1
$Section$	Useful conductor cross-section (mm²)	1
SAIDI	System Average Interruption Duration Index (minutes/year/customer)	2, 5
SAIFI	System Average Interruption Frequency Index (outages/year /customer)	2, 5
$S^{transfo\ HV/MV}$	Maximum permissible power per transformer (kVA)	5
$S^{max,feeders\ MV}$	Maximum power per MV feeders in normal operation (kVA)	5
$S^{transfo\ MV/LV}$	Maximum permissible power of the transformer (kVA)	5
$S^{max,feeders\ LV}$	Maximum power per LV feeders in normal operation (kVA)	5
S^{rated}	Rated power (or "connected power" for LV customers) (kVA)	5
S^{cons}	Maximum power consumed (kVA)	5
S^{prod}	Maximum power produced (kVA)	5

Notation	Description	Chapter(s)
SOC	State of charge (%)	6
T_{iso}	Fault isolation time (in hours)	2, 5
T_{rep}	Fault repair time (in hours)	2, 5
T_{LT}	Lifetime of the structure (years)	2
T_{co}	Cut-off time (minutes)	2
T_{AM}	Amortization period (years)	4
$T_{activation}$	Curtailment activation time (minutes)	4
T_{report}	Duration of the curtailment deferral (minutes)	4
NDR	Network infrastructure development rate (%)	4
T_p	Duration of sub-periods (years)	5
T_s	Storage periodicity (hours)	6
T_{anneal}	Simulated annealing temperature (°C)	6
U	Phase-to-phase voltage (V)	1, 2, 5
V	Single phase-to-neutral voltage (V)	1, 5
NPV	Net present value (€)	
Vu	Use value (€)	2
V_N	Replacement value of the work (€)	2
V_F	Value of the reinforcement of the structure (in k€)	2
V_G	Set of graph vertices	5
v	Graph vertex	5
W_{cur}	Consumption curtailment vector	6
X	Reactance (Ω)	1, 2
X	Vector solution	5, 6
x	Linear reactance (Ω/km)	1, 2, 5
Y	Complex admittance (S)	5
y	Linear complex admittance (S/km)	5
Z	Complex impedance (Ω)	5, 1
z	Linear complex impedance (Ω/km)	5
z	Variable function of the confidence interval (central limit theorem)	5, 6

List of Acronyms

AC	Alternating Current
ACER	Agency for the Cooperation of Energy Regulators
ACM	Avoided Cost Model
ADEeF	*Association des distributeurs électriques en France*: Association of the Distributors of Electricity in France
ADEME	*Agence de la transition écologique*: Ecological transition agency (formerly *Agence de l'environnement et de la maîtrise de l'énergie*: Agency for the environment and energy management)
ADMS	Advanced Distribution Management Systems
AEMC	Australian Energy Market Commission
AEMO	Australian Energy Market Operator
AFNOR	*Association française de normalisation*: French standards association
AI	Artificial Intelligence
AMS	Active Management System
ANEEL	*Agência Nacional de Energia ELétrica*: Brazilian Electricity Regulatory Agency
ANM	Active Network Management

ANROC	*Association nationale des régies de services publics et des organismes constitués par les collectivités locales*: National association of public service boards and local government organizations
ANSI	American National Standards Institute
API	Application Programming Interface
BEV	Battery Electrical Vehicle
BTB	Back-To-Back
CACM	Capacity Allocation and Congestion Management
CAPEX	CAPital EXpenditures (investment costs)
CBA	Cost–Benefit Analysis
CBR	Cost–Benefit Ratio
CCP	Climate Change Plan
CCRC	*Communauté de communes de la région de Condrieu*: Community of communes of the Condrieu region
CEER	Council of European Energy Regulators (a non-profit association that allows a group of European regulators to exchange best practices and it produces comparative reports on reliability)
CEP	Clean Energy Package (document proposing measures to improve energy efficiency and the integration of renewable energy by 2030)
CIGRE	*Conseil international des grands réseaux électriques*: International council on large electric systems (global community committed to collaborative development and sharing of power system expertise)
CIM	Common Information Model
CIRED	*Conférence internationale des réseaux électriques de distribution*: International conference on electrical distribution systems
Co-Op	Cooperatives whose shareholders are the consumers themselves

CPP	Critical Peak Pricing (daily peak clearance)
CPUC	California Public Utilities Commission (Californian regulator)
CRE	*Commission de régulation de l'énergie*: France's energy regulatory commission
CSP	Curtailment Service Provider (aggregator)
CU	Coefficient of Use
CVRC	*Centrales villageoises de la région de Condrieu*: Village power plants in the Condrieu region
DC	Direct Current
DCC	Demand Connection Code
DCP	Distribution Capacity Program
DER	Distributed Energy Ressource
DERMS	Distribution Energy Resource Management System (flexibility platform)
DGEC	*Direction générale de l'énergie et du climat*: France's energy and climate directorate
DIDF	Distribution Investment Deferral Framework (methodology to identify DER connection areas, evaluate their interest and select the most suitable ones)
DNDP	Distribution Network Development Plan
DOE	Department Of Energy
DRAM	Demand Response Auction Mechanism (mechanism allowing flexibility aggregators to offer their service to the DSO by going directly through the wholesale market and the CAISO)
DRP	Distribution Resources Plan
DSO	Distribution System Operator

DSR	Demand Side Response
EB	Electricity Balancing
EC	European Commission
EDSO	European Distribution System Operators
EFSI	European Fund for Strategic Investments (supports private investment projects in the fields of infrastructure, research and innovation, education, health and information and communication technologies)
EN	European Norm
ENS	Energy Not Supplied
ENTSO-E	European Networks of Transmission System Operators for Electricity
EPRI	Electric Power Research Institute
ER	Emergency and Restoration
ESCO	Energy Service Company (additional service providers such as gas and oil)
EU	European Union
EV	Electric Vehicle
FACTS	Flexible Alternating Current Transmission System
FCA	Forward Capacity Allocation
FCR	Frequency Containment Reserve (primary frequency control)
FERC	Federal Energy Regulatory Commission (US national regulator)
FP7	Seventh Framework Programme of the European Community
FPL	Flexible Power Link

FRR	Frequency Restoration Reserve or Regulating Reserve
FSP	Flexibility Service Provider (aggregator)
GDP	Gross Domestic Product
GDPR	General Data Protection Regulation
GIS	Geographical Information System
GOPACS	Grid Operators Platform for Congestion Solutions
HC	Hosting Capacity
HDI	Human Development Index
HEV	Hybrid Electric Vehicle
HV	High Voltage
HVA	High Voltage level A
HVB	High Voltage level B
HVDC	High Voltage Direct Current
ICA	Integration Capacity Analysis
IEC	International Electrotechnical Commission (international organization for standardization in the fields of electricity, electronics, electromagnetic compatibility, nanotechnology and related technologies)
IEEE	Institute of Electrical and Electronics Engineers (international association with more than 423,000 members with activities such as journal publishing, conference organization and standards writing)
IFA2	France–England Interconnection 2
IoT	Internet of Things
IOU	Investor-Owned Utilities

IPP	Independent Power Producer
IREC	Interstate Renewable Energy
IRENA	International Renewable Energy Agency
ISGAN	International Smart Grids Action Network (international platform to support governmental concerns and actions around the world to accelerate the development of clean and smart power systems)
ISO	Independent System Operator
IT	Information Technology
IT	earthing system – "I" means that the neutral point of the transformer is isolated from the earth and "T" means that the electrical device of the installation are directly connected to the earth but independently from the one of the transformer.
LBC	Loop Balance Control
LCOE	Levelized Cost Of Energy
LDC	Local Distribution Company
LED	Light-Emitting Diode
LIFO	Last In First Out
LNBA	Locational Net Benefit Analysis (optimal location of DER to defer or cancel an investment)
LTECV	*Loi relative à la transition énergétique pour la croissance verte*: Law on the energy transition for green growth
LV	Low Voltage
LVAC	Low Voltage Alternating Current
LVDC	Low Voltage Direct Current
MAEP	Multi-Annual Energy Program
MILP	Mixed Integer Linear Programming

MINLP	Mixed Integer NonLinear Programming
MV	Medium Voltage
MVDC	Medium Voltage Direct Current
MYIP	Multi-Year Investment Plans
NF	*Norme française:* French standard
NIZ	Non-Interconnected Zone (off-grid isolated network)
NP	Non-Polynomial
NPV	Net Present Value
NRA	National Regulation Authorities
NWA	Non-Wire Alternatives
OFGEM	Office of Gas and Electricity Markets
OLTC	On-Load Tap Changer
OPEX	OPerational EXpenditure
OPF	Optimal Power Flow
ORI	*Offres de raccordement intelligentes*: Flexible non-firm connection
OT	Operational Technologies
PAC	*Préfabriqués à couloir de manœuvre*: Walk-in compact substation
PEP	Product Environmental Profile
PFR	Primary Frequency Response
PHES	Pumped Hydro Energy Storage
PHEVs	Plug-in Hybrid and Electric Vehicles

PL	Product of the total power of a given area as per the cumulative total length of conductors in that area
POU	Publicly Owned Utilities
PRCS	*Préfabriqués ruraux compacts simplifiés*: Simplified compact rural substations
PS	*Poste source*: Primary substation or HV/MV substations
PSS A	*Préfabriqués au sol simplifié de type A*: Ground-mounted substation type A
PSS B	*Préfabriqués au sol simplifié de type B*: Ground-mounted substation type B
PUIE	*Postes urbains intégrés à son environnement*: Low visual impact urban compact substation
PURPA	Public Utility Regulatory Policies Act
PV	Photovoltaic
PWM	Pulse Width Modulation
QoS	Quality of Service
RCS	Remote-Controlled Switch
RE	Renewable Energy
REMIT	Regulation on wholesale Energy Market Integrity and Transparency
RESCoops	Renewable Energy Source Cooperatives
RFG	Request For Generators
RFO	Request For Offers
ROI	Return On Investment
RR	Reserve Replacement or Contingency Spinning Reserve

RSC	Regional Security Coordinator
RTO	Regional Transmission Organizations
S3RENR	*Schémas régionaux de raccordement au réseau des énergies renouvelables*: Regional grid connection schemes for renewable energies
SAIDI	System Average Interruption Duration Index (average annual outage duration per LV or MV customer)
SAIDI $_{EPXE}$	SAIDI *hors événements planifiés et exceptionnels*: SAIDI excluding planned and exceptional events
SAIFI	System Average Interruption Frequency Index (average annual outage frequency per LV or MV customer)
SAPS	Stand-Alone Power Systems
SB	Standard Bill (California laws)
SCADA	Supervisory Control and Data Acquisition
SCORE	*Schéma d'orientation des réseaux électriques*: Power system master plan
SDR	System infrastructure Development Rate
SFCL	Superconducting Fault Current Limiter
SICAE	*Société d'intérêt collectif agricole d'électricité*: Agricultural electricity collective interest company
SMAP	SMArt grid in natural Parks
SMEs	Small and Medium-sized Enterprises
SMIs	Small and Medium-sized Industries
SMOP	Soft Multi-state Open Point
SNOP	Soft Normally Open Point
SOP	Soft Open Point

SRCAE *Schémas régionaux du climat de l'air et de l'énergie*: Regional climate, air and energy plans

SWER Single Wire Earth Return

SYS SYStem

TCFM *Télécommande à fréquence musicale*: Ripple control communication system

TEPCV *Territoire à énergie positive pour la croissance verte*: Positive energy territory for green growth

TN-C earthing system diagram – "T" means that the neutral point on the transformer is directly earthed, and "N-C" means that the earth of the installation is directly connected to the earthed point of the supply via the neutral conductor

TN-S earthing system diagram – "T" means that the neutral point at the transformer is directly earthed, and "N-S" means that the earth of the installation is directly connected to the earthed point of the supply via a conductor that is separate from the neutral conductor

TOTEX TOTal EXpenditures (the sum of CAPEX and OPEX)

TSO Transmission System Operator

TSP Traveling Salesman Problem

TT earthing system diagram – the first "T" means that the neutral point on the transformer is directly earthed and the second "T" means that the earth of installation is directly connected to an earth connection that is electrically independent of the supply

TURPE *Tarifs d'utilisation des réseaux publics d'électricité*: Tariffs for the use of public electricity networks

TYNDP Ten-Year Network Development Plan

UCTE Union for the Coordination of Transmission of Electricity

URD Underground Residential Distribution

VSC	Voltage Source Converters
VVC	Volt Var Control
WEEE	Waste Electrical and Electronic Equipment

Introduction

The first power system appeared in 1882, when Edison started the public lighting system in a New York district. As technological discoveries and advances were made, the power system continued to develop as a global electrification; however, as of 2017[1] there were almost 1 billion people still without access to electricity. Each country has developed planning rules for its systems according to a technical and regulatory framework depending on its history, its geographical specificities, and the socioeconomic and political context. In most cases, electricity is produced from high-powered power plants, most of which are polluting (64% of the world's electricity is based on the fossil fuel power plant[2]). It is then exported to consumer areas through the transmission and distribution system. The definitions and roles for the generation, transmission, distribution and consumption sectors were clearly defined, and the solutions applied for the development of the systems were based on technical and economic optimizations of the investments throughout the entire chain. In recent years, this highly structured electricity landscape has become more blurred as it is at the heart of the energy and societal transition, sustainable development, technological developments and deregulation. Operation and investment rules must be reviewed to keep up with and support these developments. We can mention, in a non-exhaustive way, the multiplication of small power production, generally renewable, and generally intermittent (from 1 kW, or even 1 W in some non-electrified regions), the electrification of means of transport, heat and cooling technology, the will to consume clean and ethical energy locally (the development of local energy communities, self-sustaining mechanisms), the development of information and communication technologies, access to electrical data (development of smart meters and other sensors) and the development of flexibility services by different actors (modulation of production and/or consumption, storage, etc.).

1 IEA (2019). World energy outlook 2019. Report, International Energy Agency, Paris.

2 EDF (2018). Global electricity generation in 2018 [Online]. Available at: https://www.edf.fr/media/display/35398.

Power systems are at the heart of all of these changes. However, the power distribution system is undergoing particularly important changes as its role evolves, from supplying end consumers, to a multi-directional logic with new services. Planning tools have not been systematically used for distribution systems up until now, and traditional investment methods are reaching their limitations within these new paradigms. This book specifically focuses on these systems since they are the ones from which the most diverse and innovative solutions are emerging.

The work is intended for readers with some basic knowledge of electricity, but also for researchers working in the field of power system planning. It provides recommendations on the evolution of planning methods being applied now. It is divided into seven chapters and offers two levels of reading: an intermediate level (Chapters 1–4 and 7) and an advanced level (Chapters 1–6) for those seeking to study the theoretical aspects further, either for their scientific curiosity or to replicate in their own case study. We have tried, as far as possible and depending on the information available, to deal with planning from a general point of view, focusing on certain countries.

– Chapter 1 details the historical context for the development of power systems, and describes the fundamental notions of electrical engineering necessary to understand the choices made for this development and their operating modes, as well as their regulatory frameworks. Next, this chapter focuses on the distribution system (description of the distribution mode and its components) and presents the main factors motivating the evolution of planning methods.

– Chapter 2 defines the concept of distribution system planning, the commonly used indicators and typical architectures.

– Chapter 3 details the impact of the development of decentralized energy resources on distribution systems, as well as the latest regulatory developments to promote their integration. Alternative solutions to the classical solutions called flexibilities or *non-wire alternatives* will be listed as well as considerations for their application.

– Chapter 4 proposes examples for the evolution of planning methods through a selection of case studies: dense urban interconnected systems; rural interconnected systems with low load density; and isolated microgrids in geographical areas where the power system does not yet exist (developing countries), in areas that are difficult to access (such as islands) or in areas where electrical access would be very expensive (mountains, national parks).

– Chapter 5 proposes a comparatively exhaustive mathematical modeling of the problem and the possible methods for its resolution, for readers wishing to delve into the theoretical aspect of planning.

– Chapter 6 details the algorithms for solving some of the case studies presented in Chapter 4.

– Chapter 7 provides an overview of the new trends and challenges that are currently emerging, the economic models of which have yet to be confirmed: the evolution of performance indicators and objectives, such as the notion of environmental impact and recyclability; the renewed interest in direct current; the need for multi-objective and multi-sectoral systemic modeling tools; the evolution of data collection and processing tools; and the proliferation of actors and their interactions.

A conclusion summarizes the final recommendations made by the authors on the necessary evolutions in the planning of distribution systems so that they can be sufficiently flexible in order to integrate existing and future changes, whose dynamic shifts are increasingly rapid.

1

Power Systems

1.1. Electricity: an essential and complex product

Electricity, which in 2018 accounted for 19.3% of the world's total energy consumption (IEA 2020), has become a real "product" synonymous with progress and as such essential for the economic and social development of a country. As an example, in 1990 the United Nations Development Program defined an indicator to assess the level of human development of a country: the Human Development Index (HDI) (UNDP 2015). The HDI is calculated using three indicators: health, education, and living conditions. A significant correlation between HDI and a country's electricity consumption was demonstrated (UNDP 2015). Figure 1.1 shows HDI as a function of electricity consumption per capita (log scale on the x-axis) (World Bank 2018).

We can see that below about 5,000 kWh/capita, a small increase in electricity consumption greatly improves HDI. For example, for countries like Ethiopia, increased access to electricity would accelerate the deployment of uses such as street lighting, refrigeration and media equipment. On the other hand, above 5,000 kWh/capita, we can observe a saturation effect, that is, increasing electricity consumption no longer improves HDI. However, our dependence on electricity is double-edged as the economic and social consequences can be significant. As an example, we can cite the blackout of August 14, 2003, which affected the northeastern United States and part of Canada and caused an estimated loss of $6.4 billion for the US (Anderson and Geckil 2003) and between 1 and 2 billion dollars for Canada (Government of Canada 2006). Therefore, the main challenge is to ensure the continuity of the power supply to end-users at all times, knowing that they are generally not located where the electricity is produced. One possible way to do this

---

Throughout this book, the term "system" has been used in place of "network" or "grid".

is to use storage facilities, such as pumped hydroelectric energy storage (PHES). However, these sources of flexibility require a high capital investment and a very long construction time and are only developed in very specific cases. For example, in countries with a large number of nuclear power plants and gas turbines, their capacity to control network flow is sufficient to ensure the safety of the system, and as such the development of PHES is not economically justified. This is the case for France, which in 2015 only had 5 GW of installed PHES (or 3.8% of the total installed generation capacity of the country) compared to 27.4 GW for Japan (about 10% of the total installed generation capacity of the country; JEPIC (2018)) where the structuring of the electricity market is favorable to heavy and long-term investments (Yang 2016). Electricity must therefore be controlled at all times to ensure a balance between production and consumption. It is for all these reasons that the power system was developed in the 19th century. It is a complex system consisting of power lines, transformers, measurement and protection devices, automation and control equipment, which allows the connection of production plants to end-users in an economical, reliable and safe way. In this chapter, the key historical steps in the development of power systems are detailed in order to understand the reasons why different technological choices were made, such as the predominance of low-frequency alternating current (AC) and the different voltage levels. The electrotechnical elements needed to understand the design and operation of power systems are then detailed, before focusing on the distribution systems that are the subject of this book. Finally, we conclude with the regulatory context in which power systems are developed and the major changes that are occurring and which may lead to a reconsideration of certain historical choices.

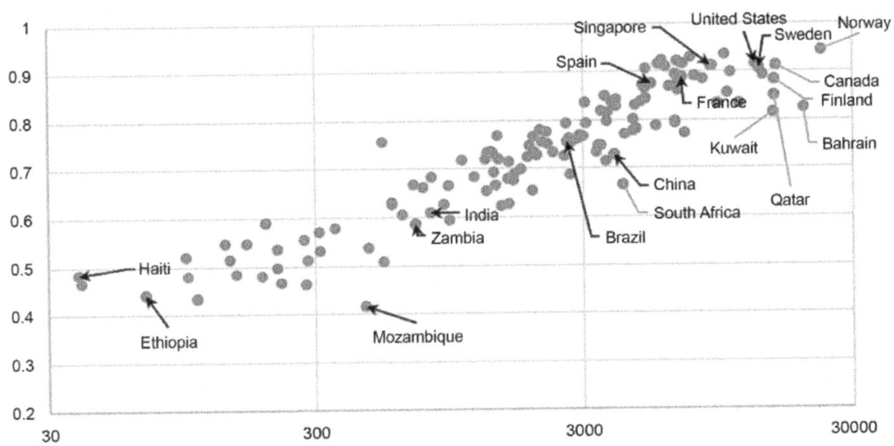

Figure 1.1. *HDI by electricity consumption (kWh per capita) in 2014 (abscissa in logarithmic scale) (sources: UNDP; World Bank)*

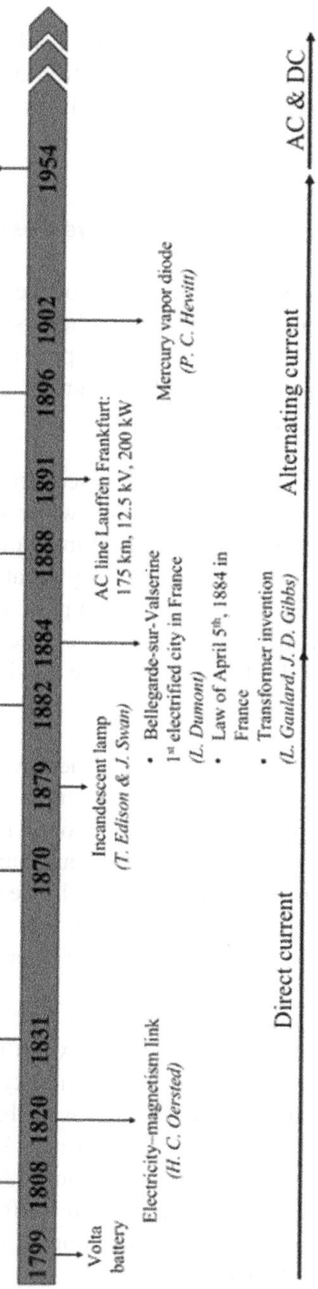

Figure 1.2. *Milestones in global electrification*

1.2. History of industrial power systems

The electrification of countries around the world has developed gradually over the past 200 years because of numerous scientific discoveries, some of which have marked important turning points and are illustrated by the timeline shown in Figure 1.2.

1.2.1. *Discovery of direct current and the design of the first generators*

The first to be developed at the beginning of the 18th century was direct current (DC), in particular through the invention of the first electric battery by Alessandro Volta in Italy in 1799. By stacking sheets of copper and zinc, separated by cardboard soaked in salt water, an electrochemical reaction generates the circulation of an electric current (BBC 2018).

Then in 1820, Hans Christian Oersted discovered, by chance, the link between electricity and magnetism. During a lecture, while illustrating the phenomenon of incandescence through an experiment connecting a platinum wire to an Auge battery (a variant of the more powerful Volta battery created by William Cruickshank in 1802), he noticed that the needle of a compass located near to the equipment deviated (Blondel and Wolff 2021). He then resumed his experiment using a power battery and demonstrated the link between the closing of a circuit on a battery and the magnetic field.

A few months later, in France, this experiment was taken up and explored further by André-Marie Ampère, who defined the notion of electric current for the first time. It also inspired Michael Faraday in London, who tried to illustrate the opposite phenomenon: generating current from a mechanical movement, now called electromagnetic induction. It was in 1831 that he demonstrated the concept of a DC electrical generator by creating a DC current and rotating a copper disc inside a magnetic field created by a U-shaped magnet (Usborne et al. 2011).

The Frenchman Hippolyte Pixii was the first to design a manual generator providing an AC to which Ampère added a switch in order for it to provide a DC, since AC was not known at that time. This first type of generator was called the dynamo, and its performance gradually improved thanks to the contributions of the Italian Antonio Pacinotti in 1860, the German Werner von Siemens, and the Englishmen Charles Wheatstone and Samuel Alfred Varley in 1867 (they simultaneously created the self-excited dynamo), followed by the Belgian Zenobe Gramme in 1870 who expanded its application to the supply of motors with high power ratings (Harvey et al. 2017).

1.2.2. *Birth of the first power systems: public lighting systems*

The interest in electricity as a means of power, in particular as a source of lighting, began to take hold of scientific minds across the world. First of all, public lighting, initially provided using gas as a resource, gradually became electric because of arc lamps, followed by incandescent lamps. Developed in 1808 by the Englishman Humphry Davy, the principle of the arc lamp consists of passing a direct electric current between two lumps of charcoal using an electric battery. The electric arc generated is suitable for public lighting; however, it is too intense to be used for indoor lighting. The use of arc lamps is underdeveloped given that they require batteries that are unreliable, very expensive and have a short life span. In the 1870s, the dynamo was created and adapted to arc lamps. In 1878, the first lit street was the *Rue de l'Opéra* in Paris, by 32 street lamps, each equipped with six lamps, each with a life of 1.5 h (Expositions universelles 2021). The current in this demonstration was alternating since the two coals used by the arc lamps did not wear out at the same rate (Owen 1997). Thereafter, the development of electric street lighting accelerated with the development of the incandescent electric lightbulb, simultaneously by the American Thomas Alva Edison and the Englishman Joseph Swan in 1879, who later signed an agreement for their commercialization in the UK (Usborne et al. 2011). The principle of the incandescent lightbulb is that an electric current passes through a filament that has a higher resistance than the rest of the circuit. The extra energy needed for the electrons to circulate across the filament causes it to heat up, increasing its resistance, until it becomes incandescent. In order to make the system more economical, Edison decided to create a production plant that would allow electricity to be sent to several consumers: this was the birth of the concept of power systems. In 1882, Edison lit up the 1st district of New York with a 110 V DC production plant located on Pearl Street, which supplied consumers spread over the 0.65 km^2 surrounding the power plant as a result of 24 km[1] of buried cables (Sulzberger 1882). The power plant consisted of six dynamos, each with a capacity of 100 kVA, driven by steam engines and capable of lighting up to 1,200 lamps each (Sulzberger 1882). The cost of this lighting system was $0.5 million and required 4 years of construction. On the day it was put into operation, 400 lamps were connected to Edison's system, and after 1 year, about 10,000 lamps were connected. However, this power system had two major drawbacks: since it is not possible to change the voltage, an electrical circuit is required for each type of equipment (lighting and motors, for example), and the electricity must be generated locally. Indeed, with the low voltage level and the high resistance of the lines, the voltage drops were significant. Edison had calculated that given the cross-section of the lines he could build at the time, for anything beyond 1.5 km, the voltage at the end of the line was no longer sufficient to allow the operation of lamps (Usborne et al. 2011). Replicating this system in other parts of the city therefore required the development of multiple lighting systems, each with a generating station, which was

1 Unit of conversion used: 1 foot = 0.3048 m.

too expensive. Another example is the city of Bellegarde-sur-Valserine, which was the first city in France with electric street lighting, due to the hydroelectric dam built by Louis Dumont in 1884 (Guinet 2015). The first long-distance DC line was put into service in 1882 by the Frenchman Marcel Deprez. It was a 57 km-long iron line with a voltage of 1.34 kV, with a diameter of 4.5 mm, which connected Miesbach to Munich (in Germany) and operated an artificial waterfall through a pump connected to a motor. However, the electrical efficiency of the transmission was only 60%, with a voltage drop of 37% along the line (Multon 2013).

1.2.3. *The expansion of AC*

The invention of the transformer by the Frenchman Lucien Gaulard and the Englishman John Dixon Gibbs in 1884 marked a turning point in the history of electricity. Indeed, this device, which only works on AC, allows for a change in voltage levels. As the power is the product of the voltage and the current, in order for the same power to be transported, increasing the voltage allows for the decrease of current and thus reduces the power losses due to the Joule effect occurring in the lines. The transformer can therefore transport electricity at high voltages before distributing it to consumers at lower voltages, thus solving the problem of transporting electricity via DC. A wealthy American industrialist, George Westinghouse, came across this breakthrough and bought the patent from Gaulard and Gibbs in 1886 (Allerhand 2017).

However, given that AC does not allow motors to operate, it was Nikola Tesla who in 1888 filed patents on polyphase motors and dynamos (Allerhand 2017), which were once again bought shortly after by Westinghouse. At the same time, Edison did not want to lose the electricity market and a "war of the currents" broke out between Edison on one side and Westinghouse and Tesla on the other. Edison tried to play on people's fear of alternative electricity, which was then little understood and "invisible", through various demonstrations that showed DC to be safer than AC. Not only did he publicize every accident that occurred during the work of Westinghouse and Tesla, but he was also helped by Harold P. Brown, who carried out electrocution experiments on animals in front of a small audience of journalists and electrical engineers in order to demonstrate that AC could kill, convincing politicians that AC was the quickest way to kill a man (giving birth to the electric chair in 1891) (Usborne et al. 2011). However, this was not enough to prevent AC from rapidly gaining a foothold in the world. In 1890, there were 202 Edison DC generating stations in the US as compared with 323 Westinghouse AC generating stations (Allerhand 2017). We can cite Usborne et al. (2011) and Allerhand (2017) in particular:

– *the Lauffen Hydroelectric Power Station (1891, Germany)*: three-phase generators at 50 V transported via a 4 mm diameter line over 175 km at 12.5 kV to Frankfurt, delivering 200 kW of power;

– *Niagara Falls Hydraulic Power and Manufacturing Company (1896, US)*: two-phase generators (20 MW in 1898) at 2.2 kV, with three-phase transmission of 11 kV to the city of Buffalo in New York State via a 42 km line system.

Three-phase current to transport electricity over long distances quickly became the standard for the vast majority of power systems in the world.

Today, the majority of power systems operate with AC, in order to be able to use transformers to adapt the voltage level to the distance the electrical energy needs to be transported. Some industrial systems are still using DC, such as in the transportation sector (trains and subways). In some cities, DC parts have coexisted with AC for a long time, such as in New York City, where the last DC line was retired in December 2007 (Spectrum 2012). In San Francisco and Chicago, we still find some DC areas, whose existence is generally justified by collateral replacement costs that are too high (Spectrum 2012).

1.2.4. *The revival of DC*

The beginning of the 20th century marked the development of power electronics with the invention of the mercury vapor diode by Peter Cooper Hewitt in 1902, which made it possible to rectify a current or a voltage, that is, to perform an AC to DC conversion.

The constant and growing technological progress in this field will increase the number of static converters at different levels of the power system:

– *at the transmission system level*: HVDC (high-voltage DC) connections;

– *at the distribution system level*: the development of optimized areas in DC in order to limit losses linked to converters as well as reduce costs:

- in medium voltage: connection of renewable energy and solar production zones,

- in low voltage: integration of renewable generation, storage facilities and charging stations for electric vehicles and rural electrification.

In particular, HVDC connections have become essential for the combining of two non-synchronous alternating systems, transport via buried or submarine cables over long distances, and the connection of offshore wind turbines. When compared to DC, the inductive and capacitive nature of AC lines in fact limits the power that can be transmitted. Currently, there are more than 200 HVDC systems around the world (Wolf and Adapa 2018), some examples are as follows:

– *underground HVDC line*: the line connecting Kashira to Moscow in Russia was the first HVDC line to be put into service in 1950 and was built by the German companies Siemens and AEG. It operated at 100 kV (a potential difference of 200 kV between the forward and return conductor) and allowed the transport of 30 MW of power over 120 km of underground cable (Travin and Khudyakov 2013). In France, the longest underground line is the 190 km France–Italy (Savoie-Piémont) line, which began construction in 2015 and is expected to be commissioned during 2022. It will allow for the transmission of 1,200 MW of power at 320 kV, thereby increasing the power exchange capacity between the two countries by 60% (RTE 2021c);

– *first commercialized HVDC line*: the Gotland 1 HVDC line in Sweden built in 1954 (and dismantled in 1986 when Gotland 3 was built) had a capacity of 20 MW at 100 kV using mercury vapor diode converters (ABB 2021);

– *overhead HVDC line*: the longest HVDC line in the world is the Rio Madeira line in the state of São Paulo, Brazil, connecting Santo Antônio to the Jirau Hydroelectric Power Plant. Commissioned in 2013, it can carry up to 7.1 GW of power over a distance of 2,385 km at 600 kV (Gupta 2014);

– *submarine HVDC line*: the 250 km IFA2 line (of which about 200 km is under the sea), which links France to England with a capacity of 1,000 MW at 350 kV, was commissioned in 2021.

1.2.5. *Development of power systems*

1.2.5.1. *The French context*

The development of electricity systems in France really began with the law of April 5, 1884, which entrusted the maintenance and distribution of electricity to municipalities via local distribution companies (LDC) or via private companies because of the law of June 15, 1906, which established management in the form of a concession. Municipalities could therefore delegate the management of electricity distribution to private companies on a long-term basis through a concession contract. However, electrification in rural areas did not develop as well as in urban areas since the profitability of investments was not compensated for by the sale of electricity. Moreover, at the end of World War II, 5.7% of urban households and 17.5% of rural households were without electricity (Massé 2002). For these various reasons, on April 8, 1946, the State decided to nationalize the private companies that operated the electricity and gas networks into two companies: *Électricité de France* (EDF) and *Gaz de France* (GDF). Note that 154 private production companies, 86 private transmission companies and 1,150 private distribution companies were thus merged into EDF, with the exception of 300 LDCs (260 local distribution companies and 40 SICAEs (*Société d'intérêt collectif agricole d'électricité*:

Agricultural electricity collective interest company)) (SDE35 2011). A large number of “microgrids” that had developed with their own technical reference systems are now united and managed by a single entity, with the price of electricity regulated by the State. The current French power system that has interconnections with neighboring countries only really began to develop at the end of the 20th century, because of the creation of the Union for the Coordination of Production and Transmission of Electricity (UCPTE 2018) on May 23, 1951; the initial objective of which was to ensure the optimal operation of electricity production plants, in particular by giving priority to less expensive production plants such as hydroelectric plants (UCPTE 2018).

1.2.5.2. *Other European contexts*

Today, the configuration of electricity system operators in Europe differs from one country to the next. In Germany, unlike France where a single operator (RTE[2]) operates the transmission system, the system is divided into four zones, each managed by a separate operator (Tennet, 50 Hertz, Amprion and TransnetBW). These operators are grouped into a coordinating body (Netzregelverbund). The distribution of electricity has been carried out by 900 operators (Verteilnetzbetreiber) since the end of the 19th century, that is, about six times more than in France. These operators perform other functions in addition to managing the system (electricity production and its transport, water treatment, waste management, urban transport, etc.). These operators are very often at the municipal level (*Stadtwerke*), a characteristic that, as of recently, is tending to become more pronounced with the takeover of systems by several cities, citing in particular the case of Hamburg in September 2013. This movement comes in opposition to the oligopoly composed of the four companies EON, EnBW, RWE and Vattenfall, which account for 82% of production and about 45% of electricity sales in Germany (Häusserman 1992). Between 2007 and 2015, 63 new *Stadtwerke* were created and 192 municipalities re-municipalized their electricity distribution.

In Belgium, the high-voltage system is operated by a single company, Elia, like RTE in France, and there are about 10 distribution system operators whose territory corresponds to intermunicipal areas, such as Sibelga for the Brussels region (19 municipalities) or ORES for an area of 200 municipalities representing 75% of the Wallonia region. These companies generally distribute gas along with electricity but do not perform other functions such as electricity generation and supply.

In Portugal, the production and distribution of electrical energy was carried out at the beginning of the electrification of the country by 12 companies at a regional level. In 1975, during the revolutionary period that ended the dictatorship, these

2 *Réseau de Transport d'Electricité*: this is the name of the TSO in France.

12 companies were merged and nationalized into a single entity: EDP (*Energias de Portugal*). The EDP group was privatized in 1994, creating, in the process, a separate entity within the group for electricity transportation: REN, which left the EDP group in 2000. Within EDP, the part that handles distribution for all medium voltage is now called E-REDES (in total there are now about 10 local companies that handle low-voltage distribution).

1.2.5.3. *The American context*

The development of electricity systems in the US began with San Francisco in 1879, with the first DC arc lighting system. In 1882, New York inaugurated the first low-voltage DC system. Electricity was first used to replace the flickering flame of kerosene lamps. Until the 1900s, electricity was the reserve of the elite, with their own generators and distribution systems. By 1907, only 8% of households had access to electricity, and most were powered by private systems (distributed systems) (Bakke 2016). Different voltage levels (100, 110, 220, 500, 600, 1,200 and 2,000 V), multiple private grids and, starting in the 1890s, the beginning of AC and its different frequencies (25, 30, 331/3, 40, 50, 60, 662/3, 831/3 and 125 cycles per second) contributed to the complexity of the electrical system by increasing the number of distribution systems. In 1896, the transmission of AC from the Niagara Falls power plant to Buffalo, 35 km away, allowed electricity to become a means of separating production from consumption. In 1907, following the example of Samuel Insull's success in Chicago, with the takeover of more than 20 distribution companies, the LDCs organized themselves as monopolies and became regulated by the states in exchange for protection against competition. The mass consumption of electricity through household appliances did not arrive until the 1940s, when electricity was no longer reserved for the elite, which allowed LDCs like General Electric to obtain a double income by also selling household appliances. In the countryside in 1930, only 10% of American farms were connected to the grid. Fifteen years later, almost 45% of farms were connected because of public works (the Rural Electrification Administration, 1935) (Smithsonian Institution 2019). By 1970, 77% of kilowatt-hours were generated and distributed by privately owned LDCs, while municipal and governmental LDCs produced the rest (Smithsonian Institution 2019). The first oil shock in 1973, along with the need to open up the LDC monopoly to competition, led to the Public Utility Regulatory Policies Act of 1978 (PURPA), requiring LDCs to purchase energy output from independent generators at a regulated rate. As soon as PURPA was implemented in California, many permit applications arrived: 1,800 hydro dams and 16,000 wind turbines, as well as solar projects (Bakke 2016). In 1992, the Energy Act opened the door for the deregulation of the energy market by distinguishing between production and distribution. In California, an energy market went live in 1998 and served 80% of the state. LDCs were then required to go through this wholesale market to source their energy (FERC 2019). The retail market for LDCs is also evolving; in 2016, the State of New York initiated a restructuring of LDCs toward a distribution system operator model, facilitating the link between consumer and energy provider (State of New York Public Service Commission 2016).

1.2.6. *The frequency choice for power systems*

1.2.6.1. *Two preferred frequencies: 50 Hz and 60 Hz*

An alternating quantity is characterized by the shape of the wave (sinusoidal, triangular, etc.) and by its periodicity, that is, the duration of the elementary pattern that is perpetually repeated. The unit typically used is the frequency, expressed in Hz, defined as the number of repetitions of the elementary pattern over 1 s. Electrical machines can only operate with a sine wave whose amplitude and frequency can vary according to its characteristics. The choice of operating frequency for a power system is the result of a compromise between the cost and technical performance for the various uses of the power system. Indeed, if the frequency is too low, the filament of incandescent lamps and the carbon of arc lamps have time to cool down between two alternations and an unpleasant phenomenon of flicker can appear. Concerning the transformers, at equal power, their dimensions are inversely proportional to the value of the frequency and therefore the high frequencies make it possible to reduce the size of the transformers, and thus their cost and their dimensions. On the other hand, if the frequency is too high, then for a given machine rotation speed, the number of poles will have to be increased, which is technically complicated to achieve in general. Indeed, the relation between the frequency (f) of the current generated by a generator, its number of poles (p) and its speed of rotation in rpm (N) is given by the relation: $f = \frac{N \times p}{120}$. Moreover, too high a frequency increases the inductive character of rotating machines (generators and motors) and transmission lines, degrading the efficiency and the amplification of voltage drops. In conclusion, high frequency is preferred for lighting, transformers and lower frequency for the electromechanical characteristics of rotating machines and power lines. At the genesis of AC systems, more than 10 frequencies were used in the world depending on the system developer, they gradually converged to two preferred frequencies: 50 Hz (Europe and countries under European influence) and 60 Hz (North America). A special case is Japan where both frequency systems coexist.

The generalization of these two frequencies began in 1891, that is, with the development of AC:

– *North American continent*: in 1886, William Stanley designed the first AC distribution system for lighting in Great Barrington, Massachusetts, financed by

Westinghouse, with a generator consisting of 16 poles at 1,000 rpm, thereby creating a current of 133 Hz. Another important American manufacturer, the Thomson-Houston Company, favored a frequency of 125 Hz (Owen 1997). Frequencies around 130 Hz became more common in North America but have some disadvantages. The first is that at the time, generators were driven by low-speed steam engines and therefore a large number of poles were needed to increase the frequency (Neidhofer 2011). In addition, the use of high frequencies causes resonance in transmission lines. For this reason, the two important American manufacturers, Westinghouse and General Electric (a merger of the Thomson-Houston Company and the Edison General Electric Company) both chose to adopt 60 Hz, in 1890 and 1894, respectively (Owen 1997). 25 Hz also experienced a boom for generation and transmission following the construction of the 20 MW Niagara Falls generator and remained popular until 1919.

– *European continent*: in England in the 1900s, we mostly find 100 Hz and already a tendency to go to lower values such as 50 Hz, while in the rest of Europe, low-frequency values have always been used in majority (between 30 Hz and 50 Hz). In Italy, for example, five different frequencies coexisted for several years: 42 Hz, 50 Hz, 46 Hz, 16 Hz and 25 Hz (taken from an assessment made in 1918 by Garver Lamme, then Chief Engineer for Westinghouse). The generalization of 50 Hz was launched by the companies AEG and MFO, after noticing a flicker phenomenon in the lamps connected to the Lauffen–Frankfurt line that they had developed and which operated at 40 Hz.

– *The case of Japan*: Japan has been influenced by both the US and Europe and is therefore currently divided into two operating areas: the northeast area of Mount Fuji operates at 50 Hz while the southwest area is at 60 Hz. In 1889, two Japanese returned from the US with Westinghouse equipment and know-how to electrify the eastern part of Japan. In the years 1896–1897, AEG and Siemens sold generators operating at 50 Hz to the Tokyo Electric Company.

From 1925 onwards, the various countries understood the importance of interconnection, for the commercial exchange of electricity but also to increase the reliability of the system. These interconnections and the standardization of industrial production therefore required the use of a single frequency. This process of frequency harmonization was completed in the 1950s. It should be noted, however, that despite being interconnected and at the same frequency, system zones can have different phase rotation directions, which can make interconnection impossible, as in Mauritania, Corsica, Sardinia and Texas. On-board networks such as airplanes and networks supplying aircraft on land often use 400 Hz, as this increases the power density of transformers and high-frequency converters. The Swiss railway system is also 16 2/3 Hz.

1.2.6.2. *Frequency: a key element of system stability*

Frequency depends on the balance between generation and consumption, which must be maintained at all times. Moving out of this state of equilibrium can cause widespread instability across the system. Indeed, the very strong dynamic coupling between generation resources and loads, due to the low mechanical inertia of some power system components, means that a major disturbance in Northern Europe can be felt by Southern Europe in a matter of seconds. A frequency variation can cause material damage (reduction in the efficiency of rotating machines, variations of the magnetic flux in the transformers, which in turn can increase saturation and thus iron losses, and the production of harmonics). Moreover, the most significant impact is the de-energizing of part of the system by a cascade tripping of the protection mechanisms. This was the case in Italy, for example, in September 2003 (RTE 2004), when all the lines interconnecting Italy and its neighbors were disconnected by a cascade following the loss of two 400 kV lines located in Switzerland, which led to a widespread blackout. Three levels of action exist to ensure the production–consumption balance (RTE 2004):

– *primary frequency control* (frequency containment reserve (FCR) or primary frequency response (PFR)): this is local and automatic control of the generation of each group participating in this control. Within each synchronous zone of an interconnected system, the control capacities are defined between the different transmission system operators;

– *secondary frequency control* (frequency restoration reserve (FRR) or regulating reserve): this is carried out by a centralized unit located at the regional dispatching center. It adapts the control laws of the generation units that are based on remote frequency measurements and transits along the interconnection lines, with a response time of about 5–15 min;

– *tertiary control* (reserve replacement (RR) or contingency spinning reserve): this is the optimization of a generation margin of maneuver called the spinning reserve, which can be mobilized almost instantaneously and is available for a guaranteed period of time. In general, this is a lightly loaded conventional generator or PHES-type storage that can respond quickly, and has a capacity corresponding to that of a large generator, or which can cover the load of a transmission line (DOE 2016). This is a manual adjustment made at the national level, with a response time of about 15 min. If necessary, other types of generation capable of starting quickly can be called into service.

The specifications in France allow for the frequency to fluctuate by 1% (RTE 2009) for 99.5% of the time over the year, for systems linked by synchronous connections to an interconnected system, according to the NF EN 50160 standard.

However, with interconnections, the deviation is more typically 0.1%. Other variation ranges are defined for exceptional conditions. For example, the frequency may drop to 47–47.5 Hz for 1 min once every 5–10 years (RTE 2009). It is important to remember that frequency control, which is an essential element of power system stability, concerns large generating units and transmission system management, which is presented later in this chapter.

1.2.7. *Choosing voltage levels for power systems*

The choice of voltages encountered in power systems is the result of a techno-economic trade-off between power quality, safety, investment costs, operational costs and space constraints. This optimum varies according to the total load to be supplied and the distance to be covered to connect the customer. This is why the power system has several voltage levels. Figure 1.3 shows the different voltage levels standardized according to AFNOR NF C 18-510 (France), IEC (Europe) and ANSI C84.1 (USA). High voltage corresponds to voltages above 50 kV according to AFNOR and is called HVB, voltages above 35 kV according to IEC 60038, and above 69 kV according to ANSI and is called HV. It is used in transmission systems to transport large volumes of electrical energy over long distances, from major production centers to consumer regions.

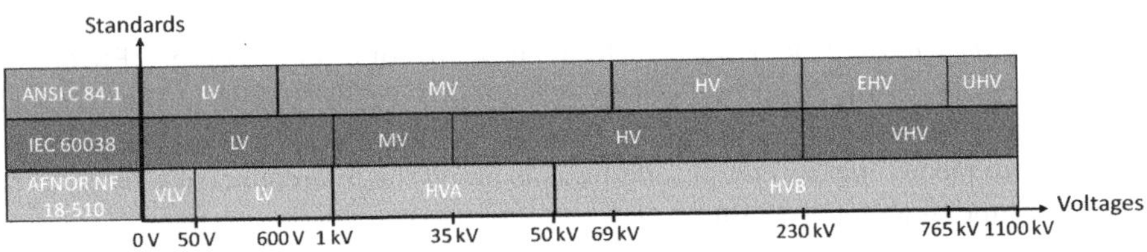

Figure 1.3. *Voltage levels according to different standards. For a color version of this figure, see www.iste.co.uk/alvarezherault/distribution.zip*

In France, we can find high voltage lines operating at 63, 90, 225 or 400 kV and at 115, 138, 161, 230, 500 and 765 kV in the US. The medium voltage corresponds to voltages between 1 kV and the threshold values previously mentioned. It is called HVA according to the NF C 18-510 standard and MV according to the IEC 60038 and ANSI standards and is used in distribution systems to distribute electricity to consumer areas. These lower voltages make it possible to reduce the size of the equipment and therefore their cost, which is not justified over shorter distances. In France, the voltage level is 20 kV for the HVA level, a value resulting from a long historical evolution (see Table 1.1). There are still some 5.5 kV and 15 kV systems that tend to disappear. The typical values we can find in the US are 34, 46 and 69 kV. Finally, the European and French standards define low voltage as being lower than 1 kV (600 V for the ANSI standard) and corresponding to inhabited areas. For the sake of homogeneity, we use the terminology of IEC 60038, that is, LV, MV and HV.

After the Second World War	Eight voltage levels coexist: 10; 11; 13.5; 15; 16.5; 17.3; 18; 22 kV.
First standardization guidelines	Five possible values: 5.5; 10; 15; 20 and 30 kV, this last value is used for sub-distribution systems.
Early 1960s	85% of MV systems are operated at standard voltage. The 15 kV level represents 52% of MV systems.
1991	Systems operating at 20 kV represent 75% of MV systems.
End of 2004	The remaining 11% of the MV system is operated at 15 kV and the rest at 20 kV.

Table 1.1. *Historical evolution of MV voltage levels in France*

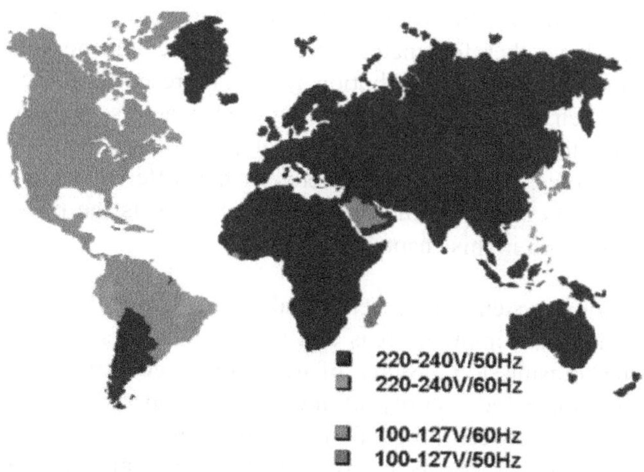

Figure 1.4. *Operating frequency and voltage of the LV power system across the world (source: Wikitravel 2021). For a color version of this figure, see www.iste.co.uk/alvarezherault/distribution.zip*

The LV voltage level in France is 400 V between phases or 230 V between a phase and a neutral. This ratio of √3 allows for the provision of two different voltages through a three-phase system and has been used since the 1950s. At that time, the 127/220 V and 220/380 V levels existed. The second pair of values was standardized in 1956, then replaced in 1986 by the 230/400 values following the recommendations of the IEC. The generalization to 230/400 was completed in 1996. In addition to 230/400 V, the norm across most of Europe, another LV voltage pair, 115/220 V, is widely used in North America, most often in the form of two 120 V single-phase bridges. Note that 230/400 V is justified for LV systems with large MV/LV substation service areas and long lines in order to save on technical losses. In North America, LV systems are reduced to simple sidings of two to four customers, while the main service is provided by the MV system in 4 kV (12.8% of the departures), 12 kV (70.2%), 17 kV (1.5%) and 21 kV (15.5%). Figure 1.4 shows the distribution of the use of these two voltage ranges around the world. It also shows the 50 and 60 Hz operating areas as presented in section 1.2.6.

1.2.8. *Structuring the power system*

In order to achieve economies of scale and to adapt to the constraints of locating hydraulic reserves and cold sources for nuclear reactors, production units have been built in limited and concentrated geographical areas. On the other hand, consumption is spread over the whole territory in a very dispersed manner. Three types of systems have therefore naturally emerged: the transmission system, the subtransmission system and the distribution system, represented in Figure 1.5. These systems differ in their voltage levels, operating modes, actors (managers, regulators and owners), architectures, types of users, equipment used and the volumes of electrical energy flows. The connection between the different types of systems is made via transformers and a set of equipment that protects the power system and that is described later on in this chapter.

The transmission system carries electrical energy from the major production centers to the main consumption areas and major customers. This system is also connected to the transmission systems of neighboring countries in order to share resources and optimize the security of the system's operation, particularly with regard to the balance between production and consumption. In the case of France, the transmission system is connected to Germany, Belgium, Spain, Great Britain, Italy and Switzerland through 17 interconnection lines. As an example, it was expected that for 2017 there would be a net transfer capacity varying between 10,923 and 14,901 MW for export and 6,155 and 9,910 MW for import, depending on the month of the year (RTE 2021a) with the knowledge that the total installed generation capacity in 2017 was 130 GW (RTE 2017).

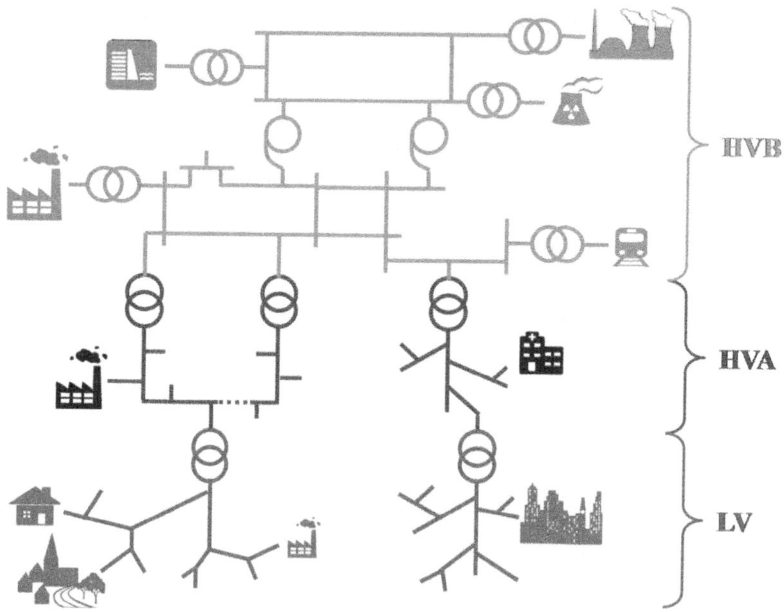

Figure 1.5. *Structure of the French power systems. For a color version of this figure, see www.iste.co.uk/alvarezherault/distribution.zip*

North America is divided into four interconnected areas:

– Eastern interconnection: two-thirds of the US and part of Canada;

– Western interconnection (the western third of the US, Alberta and British Columbia in Canada and the Baja California Norte part of Mexico);

– Texas interconnection (a large part of the state of Texas);

– Quebec.

The subtransmission system supplies the distribution systems and high-power industrial consumers (e.g. metallurgy, steel, chemical, automotive, paperboard and railroad plants) at lower voltages than those used for transmission. In France, the distribution system represents a volume of approximately 30.2 TWh of energy consumed per month, or 76% of the total energy consumed by the transmission system. The rest of the energy is consumed by major industries, accounting for approximately 14% of total energy consumption.

The distribution system carries electrical energy from the transmission and subtransmission system to end-users, either at medium voltage (e.g. industrial, commercial, hospitals) or at low voltage (residential, small businesses). In 2017, the French distribution system supplied 151.1 TWh to residential customers, 45.6 TWh to SMEs, 112.3 TWh to businesses and 41.9 TWh to professionals (RTE 2017).

System	Transmission	MV distribution	LV distribution
Lines	Overhead[2]: 99,964.3 km Underground[2]: 5,996.8 km Continuous[2]: 183 km Total[2]: 106,144.1 km	Naked overhead[3]: 326,712 km Twisted overhead[3]: 1,395 km Underground[3]: 312,580 km Total[3]: 639,687 km	Naked overhead[2]: 65,689 km Twisted overhead[3]: 328,270 km Underground[3]: 323,129 km Total[3]: 717,088 km
Annual growth	On average 231.5 km/year since 2011[2]	On average 4,594 km/year since 2011[3]	On average 4,187 km/year since 2011[3]
Substations	2,740 substations[2]	4,389 substations HV/MT[3] 783,262 substations MV/BT[3]	
Annual growth	22 new substations in 2017[3]	On average 5,402 substations LV/MV/year since 2011[3]	
Architecture	Meshed	Looped Operated radial	Tree structure
Voltage level (phase-to-phase)	225/400 90/63	20 kV mainly and some 5.5 kV (Grenoble) and 15 kV	400 V
Neutral regime	Three-phase	Three-phase	Three-phase + neutral

Table 1.2. *Comparison of French transmission and distribution systems*

3 Source: RTE – Electrical Balance Sheet 2017.

The main differences between the transmission and subtransmission systems and the distribution systems summarized in Table 1.2 for France (Enedis, 36 million people, 543,965 km^2) and Table 1.3 for part of California (PG&E, 5.4 million people, 181,300 km^2) are as follows:

– *The amount of energy to be transported*: this affects the criticality of the system and the need for redundancy, and therefore its architecture. The loss of a line or substation on the transmission system has a much greater impact than on the distribution system, where the number of MV/LV and HV/MV substations is 290 times greater than the number of substations on the transmission system in France and eight times greater on the PG&E system. This is why a meshed architecture is chosen for the transmission system in order to guarantee operation in the event of the loss of an element (generation facility, line, source substation). The subtransmission system is less meshed than the transmission system for economic reasons and because the level of criticality is less important. Finally, the distribution system has a loopable architecture but is operated in a radial manner. The few redundancies allow for the resupply of customers in the event of a fault (see Chapter 2).

– *Total distance*: we can see that the distribution system has 13 times as many lines as the transmission and subtransmssion systems in France (and almost six times as many on the PG&E system), and increases by several thousand kilometers each year, as compared to a few hundred kilometers each year for the transmission system. It is therefore much more expensive to mesh the distribution system. In rural areas where load density is low and the line length is long, the architecture will more often be radial with a tree structure. In denser urban areas, backup is made possible by adopting architectures such as source-to-source secured feeder or double feeder supply (see Chapter 2).

– *Type of line*: the transmission system is mainly overhead, while the French distribution system is currently 53% overhead (80% for the PG&E system) or twisted and 47% underground. The proportion of overhead lines is not likely to change for the transmission system because of the very high cost of underground cables, strong topological constraints and the volume of materials and insulation requirements proportional to the voltage level. For distribution systems, the choice between underground and overhead depends more on the type of area to be supplied (urban, rural or semi-rural), which influences the type of topological constraints (e.g. buildings in urban areas) and the power quality policy, as underground cables are less prone to faults.

System	Transmission	MV distribution	LV distribution
Lines	Total: 29,430 km[4]	Total: 171,686 km (of which 75% are overhead)	
Substations	91 substations (Global Transmission 2017)	622 substations HV/MV (PG&E 2020)	
Architecture	Meshed	Looped Operated radial	Tree structure
Voltage level (phase-to-phase)	60, 115, 230 or 500 kV	4, 12, 17 or 21 kV	120 or 480 V
Neutral regime	Three-phase	Three-phase or single-phase + neutral	Single-phase or three-phase (when MV is also three-phase) + neutral

Table 1.3. *Comparison of PG&E's transmission and distribution systems (California)*

1.3. Technical description of the power system

The power system has a structure inherited from a series of significant historical technical choices to ensure the transmission, dispatch and distribution of electrical energy while respecting the technical constraints inherent to the materials constituting it. In this section, we introduce the preliminary elements of modeling and theoretical calculations that allow us to understand the choices that led to the current structure of the power system.

1.3.1. *The three-phase system*

1.3.1.1. *Balanced three-phase system*

Currently, with the exception of HVDC lines, most power systems are three-phase AC with some parts in single- and two-phase systems. A three-phase system is a set of three AC quantities of the same frequency and nature. A balanced three-phase AC voltage source consists of three sinusoidal voltage sources, of the same frequency and amplitude, and phase-shifted by an electrical angle of $2\pi/3$. These voltages can be represented graphically by three sine waves or three vectors as shown in Figure 1.6 and mathematically by three equations in time-based representation (equation [1.1]) or in complex format (equation [1.2]).

4 Unit of conversion used: 1 mile = 1.60934 km.

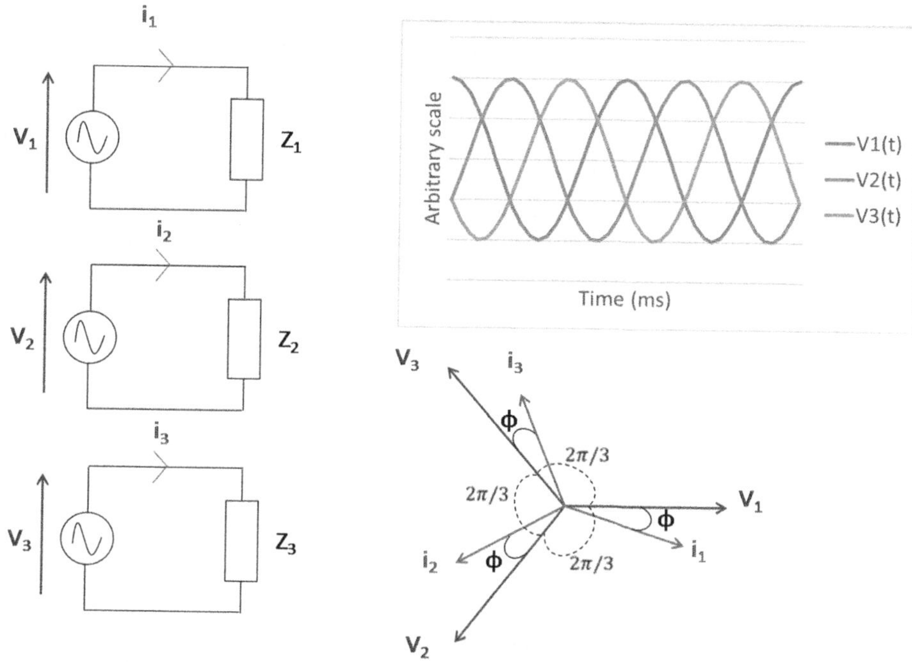

Figure 1.6. *Representations of the balanced three-phase electrical system (source: Escane and Bastard 2004). For a color version of this figure, see www.iste.co.uk/alvarezherault/distribution.zip*

$$
\begin{aligned}
V_1(t) &= V \times \cos(\omega.t) \\
V_2(t) &= V \times \cos(\omega.t - 2\pi/3) \\
V_3(t) &= V \times \cos(\omega.t + 2\pi/3)
\end{aligned} \qquad [1.1]
$$

$$
\begin{aligned}
\underline{V_1}(t) &= Ve^{j0}.e^{j\omega.t} \\
\underline{V_2}(t) &= Ve^{-2\pi/3}.e^{j\omega.t} \\
\underline{V_3}(t) &= Ve^{j2\pi/3}.e^{j\omega.t}
\end{aligned} \qquad [1.2]
$$

with:

– ω: electrical pulsation of the three-phase sinusoidal system equal to 2πf;

– f: power system frequency;

– V_1 chosen as a reference for the phases.

NOTE.– For studies the in static regime, we are interested in the phase shift of the voltages between them, and equation [1.2] is simplified by placing ourselves at t = 0.

Consider that the three-phase AC voltage source is connected to a balanced load Z. Each of the sources V1, V2 and V3 is connected to three equivalent loads Z1, Z2 and Z3. Considering that these three loads are balanced and in star-connection, the currents flowing through these three loads are sinusoidal and given by equation [1.3] in time-based representation and equation [1.4] in complex format.

$$\begin{aligned} I_1(t) &= I \times \cos(\omega.t - \varphi) \\ I_2(t) &= I \times \cos(\omega.t - 2\pi/3 - \varphi) \\ I_3(t) &= I \times \cos(\omega.t + 2\pi/3 - \varphi) \end{aligned} \quad [1.3]$$

$$\begin{aligned} \underline{I_1} &= Ie^{-j\varphi}.e^{j\omega.t} \\ \underline{I_2} &= Ie^{j(-2\pi/3-\varphi)}.e^{j\omega.t} \\ \underline{I_3} &= Ie^{j(2\pi/3-\varphi)}.e^{j\omega.t} \end{aligned} \quad [1.4]$$

The phase shift φ is related to whether the load Z is inductive or capacitive. The three single-phase electrical circuits connecting the three-phase voltage source to the three balanced loads require six conductors: three "forward" conductors and three "return" conductors. Therefore, it is more interesting to couple together the three phases, which can be done in two ways, as shown in Figure 1.7.

Delta connection: $V_1(t) + V_2(t) + V_3(t) = 0$ Star connection: $I_1(t) + I_2(t) + I_3(t) = 0$

a)

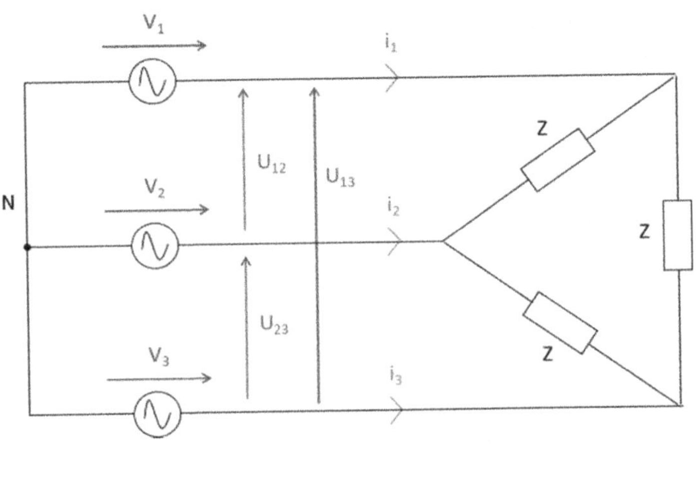

b)

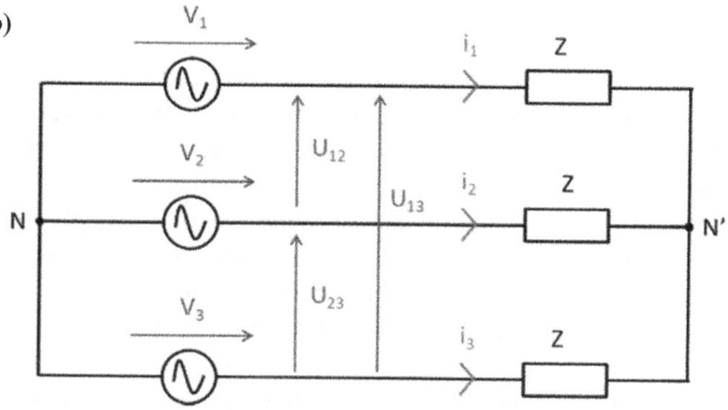

Figure 1.7. *Two possible connections: a) delta and b) star. For a color version of this figure, see www.iste.co.uk/alvarezherault/distribution.zip*

In the case of the star connection, the coupling of the three phases reduces the number of return conductors from 3 to 1. By pooling a single return conductor, we obtain a three-phase system with four conductors. If the system is also balanced, the return current will be zero. It is also possible to do without a return conductor. In the case of a delta connection, since there is no return conductor, we go directly from a six-conductor system to a three-conductor system.

On the diagrams in Figure 1.7, we can see two types of voltages: the phase-to-phase voltages U_{12}, U_{23} and U_{31} which are the voltages between two phases and the single-phase voltages V_1, V_2 and V_3 which are the voltages between the phases and neutral. The instantaneous relationship between single and phase-to-phase voltages is described by equation [1.5].

$$U_{ij}(t) = V_i(t) - V_j(t) \; \forall\{i,j\} \in \{1;2;3\} \qquad [1.5]$$

$\forall\, i \in \{1;2;3\}$, by taking V_i as the phase reference, then the phase-to-phase voltage between phases *i* and *j* is $\pi/6$ out of phase, in the balanced state, with respect to the single voltage of phase *i* as shown in equation [1.6]. The ratio between the modulus of the phase-to-phase voltage and the modulus of the single-phase voltage is $\sqrt{3}$: thus, in France, the low voltage phase-to-phase voltage is 400 V while the low-voltage single-phase is ${}^{400}/_{\sqrt{3}} = 230.94$, or about 230 V.

$$\underline{U_{ij}} = \underline{V_i} - V_i \cdot e^{j2\pi/3} = V_i \times \left(1 - e^{j2\pi/3}\right) = \sqrt{3} \cdot V_i \cdot e^{j\pi/6} \qquad [1.6]$$

1.3.1.2. *Advantage of the three-phase system*

The advantage of three-phase current is that in balanced mode, the instantaneous power is constant and smooth (zero fluctuating power), which causes less noise and vibration and therefore less material fatigue. With single-phase current, the resulting fluctuating power would destroy the connection between a very high-powered alternator and its mechanical energy source. The use of three-phase current also allows the creation of a rotating magnetic induction, which is the basis of all alternating machines. These machines can be connected directly to the grid as generators. In addition, a three-phase system allows for a 50% reduction in the total conductor cross-section, as compared to a single-phase system, in order to transmit the same power (Carrive 1991). In a single-phase system, the load is connected between a phase conductor and a neutral conductor. For each conductor cross-section, there is a maximum ampacity beyond which the conductor deteriorates. Let S be the cross-section of the phase and neutral conductors and I_{max} the maximum current that can flow. The maximum power that can flow is equal to $V \times I_{max} \times \cos\varphi$ for an equivalent conductor cross-section of 2S (V being the single-phase-neutral voltage and $\cos\varphi$ the power factor of the load). In a three-phase system, the maximum power that can be transmitted by three phases of cross-section S (equivalent conductive cross-section of 3S) will be equal to $3 \times V \times I_{max} \times \cos\varphi$. To pass $3 \times V \times I_{max} \times \cos\varphi$, an equivalent conducting cross-section of 6S in single-phase is therefore required against 3S in the three-phase system. The change from single-phase to three-phase systems therefore allows for a reduction in the conductive cross-section by 50%. Finally, the three-phase electrical system offers the same system two different voltages of use, either between phase and neutral (single-phase voltage), or between two phases (phase-to-phase voltage). In the various standards and technical documents of power systems, voltage refers to phase-to-phase voltage. Indeed, phase-to-phase voltage does not require the use of neutral voltage, which does not always correspond to an existing physical element. In the case of a star connection, the neutral point is accessible. In the case of a delta connection, it is possible to artificially recreate a neutral point by connecting a balanced three-phase load. The phase-to-phase voltage is suitable for the various three-phase systems we encounter, with three or four conductors as described below. However, a single-phase system can be advantageous in the service of small dispersed loads because of a weak flow, as compared with the capacity of the conductors, instead of using a three-phase system whose infrastructure costs are too significant, as is predominantly the case in rural areas.

1.3.1.3. *Unbalanced three-phase system*

Voltages and currents are unbalanced when their amplitudes and the phase shift between them are no longer identical. In this case, the vector sum of voltages and currents is no longer zero. In general, multiphase systems, whether balanced or not, can be reduced to the simplified study of balanced systems because of the Fortescue transformation, the principle of which we briefly recall here. As shown in Figure 1.6, a balanced three-phase system can be represented by three phase-shifted vectors $\frac{2\pi}{3}$. Let "a" be the rotation operator equal to $e^{j\frac{2\pi}{3}}$. Then, provided the quantities $\left(\overrightarrow{\underline{G_1}}, \overrightarrow{\underline{G_2}}, \overrightarrow{\underline{G_3}}\right)$, where G can represent voltage or current, form a balanced three-phase system, we obtain the following relations:

$$\overrightarrow{\underline{G_2}} = a^2 \times \overrightarrow{\underline{G_1}} \text{ and } \overrightarrow{\underline{G_3}} = a \times \overrightarrow{\underline{G_1}} \tag{1.7}$$

Let $\left(\overrightarrow{\underline{G_1}}, \overrightarrow{\underline{G_2}}, \overrightarrow{\underline{G_3}}\right)$ be any voltages or currents that are not necessarily balanced. Then, since $1 + a + a^2 = 0$, we can write the system of equation [1.8], which can be written as equation [1.9].

$$\begin{cases} \underline{G_1} = \dfrac{3\underline{G_1}}{3} + \dfrac{\underline{G_2}}{3} \times (1 + a + a^2) + \dfrac{\underline{G_3}}{3} \times (1 + a + a^2) \\ \underline{G_2} = \dfrac{3\underline{G_2}}{3} + \dfrac{\underline{G_1}}{3} \times (1 + a + a^2) + \dfrac{\underline{G_3}}{3} \times (1 + a + a^2) \\ \underline{G_3} = \dfrac{3\underline{G_3}}{3} + \dfrac{\underline{G_1}}{3} \times (1 + a + a^2) + \dfrac{\underline{G_2}}{3} \times (1 + a + a^2) \end{cases} \tag{1.8}$$

$$\begin{cases} \underline{G_1} = \dfrac{\underline{G_1} + \underline{G_2} + \underline{G_3}}{3} + \dfrac{\underline{G_1} + a\underline{G_2} + a^2\underline{G_3}}{3} + \dfrac{\underline{G_1} + a^2\underline{G_2} + a\underline{G_3}}{3} \\ \underline{G_2} = \dfrac{\underline{G_1} + \underline{G_2} + \underline{G_3}}{3} + \dfrac{a^2\underline{G_1} + \underline{G_2} + a\underline{G_3}}{3} + \dfrac{a\underline{G_1} + \underline{G_2} + a^2\underline{G_3}}{3} \\ \underline{G_3} = \dfrac{\underline{G_1} + \underline{G_2} + \underline{G_3}}{3} + \dfrac{a\underline{G_1} + a^2\underline{G_2} + \underline{G_3}}{3} + \dfrac{a^2\underline{G_1} + a\underline{G_2} + \underline{G_3}}{3} \end{cases} \tag{1.9}$$

By defining the homopolar (or zero sequence), direct (or positive sequence) and inverse (or negative sequence) components as the three symmetric systems of equations [1.10], [1.11] and [1.12], we can express any three-phase voltage or current as the sum of the direct, inverse and homopolar components, as shown in equation [1.13].

– Homopolar system:

$$\begin{cases} \underline{G}_{h1} = \dfrac{\underline{G}_1 + \underline{G}_2 + \underline{G}_3}{3} \\ \underline{G}_{h2} = \dfrac{\underline{G}_1 + \underline{G}_2 + \underline{G}_3}{3} \\ \underline{G}_{h3} = \dfrac{\underline{G}_1 + \underline{G}_2 + \underline{G}_3}{3} \end{cases} \qquad [1.10]$$

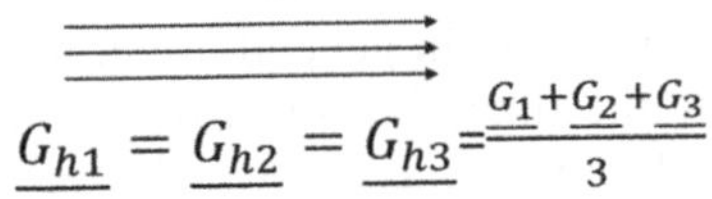

– Direct system:

$$\begin{cases} \underline{G}_{d1} = \dfrac{\underline{G}_1 + \underline{a}\underline{G}_2 + \underline{a}^2\underline{G}_3}{3} \\ \underline{G}_{d2} = \dfrac{\underline{a}^2\underline{G}_1 + \underline{G}_2 + \underline{a}\underline{G}_3}{3} \\ \underline{G}_{d3} = \dfrac{\underline{a}\underline{G}_1 + \underline{a}^2\underline{G}_2 + \underline{G}_3}{3} \end{cases} \qquad [1.11]$$

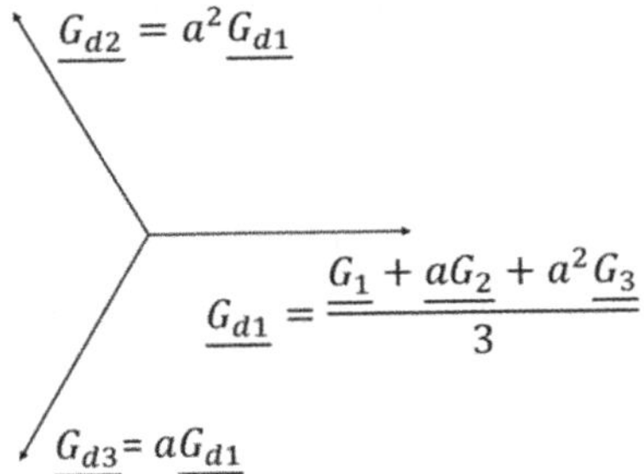

– Inverse system:

$$\begin{cases} \underline{G}_{i1} = \dfrac{\underline{G}_1 + \underline{a}^2\underline{G}_2 + \underline{a}\underline{G}_3}{3} \\ \underline{G}_{i2} = \dfrac{\underline{a}\underline{G}_1 + \underline{G}_2 + \underline{a}^2\underline{G}_3}{3} \\ \underline{G}_{i3} = \dfrac{\underline{a}^2\underline{G}_1 + \underline{a}\underline{G}_2 + \underline{G}_3}{3} \end{cases} \qquad [1.12]$$

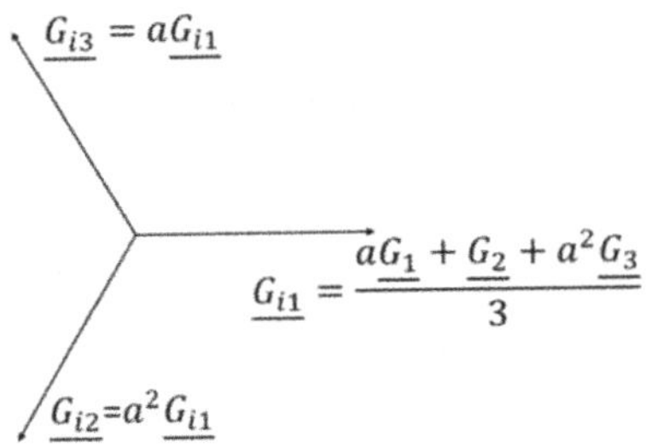

$$\begin{cases} \underline{G}_1 = \underline{G}_{h1} + \underline{G}_{d1} + \underline{G}_{i1} \\ \underline{G}_2 = \underline{G}_{h2} + \underline{G}_{d2} + \underline{G}_{i2} \\ \underline{G}_3 = \underline{G}_{h3} + \underline{G}_{d3} + \underline{G}_{i3} \end{cases} \qquad [1.13]$$

Thus, even an unbalanced three-phase system can be decomposed into three balanced subsystems, thus simplifying the solution method. This method can also be generalized to a polyphase system with n phases defined by equation [1.13]. Fortescue (1918) provides more details concerning the calculations of polyphase systems.

$$\forall\, 1 \le i \le n, \overrightarrow{\underline{G_i}} = \underline{a}^{i-1} \times \overrightarrow{\underline{G_1}} \text{ with } \underline{a} = e^{j\frac{2\pi}{n}}$$
$$1 + \underline{a} + \underline{a}^2 + \cdots + \underline{a}^{n-1} = 0 \quad [1.14]$$

1.3.2. *Connection mode for components of the power system*

As seen above, the power system is structured into three sub-systems: transmission, subtransmission and distribution, which have different roles. Similarly, the way in which the different components (generators, transformers, loads) are connected depends on a set of technical and economic constraints that vary from one country to another. For generation, transmission and subtransmission, the system used is the three-phase system without a distributed neutral. It is indeed easy to ensure the balance between phases at the generation level. At the transmission and subtransmission system level, since the load and the number of customers is significant, a phenomenon of diversity (the consumption fluctuations of each customer are statistically attenuated) allows for a balanced system, that is, the vectorial sum of the currents and voltages is zero. Installing a neutral conductor would therefore serve no purpose other than to incur unnecessary additional costs.

Regarding MV distribution systems, all countries have a principal three-phase backbone and the choice of whether or not to distribute the neutral as well as the number of distributed phases will strongly depend on the load density (urban or rural area) and geographical constraints (e.g. mountains, deserts, natural parks). As shown in Figure 1.8, for European and European-influenced countries, the neutral is not distributed in MV, unlike North American and North American-influenced countries. Branches then start from this three-phase backbone. They can be:

– Case 1: three-phase as in France;

– Case 2: two-phase or three-phase system as in the UK;

– Case 3: single-phase with single wire earth return (SWER), in some very isolated areas in Australia, Canada, the US, Brazil and Africa;

– Case 4: single-phase + neutral, two-phase + neutral or three-phase + neutral in North American and North American-influenced countries.

The SWER system is reserved for cases where the length of the lines is very important to the supply of a few very dispersed consumers (large areas with very few inhabitants). Moreover, the earthing resistivity must be quite homogeneous and very low. This solution is not generalizable for areas where earthing resistivity varies according to the seasons because it would be necessary to create an underground "copper plate" to overcome this problem, which would increase costs considerably. Moreover, it is a dangerous solution for people because of the step voltage that may exist as the earthing potential may not be homogeneous everywhere.

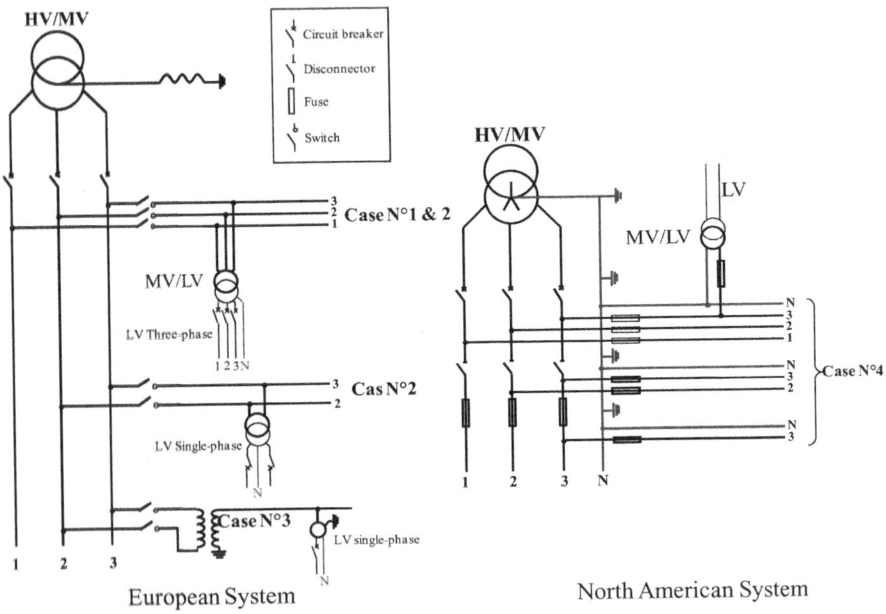

Figure 1.8. *The two distribution systems in the world. For a color version of this figure, see www.iste.co.uk/alvarezherault/distribution.zip*

The majority of LV residential and commercial customers are single-phase systems, so they need to be connected to the power system via two conductors only: a forward and a return conductor. The forward conductor is one of the phases of the three-phase system and the return conductor is usually the neutral. For European-type systems, the neutral is created in LV because of the MV/LV transformer. The diversity effect does not intervene anymore in LV and this part of the system is more or less unbalanced according to the load density and their distribution on each phase. In general, distribution system operators try to distribute single-phase loads evenly over the three phases of the system in order to minimize system imbalance. However, if there are too few consumers with different profiles, phase balancing is impossible. Industrial consumers are rarely supplied with single-phase power because fixed or variable speed induction motors represent about 80% of the load (pumps, fans, etc.), and starting a single-phase induction motor is more complex because it requires a specific starting device in the absence of a rotating magnetic field.

Table 1.4 summarizes the rural service solutions used around the world based on the expected load density on the branch.

Expected load on the branch	North American system	European system
Strong	3 phases + neutral	3 phases
Average	2 phases + neutral	3 or 2 phases
Weak	1 phase + neutral	2 phases
Very weak	1 phase (earth return)	1 phase (earth return)

Table 1.4. *The different modes of distribution in rural areas (Persoz 1984)*

1.3.3. *Electrotechnical imperfections of power systems*

The choice of the number of voltage levels and their value is essential because it is linked to standardization, the sizing of the equipment (to limit investment and operating costs) and the protection of the system. The impact of the voltage level on equipment sizing is directly linked to the physical relationships between the different quantities: the maximum admissible current in the cables, the skin effect, the losses through the Joule effect and the voltage drops and rises.

1.3.3.1. *Maximum admissible current in conductors*

The circulation of a current in a conductor creates heat. If this heat becomes too high, there is a risk that the insulation and/or the conductor will deteriorate and result in fire. Let S be the apparent power of a single-phase load supplied by a conductor. The relationship between the current flowing in the conductor, the apparent power S and the supply voltage is given by equation [1.15]. Therefore, for a given conductor (material, cross-section, insulation, overhead or underground technology, type of installation), at constant apparent power S, the intensity of the maximum current flowing in the conductor is inversely proportional to the voltage level. For very large loads to be supplied (e.g. transmission systems), the voltage level will be increased in order to reduce current stress in the conductors.

$$I = \frac{S}{\sqrt{3} \times U} \quad [1.15]$$

1.3.3.2. *The skin effect*

This is a non-uniform distribution of current that occurs in AC only. The higher the frequency, the more the current tends to spread over the surface of the conductor. As a result, the "useful" cross-section is smaller than the actual cross-section of the conductor. As the resistance is inversely proportional to the cross-section, the resistance in AC is higher than the resistance of a conductor in DC. To overcome this problem, different technological solutions exist such as the use of several smaller cross-section conductors, called strands, instead of a single large cross-section conductor.

1.3.3.3. *Losses through the Joule effect*

Under a given electrical voltage, an electrical material tends to oppose the current flowing through it. This tendency is modeled by the electrical resistance of the material and is due to the flow of charges and their interaction with the crystal network. Equation [1.16] gives the expression for the linear resistance of conductors for a given temperature θ.

$$r_{20} = \frac{\rho_{resistivity}}{Section} \times 10^3$$
$$r = r_\theta = r_{20} \times \left[1 + \alpha_{temperature}(\theta - \theta_{20})\right] \qquad [1.16]$$

with:

– r_{20}: the linear resistance of the conductors for a rated temperature of 20°C (in Ω/km);

– $\rho_{resistivity}$: the resistivity of material at 20°C (in Ω·m);

– section: the useful cross-section of the conductor (mm²);

– $\alpha_{temperature}$: the temperature coefficient α (°C^{-1});

– θ_{20}: the reference temperature (20°C).

The circulation of charges leads to ion oscillations and heating. This heating is responsible for the Joule effect losses, which are equal to the product of the resistance and the square of the current flowing in the conductor, and expressed by equations [1.17] and [1.18].

$$P_{losses} = R \times I^2 \qquad [1.17]$$

$$P_{losses} = R \times \left(\frac{S}{\sqrt{3} \times U}\right)^2 = \frac{R \times S^2}{3 \times U^2} \qquad [1.18]$$

with:

– P_{losses}: the joule losses in the conductor (in W);

– R: the conductor resistance (in Ω);

– I: the current flowing in the conductor (in A).

In a three-phase system, the total joule losses are equal to the sum of the joule losses in each conductor. If the same apparent power S flows through each of the three phase conductors, the joule losses $P_{losses,3\varphi}$ for a balanced three-phase system are defined in the following equation:

$$P_{losses,3\varphi} = 3 \times R \times I^2 = 3 \times R \times \left(\frac{S}{\sqrt{3} \times U}\right)^2 = \frac{R \times S^2}{U^2} \qquad [1.19]$$

The joule losses depend on the type of conductor, the length of the line and the current level, which in turn depends on the power demand of the load and the voltage level. Their impact is mainly economic. In a 2010 CRE report on loss coverage mechanisms (CRE 2010a), the annual charge due to all losses on the power system was estimated at 2.1 billion euros, that is, more than 13% of the charges covered by the TURPE (*Tarifs d'utilisation des réseaux publics d'électricité*: Tariffs for the use of public electricity systems). This represents 33 TWh of losses per year, of which 11% occur on the transmission systems and 22% on the distribution systems. Among the technical losses, 20% are due to the corona effect (electrical discharge caused by the ionization of the area surrounding a conductor) and 80% to the Joule effect. On the distribution system, about one-third of losses are of non-technical origin: fraud, errors, technological limits of metering, differences between the termination and commissioning indexes in cases where there has been a change of occupant of a premises. Finally, losses due to the Joule effect represent an annual cost of 1.2 billion euros, that is, 7.5% of TURPE.

1.3.3.4. *Voltage drops and rises*

An electrical conductor can classically be represented by an "Π diagram", consisting of the elements R, L and C. The resistance R represents the resistance of the conductor, the inductance L is related to the fact that a line is also a current loop, and the capacities C represent the capacitive couplings between the conductors and between the conductors and the earth. Neglecting the capacitive elements, we observe a voltage drop across this dipole as expressed by equation [1.20].

$$\Delta V(t) = R \times I(t) + L \times \frac{dI(t)}{dt} 1 \qquad [1.20]$$

Or in complex notation:

$$\underline{\Delta V} = R \times \underline{I} + j \times X \times \underline{I}$$
$$X = L \times \omega \qquad [1.21]$$

X is called reactance and is expressed in Ω.

Therefore, depending on the nature of the customers connected to the system (consumers or producers), we can observe at the different nodes a voltage different from the one imposed by the voltage source(s). An overvoltage will cause the wear and tear of electronic circuits for household appliances that are connected to the mains to happen more quickly. For example, the life of an incandescent lamp is proportional to V^{-16} (Center 2016), which can be reduced by almost 80% when subjected to a 10% increase in voltage. At undervoltage, some devices may not operate below a certain threshold. In addition, a voltage level that is too low will also increase the level of joule losses as shown in equation [1.18].

1.3.3.5. *Action levers to reduce the effects of electrotechnical imperfections*

To calculate the joule losses and voltage drops for a balanced sinusoidal three-phase electrical system, we can reduce it to an equivalent single-phase system, applicable in linear regime and for low frequencies (50 or 60 Hz in our case). The two systems, three-phase and single-phase equivalent, are shown in Figure 1.9 with the following notations:

- I_{1am}, I_{2am} and I_{3am}: the currents upstream of phases 1, 2 and 3 of the line;
- I_{1av}, I_{2av} and I_{3av}: the currents downstream of phases 1, 2 and 3 of the line;
- I_{1line}, I_{2line} and I_{3line}: the currents flowing in phases 1, 2 and 3 of the line;
- V_{1am}: the potential of phase 1 upstream of the line;
- V_{1av}: the potential of phase 1 downstream of the line;
- V_{nam}: the potential of the neutral upstream of the line;
- V_{nav}: the potential of the neutral downstream of the line.

A line can therefore be modeled by its length L (km) and the physical characteristics of the conductor: linear resistance r (Ω/km), linear reactance x (Ω/km) and linear capacitance c (F/km). In general, for medium and low voltage system calculations, the capacitive component of the line for the calculation of voltage drops is negligible. Indeed, the linear capacitance of a line varies from about 0.005 μF/km for overhead lines to about 0.5 μF/km for underground lines. With power transits of the order of MVA and voltage levels of the order of 10 kV, the currents transiting in medium voltage lines are of the order of 100 A. For 1 km of overhead line, the currents flowing in the C-capacities are therefore of the order of C × ω × V, or 0.0157 to 1.57 A, which is negligible. However, in the case of a long, lightly loaded conductor, the capacitive currents may no longer be negligible and may even cause a voltage rise at the end of the line (the Ferranti effect). In the case of low-voltage lines, since resistance is generally at least 10 times greater than reactance and 100 times greater than capacitance, the latter are commonly neglected (Nexans 2018).

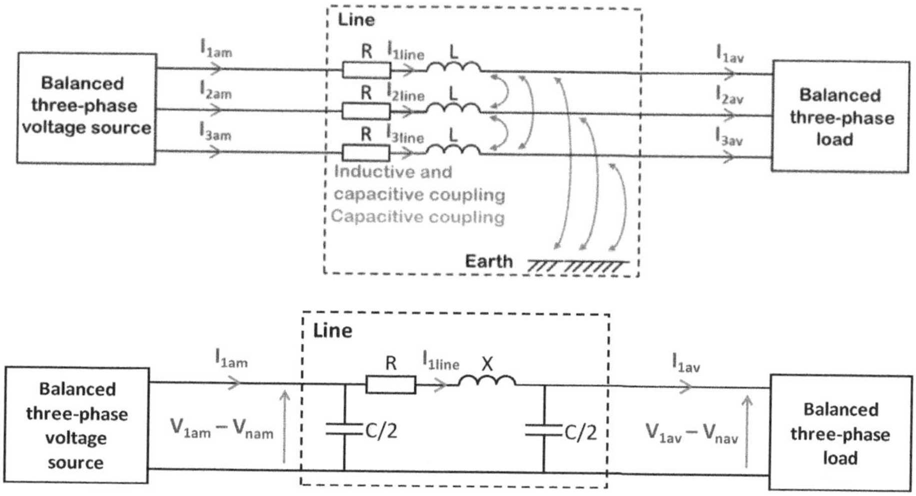

Figure 1.9. *Three-phase power line model and single-phase equivalent. For a color version of this figure, see www.iste.co.uk/alvarezherault/distribution.zip*

If the load to which the line is connected consumes (or produces) a total apparent power $\underline{S} = P + j.Q$ equal on each of the three phases, we can define the power factor noted cos(φ). The active power consumed is equal to $P = S \times \cos\varphi$ and the reactive power consumed is equal to $Q = S \times \sin\varphi$. The vector diagram associated with the equivalent single-phase diagram (with neglected capacitances) is the Kapp triangle shown in Figure 1.10. It allows us to introduce a new phase shift θ, which is called the transmission angle. To simplify the notation, we posit $\underline{V_{am}} = \underline{V_{1am}} - \underline{V_{nam}}$ and $\underline{V_{av}} = \underline{V_{1av}} - \underline{V_{nav}}$. $\underline{V_{am}}$ and $\underline{V_{av}}$ are the single-phase system voltages. In the balanced state, $V_{nam} = V_{nav} = 0$ and $V_{am} = V_{1am}$, $V_{av} = V_{1av}$. Similarly, with capacitances neglected, we have $I_{1line} = I_{1am} = I_{1av}$.

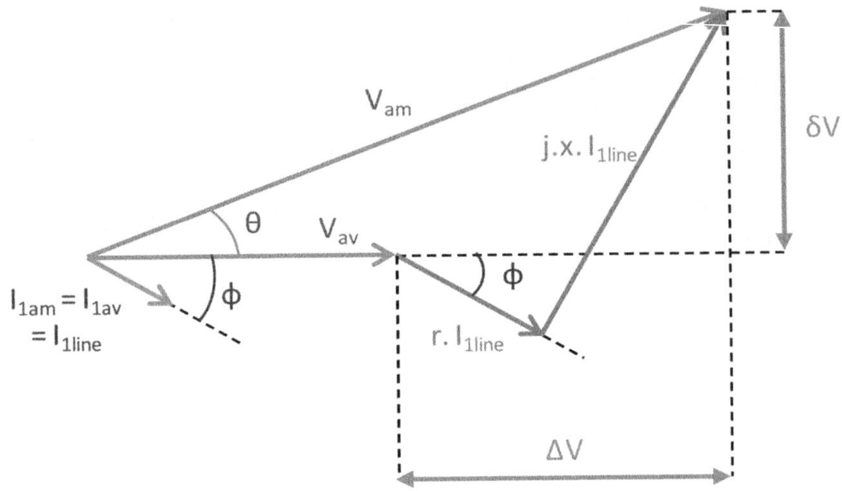

Figure 1.10. *Kapp's triangle. For a color version of this figure, see www.iste.co.uk/alvarezherault/distribution.zip*

The voltage difference across the line is calculated in equation [1.22] through equation [1.24].

$$\underline{V_{am}} - \underline{V_{av}} = (R + j.X) \times \underline{I_{1line}} \quad [1.22]$$

$$\underline{V_{am}} - \underline{V_{av}} = (R + j.X) \times \underline{I_{1av}} \quad [1.23]$$

$$\underline{V_{am}} - \underline{V_{av}} = (R + j.X) \times \frac{\underline{S}^*}{3 \times \underline{V_{av}}} \quad [1.24]$$

If $\underline{V_{av}}$ is taken as the reference of the voltages in Figure 1.10, then we find that $\underline{V_{av}} = V_{av}$ (see equations [1.25] through [1.27]).

$$\underline{V_{am}} - \underline{V_{av}} = (R + j.X) \times (\cos(\varphi) - j.\sin(\varphi)) \times \frac{S}{3 \times V_{av}} \quad [1.25]$$

$$\underline{V_{am}} - \underline{V_{am}} = [R.\cos(\varphi) + X.\sin(\varphi) + j.(X.\cos(\varphi) - R.\sin(\varphi))] \times \frac{S}{3 \times V_{av}} \quad [1.26]$$

$$\underline{V_{am}} - \underline{V_{av}} = \Delta V + j.\delta V \quad [1.27]$$

The real part of a voltage difference across a power line is called the voltage drop ΔV, while the imaginary part is called the voltage gap δV. In general, the imaginary part of the voltage drop is much smaller than the actual part, which we can verify in Figures 1.11 and 1.12.

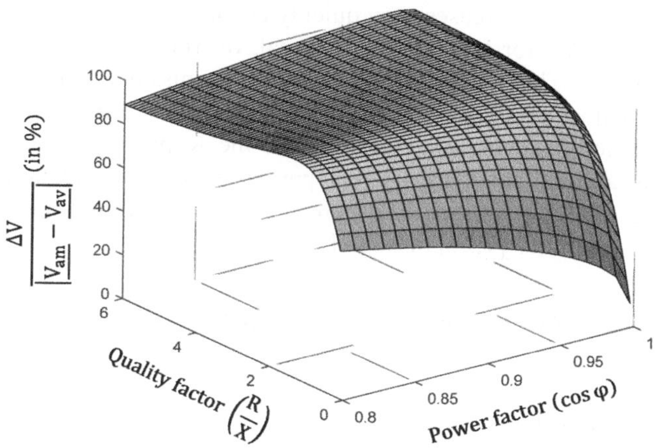

Figure 1.11. $\frac{\Delta V}{|V_{am}-V_{av}|}$ *as a function of quality and power factors. For a color version of this figure, see www.iste.co.uk/alvarezherault/distribution.zip*

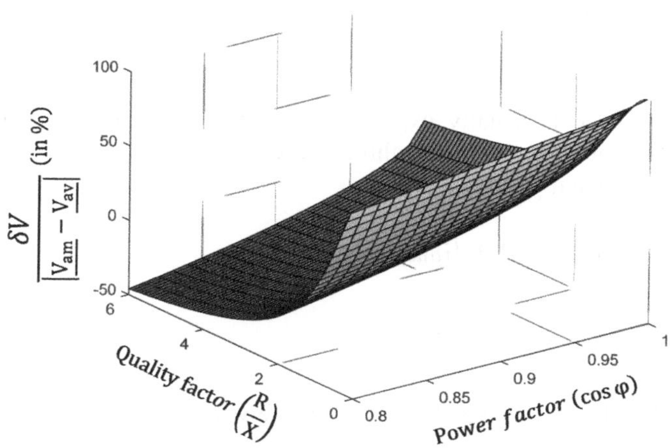

Figure 1.12. $\frac{\delta V}{|V_{am}-V_{av}|}$ *as a function of quality and power factors. For a color version of this figure, see www.iste.co.uk/alvarezherault/distribution.zip*

Two areas of operation stand out. When the quality factor is greater than 1, the assumption to study only ΔV holds since it always remains greater than 90% of the modulus of $\underline{V_{am}} - \underline{V_{av}}$. For a quality factor R/X equal to 3 and a power factor of 0.92 (common values in distribution), ΔV is 99.67% of $\left|\underline{V_{am}} - \underline{V_{av}}\right|$. When the quality factor is less than 1, ΔV decreases very quickly as the power factor increases, so δV will be larger than ΔV for large cross-sectional, overhead lines with low reactive transmission power. This is the case for the transmission system and controls implemented to adjust the transmission angle θ, via phase shifting systems, to reduce δV. For distribution systems, we therefore use the Kapp assumption, that is, the voltage drop is approximated by its real part as shown in equation [1.18]. The voltage drop is usually expressed as a percentage of the desired operating voltage (here V_2), hence equations [1.29] through [1.32].

$$\left|\underline{V_{am}} - \underline{V_{av}}\right| = \Delta V = [R.\cos(\varphi) + X.\sin(\varphi)] \times \frac{S}{3 \times V_{av}} \quad [1.28]$$

$$\frac{\Delta V}{V_{av}} = [R.\cos(\varphi) + X.\sin(\varphi)] \times \frac{S}{3 \times {V_{av}}^2} \quad [1.29]$$

$$S = \mathrm{P}.\cos(\varphi) = \mathrm{Q}.\sin(\varphi) \text{ and } {U_{av}}^2 = 3 \times {V_{av}}^2 \quad [1.30]$$

$$\frac{\Delta V}{V_{av}} = \frac{R.P + X.\mathrm{Q}}{{U_{av}}^2} \quad [1.31]$$

$$\frac{\Delta V}{V_{av}} = L \times \frac{r.P + x.\mathrm{Q}}{{U_{av}}^2} \quad [1.32]$$

The voltage drop is usually expressed as a percentage of the voltage at the reference node, that is, V_{av}. Along a three-phase line, the joule losses are expressed by equations [1.33] and [1.34].

$$P_{Joule,3\varphi} = 3 \times L \times r \times I_{1line}^2 \quad [1.33]$$

$$P_{Joule,3\varphi} = L \times r \times \frac{P^2 + \mathrm{Q}^2}{{U_{av}}^2} \quad [1.34]$$

We note that there are several ways to reduce voltage drops, rises and joule loss effects:

– the length of the line can be reduced by optimizing the route it takes to supply customers;

– the conductor can be chosen for its low resistance value, either by a lower resistivity, using copper over aluminum for example, or by increasing its cross-section;

– the last solution to reduce voltage drops is to increase the value of the rated voltage. This is possible because of the transformers that are distributed across the power system: the higher the power demand and the longer the length of line needed for routing, the higher the voltage level will be in compensation. However, a high voltage level requires more protective devices and increases the cost of the lines considerably. In addition, the size of the structures increases with the voltage level, which leads to space constraints that quickly become limited in urban areas.

1.4. Distribution systems

The medium voltage distribution system is made up of HV/MV primary substations that receive energy from the transmission system and transmit it to the various loads on the distribution system. The energy is transferred by MV lines whose architecture and conductor type vary according to the type of system (urban, semi-urban or rural), the country and the load density. The energy is then transmitted to MV/LV distribution substations (also called secondary substations) and to medium voltage consumers. The different architectures and their operating modes (normal operation, backup operations) are discussed in detail in Chapter 2.

1.4.1. *HV/MV primary substations*

A primary substation consists of one or more HV feeders (63 kV, 90 kV or 225 kV for high load density areas in France and 60 kV, 115 kV, 230 kV and 500 kV on the PG&E system) called antennas, which feed one or more HV/MV transformers via an HV busbar. These transformers are then connected to one or more MV busbars, from which MV feeders start, that is, the power lines being supplied with medium voltage that will supply important MV consumers and MV/LV transformers. Protective equipment is also placed at various locations to protect the primary substation against short circuits.

In France, depending on the load density of the area in which the primary substation is located and the desired level of continuity (the notion of "guaranteed power" substation), the three architectures shown in Figure 1.13 are encountered: single antenna, double antenna single busbar and double antenna double busbar (Carrive 1991).

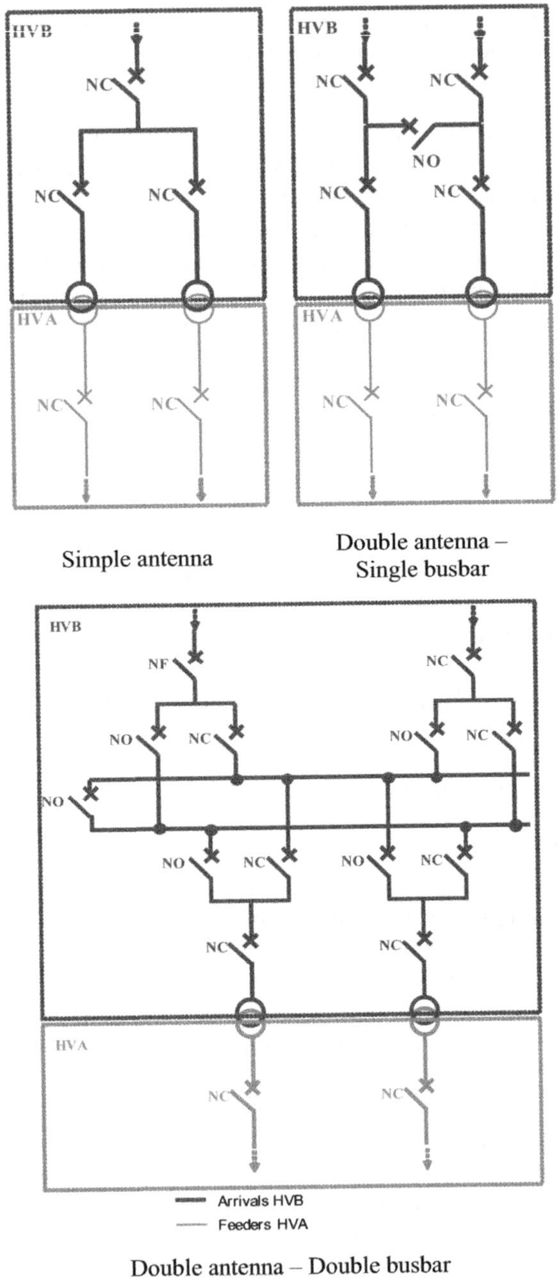

Figure 1.13. *Structure of source stations in France. For a color version of this figure, see www.iste.co.uk/alvarezherault/distribution.zip*

1.4.1.1. *Concept of "guaranteed power"*

In the case of a double antenna substation, in the event of the loss of an element of the substation, such as a fault on the HV or MV side, or transformer, there will be no interruption of power supply to downstream consumers because of the redundancies present in the primary substation. This is a guaranteed power substation. However, this only works if the transformers of the primary substation are sized to carry the entire load of the other transformers of the substation, which is not always economically justifiable. Otherwise, we speak of a substation with partially guaranteed power. In the case of a single antenna substation, there is no continuity of supply in the event of an HV fault, so it is a non-guaranteed power substation. The notion of guaranteed power is gradual: a guarantee "n-1" means that the power is guaranteed in the event of the loss of one power system element, while a guarantee "n-2" means the power is guaranteed in the case where two power system elements are lost. The maximum power of the transformers in the "n-1" regime is determined by the number of transformers and busbars, and their configuration.

1.4.1.2. *HV/MV transformers and their on-load tap changers*

The number of transformers in the primary substations depends on the load density of the area to be supplied. In rural areas (with low load density), we can find between one and two 20 MVA or 36 MVA transformers. In urban areas, the antennas are usually 63 or 90 kV and we can find up to three 36 MVA transformers. In the case of very dense urban areas, the antennas are 225 kV and there can be up to three 70 MVA transformers (Coiffier 2013). These transformers are equipped with on-load tap changers (OLTCs) which set the secondary voltage independently of the primary voltage within a certain operating range around its rated value (usually between –8% and +22%; Berseneff (2010)). The rated transformation ratio of a transformer is defined by equation [1.35].

$$r_{\text{transformation}} = \frac{V_S}{V_P} = \frac{N_{Secondary}}{N_{Primary}} \quad [1.35]$$

with:

– $r_{\text{transformation}}$: the transformation ratio;

– V_S: the rated single voltage at the secondary (in kV);

– V_P: the rated single voltage at the primary (in kV);

– $N_{Secondary}$: the number of turns of the secondary winding;

– $N_{Primary}$: the number of turns of the primary winding.

The operating principle of an OLTC is shown in Figure 1.14.

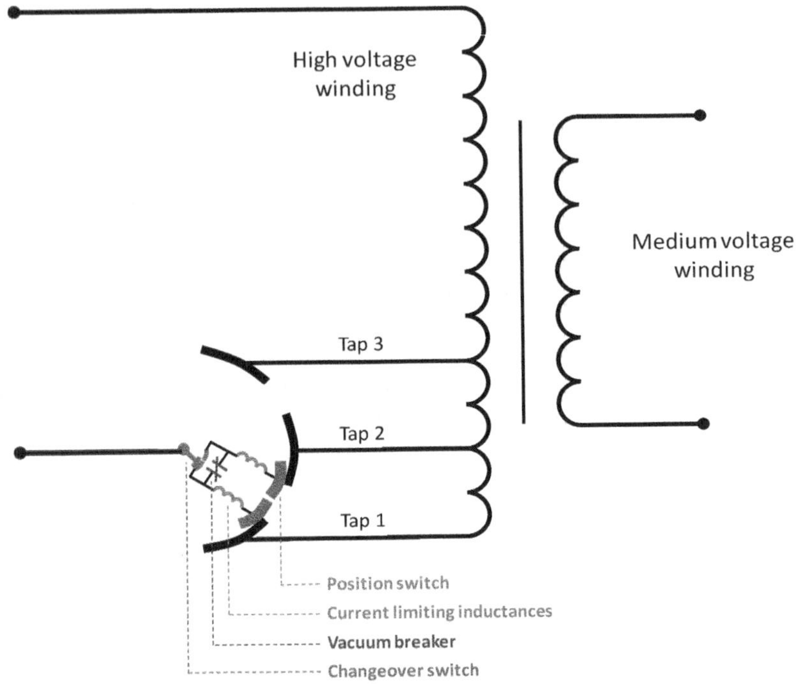

Figure 1.14. *Operating principle of the on-load controller. For a color version of this figure, see www.iste.co.uk/alvarezherault/distribution.zip*

When the secondary voltage of the transformer falls below the set voltage for about a minute, an automatic system decreases the number of taps in the transformer, allowing the secondary voltage to increase. Indeed, equation [1.35] shows that if the number of primary turns decreases then the transformation ratio increases. Therefore, at a fixed HV primary voltage, the MV secondary voltage increases. If the new voltage obtained is always lower than the set voltage, successive changes occur approximately every 10 s (Enedis 2018). The preliminary waiting time of about 1 min allows the primary and secondary controls of the transmission system to act (about 30 s) and also allows for any rapid temporary variations resulting from load connections and disconnections to be ignored (Berseneff 2010). During a tap change, for a short time two turns of the transformer are short-circuited, which generates a large current. To limit the current, a resistor is integrated. The setpoint voltage is generally set at +4% of the rated secondary voltage (Enedis 2018) unless the generation is important, in which case it is set at +2%. This is because the distribution systems downstream of the primary substations are subject to voltage drops as a result of the load. It is therefore in one's interest to "translate" the voltage plane in order to avoid undervoltage constraints in the system. The voltage that is compared to the setpoint voltage is either the voltage at the MV busbar just downstream of the transformer or the voltage at a well-chosen node in the system, which is estimated by measuring the current through the transformer. The error due to the measurement chain and the discrete operation of the OLTC is estimated at 1% (Enedis 2018).

1.4.1.3. *MV feeders*

Each transformer supplies between 1 and 50 feeders. A feeder is a line starting from a primary substation and supplying MV consumers or MV/LV distribution substations, which we describe next. There are two types of feeders: direct feeders (only one consumer is connected) and shared feeders (several consumers are connected), also called mixed feeders. Depending on the load density, these feeders generally join other primary substations; however, their operation is carried out in a radial manner because of the presence of a normally open switching device positioned at a well-chosen point. Therefore, in the event of a fault on the MV feeder, it is possible to isolate the fault and to supply the consumers via the other primary substation. Chapter 2 explains how this mechanism works in detail.

1.4.1.4. *Busbars and connected devices*

The HV busbar allows a transformer to be fed by a different antenna in the event of a fault or intervention on the antenna feeding it, and the MV busbar allows a set of feeders to be fed by another transformer in the event of a fault or intervention on the transformer that normally feeds them. Shunt circuit breakers on the MV busbars allow a faulty phase to be earthed in the event of an incident on the distribution system in order to eliminate most single-phase faults without any disruption to users. Finally, capacitor banks can be connected to the MV busbar in order to compensate for the reactive energy, to reduce Joule effect losses, and adjust the voltage.

1.4.2. *MV/LV distribution substations*

MV feeders supply MV consumers and LV consumers via MV/LV distribution substations. Table 1.5 lists the different types of distribution substations that can be found in France. The characteristics of rural substations are detailed in the NF C 11-201 standard while urban substations are detailed in a distributor's circular (Valentin et al. 1996).

1.4.2.1. *Rural distribution substations*

Rural distribution substations are supplied by a single MV feeder feeding an MV/LV transformer. Downstream of the transformer, we can find between one and four LV feeders depending on the transformer power rating. We distinguish two categories of rural distribution substations. The first one is a pole-mounted distribution substation, also called "H61" in France. This type of substation is simple and inexpensive. It is installed on a pole and can therefore be connected to the MV distribution system via a simple overhead branch. A circuit breaker placed just downstream of the transformer, on the LV side, protects the system against faults as well as protecting the transformer against overcurrent (Carrive 1991). The second type of rural substation is the PRCS (*Préfabriqués ruraux compacts simplifiés*: Simplified compact rural substations) and the PSS A (*Préfabriqués au sol simplifié de type A*: Ground-mounted substation type A), used when it is not possible to connect the substation exclusively overhead due to topological constraints (e.g. the presence of buildings) or for aesthetic reasons (Carrive 1991). This type of substation is 1.5 m high and has a footprint of 3 m² so as not to mar the landscape. The LV feeders are equipped with high breaking capacity fuses to protect against polyphase faults. Single-phase faults are eliminated because of the protection of MV systems. We also find the PSS in peri-urban areas (PSS B – *Préfabriqués au sol simplifié de type B*: Ground-mounted substation type B) to which it is possible to add MV side breaking devices. It is connected as a backbone disconnect. Figure 1.15 shows the architecture of the different types of rural distribution substations.

<table>
<tr><th>Types of distribution substations</th><th>Power rating</th><th>Zone and connection</th><th>Overhead or underground</th><th>Maximum number of feeders</th></tr>
<tr><td>Pole-mounted stations (or "H61")</td><td>50, 100 and 160 kVA</td><td rowspan="3">Rural, single branch</td><td>Overhead</td><td rowspan="2">2</td></tr>
<tr><td>PRCS: Simplified compact rural substations</td><td>50 and 160 kVA</td><td rowspan="3">Overhead or underground</td></tr>
<tr><td>PSS A: Ground-mounted substation type A</td><td>100, 160 and 250 kVA</td><td rowspan="2">4</td></tr>
<tr><td>PSS B: Ground-mounted substation type B</td><td>100, 160 and 250 kVA</td><td>Peri-urban, backbone cut-off</td></tr>
<tr><td>PAC – Préfabriqués à couloir de manœuvre: Walk-in compact substation</td><td>400, 630 to 1,000 kVA</td><td rowspan="3">Suburban and urban, secured feeder</td><td rowspan="3">Underground</td><td rowspan="3">8</td></tr>
<tr><td>PUIE – Postes urbains intégrés à son environnement: Low visual impact urban compact substation</td><td>400 and 630 kVA</td></tr>
<tr><td>Substations in buildings</td><td>400, 630 to 1,000 kVA</td></tr>
</table>

Table 1.5. *Different types of distribution substations (sources: Enedis (2018); Groupe Cahors (2018); ORMAZABAL (2018))*

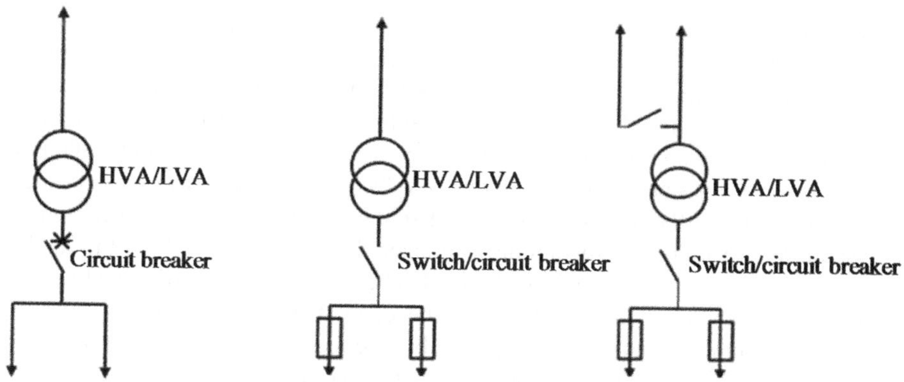

Architecture of H61 substations | Architecture of H61, PRCS and PSS A substations | Architectures of PSS B substations

Figure 1.15. *Rural distribution substations*

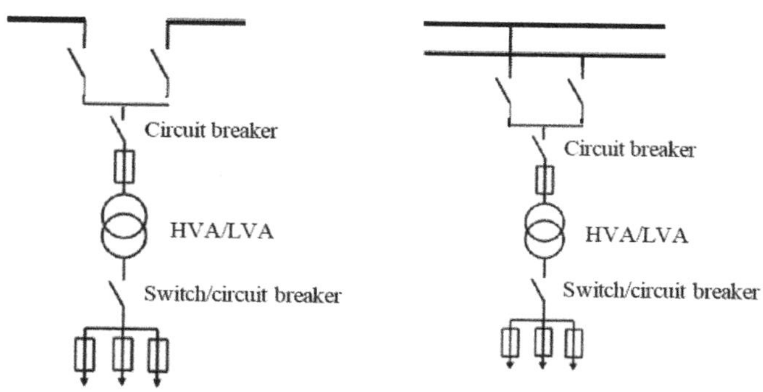

Secured feeder | Double feeder

Figure 1.16. *Urban distribution substations (PAC, PUIE, and in buildings)*

1.4.2.2. *Urban distribution substations*

In urban areas, we find PAC (*Préfabriqués à couloir de manœuvre*: Walk-in compact substation), PUIE (*Postes urbains intégrés à son environnement*: Low visual impact urban compact substation) and building-type substations (rated capacity from 400 to 1,000 kVA). The MV feeder that supplies the substation transformer is equipped with fuse switches to protect the transformer (Valentin et al. 1996). Downstream of the transformer, we can find between one and eight LV feeders that are protected by fuses. In France, these substations are usually connected with a secured feeder, and more rarely, a double feeder, as shown in Figure 1.16. In the secured feeder connection, an MV feeder starts from a primary substation and successively passes through the distribution substations before reaching the same primary substation or another primary substation. The two switches through which the distribution substations are connected to this MV feeder are all normally closed except for one, which allows the MV loop to be opened, thereby ensuring radial operation for the system. We examine this operation in greater detail in Chapter 2. In the double feeder connection, each MV feeder is doubled. The substations are connected to both feeders via two switches, one closed and one open. Under normal operations, the distribution substations are therefore supplied by one feeder, but in the event of a fault on this feeder, operations are then transferred to the other feeder.

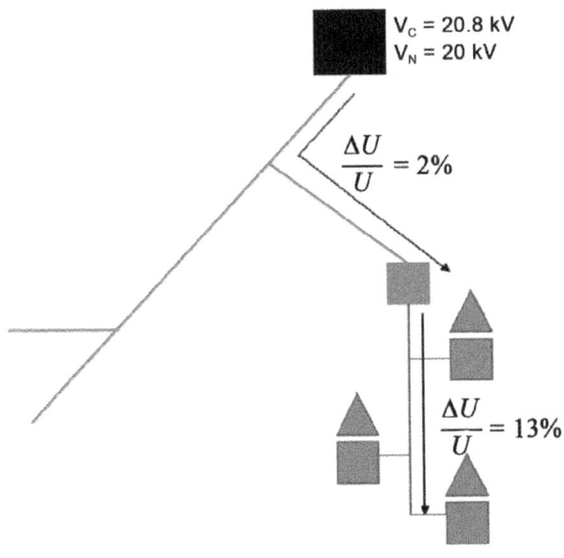

Figure 1.17. *Example setting of the off-load tap changer. For a color version of this figure, see www.iste.co.uk/alvarezherault/distribution.zip*

In France, urban distribution substations are equipped with off-load tap changers. The principle is the same as for the OLTC, that is, by increasing the number of turns at the primary, we decrease the voltage of the transformer's secondary. However, the tap change is done manually and off-load, which means consumers are temporarily cut off. The transformers are equipped with three taps (0%, +2.5% and +5%) or five taps (we add the –5% and the –2.5%) (Enedis 2018). The taps are chosen during the sizing phase and are not generally changed. The advantage is that future changes in load and/or generation may require a change of tap in the future, which is more economical than a reinforcement. In Enedis (2018), the tap choice is given by the following criterion: if $\Delta U_{tot} < \Delta U_{cons} + 10\%$, then the tap should be set to +2.5% (i.e. 410 V at the MV/LV substation's secondary) and, if not, for cases where no generation is connected to the LV system and fed by the distribution substation in question, then the tap can be set to +5% (i.e. 420 V at the MV/LV substation's secondary). ΔU_{cons} is the difference between the primary substation OLTC and the MV-rated voltage, and ΔU_{tot} is the total voltage drop between the primary substation and the furthest point from the distribution substation. Both of these parameters are expressed as a percentage of the rated voltage. In the example shown in Figure 1.17, $\Delta U_{cons} = 4\%$ and $\Delta U_{tot} = 12\%$. Therefore, $\Delta U_{tot} < \Delta U_{cons} + 10\% = 14\%$. The chosen catch will thus be 2.5%.

1.4.2.3. *MV and LV conductors*

The way in which conductors connect primary substations, distribution substations, and MV and LV consumers is what gives the distribution system its architecture. It has a significant impact on costs, system reliability and power continuity. Chapter 2 focuses in greater detail on the description of these architectures, which are not presented here. The distribution system is made up of overhead and underground lines. Although overhead lines generate more faults on the system because they are more subject to climatic hazards, they are less expensive, easier to install and faults are eliminated more quickly than those occurring along underground cables.

1.4.2.4. *Protection in distribution systems*

1.4.2.4.1. In medium voltage

A set of protection devices is deployed on the system to deal with the various incidents. They are made up of physical elements (sensors, circuit breakers, etc.) supported by a set of algorithms and automated systems forming the protection plan. It is a compromise between cost, sensitivity, selectivity, speed, robustness and simplicity. It must allow one to detect and isolate the fault in a first step, to repair it in a second step, and if necessary to switch the system from normal operation mode onto the backup mode. The objective is threefold: to protect people, protect equipment and limit service interruptions. Polyphase or single-phase faults are of four types, the first three are typically associated more with single-phase faults:

– Self-extinguishing faults (10% of faults): these disappear spontaneously in a very short time (less than 60 ms; Oddi (2011)), without triggering any protection device. An example is the straw stems that are thrown up into the air during harvesting and that then hit a power line (Oddi 2011).

– Fugitive faults (75% of faults): to eliminate these, there needs to be a very short interruption of the system supply (a few tenths of a second) or, in the case of a single-phase fault in a rural environment, a connection to earth via a shunt circuit breaker is required. We can cite as examples an animal touching the conductors of an overhead line, a tree branch touching a phase because it has not been pruned, a lightning strike.

– Semi-permanent faults (10% of faults): unlike fugitive faults, these require a longer interruption to the power supply, that is, a few tens of seconds to a few seconds. The absence of tree pruning can also cause semi-permanent faults, or a strong wind that brings two phase conductors together can cause an electric arc (Oddi 2011).

– Permanent faults (5% of faults): these cause tripping of protection devices and an intervention is necessary in order for the system to return to its normal operation mode. We can cite, for example, the breakdown of an insulator, or the strike of a shovel hitting a buried conductor due to roadworks.

Permanent or semi-permanent faults can be polyphase (short-circuit) or single-phase (earthing fault). Polyphase faults are easily detected due to their high current value, while single-phase faults are more complicated to handle because their occurrence depends on the connection of the neutral to earth. The criticality and management of single-phase faults depends on the neutral system. Indeed, the motivation to detect them is solely to minimize outage time and to increase the lifetime of the equipment, since a single-phase fault generally evolves after a certain amount of time into a three-phase fault. Eighty percent of faults are single-phase (Oddi 2011), so the decision to neutral earth HV/MV transformers is very important.

In France, the neutral is connected to earth via an impedance at the HV/MV primary substations, while in the US, it is connected directly to earth and distributed. In France, a three-phase inductance is used if the neutral is not directly accessible. MV/LV substations usually have a delta connection on the primary side, so the homopolar current in the different feeders should be zero. In reality, the homopolar current is in the order of amperes due to current leakage and imbalance. When a fault occurs, the fault current flows back through the earth to the primary substation's secondary. The aim is to limit the single-phase short-circuit current to avoid serious consequences for the equipment, but also to keep a sufficient value in order to ensure detection. A classical value of earthing impedance is 40 Ω for a limitation of the fault current at 300 A. However, with the development of underground systems, this value can be equal to 12 Ω for a limitation of 1,000 A.

The main disadvantage due to the system structure is the appearance of capacitive currents in the healthy lines feeding the fault, which can have the effect of triggering the healthy feeder protection devices. In the case of urban or mixed systems made up of a certain proportion of underground lines with significant capacitive effect, it can be prudent to connect the neutral of the transformer to the earth via compensation coils called Petersen coils. These are designed to compensate precisely for these capacitive currents. Taking into account the system topology and the type of conductors is therefore very important when considering a protection plan that strikes a compromise between fault detection and reducing nuisance tripping. Table 1.6 gives some characteristics of the different neutral systems found in the world, as well as their drawbacks.

1.4.2.4.2. In low voltage

The neutral conductor can be connected to the neutral of other system elements, namely power transformers and customer facilities. This choice is very much influenced by the urban or rural environment of the system and accessibility to the neutral. The aim is to limit the earth fault currents in order to reduce their consequences on the system, all the while ensuring there is a sufficient threshold to detect them.

In low-voltage distribution systems, at the output of the transformers, the neutral is accessible and connected to the neutral line conductor. In France, the secondary of the transformers are always in star or zigzag configuration with the output neutral. The neutral point at the transformer can be directly earthed (denoted T) or isolated from the ground (denoted I). On the customer side, the installation earths can be directly connected to an electrically independent earthing point (denoted T), or directly connected to the earthed point of the supply, either via the neutral conductor (denoted N-C) or via a conductor separated from the neutral conductor (denoted N-S) (Lacroix and Calvas 1998).

Unfortunately, there are no valid statistics to show that one scheme is better than the other. In Häusserman (1992), the most common scheme in the world is the TT scheme. We find the TN-C scheme mostly in the US, in Great Britain for new installations and a few in Germany, the TN-S in Ireland, Germany and France in the tertiary sector, and the IT scheme only in Norway because of its poor earthing abilities.

The TT scheme, in which the neutral and the users' earths are connected to the earth, has the advantage of not energizing the earths in the event of a neutral break with the system. The disadvantage is that there is no interconnection with the other earths through the neutral conductor. The TN scheme, where the neutral is connected to the earth and the grounds to the neutral, has the advantage of limiting the value of the earth resistance because of the interconnection with other subscribers via the neutral conductor. This limits overvoltage due to fault currents or imbalances. The disadvantage is that the earths are exposed to a higher or lower voltage in the event of a neutral break. In France, the TT scheme has been chosen to avoid the potential rise of earths in the event of a neutral break. This choice is imposed by regulations, in particular the decree of May 17, 2001 that laid down the technical conditions to which electrical energy distributions must conform.

Neutral system	Direct to earth	To earth via an impedance	To earth via a Petersen coil	Isolated
Intensity of single-phase fault current	Very important	Limited (300 A in rural and 1,000 A in urban areas)	Close to 0	Close to 0
Overvoltages	Moderate	Moderate	Yes, but lower than in isolated neutral	Important in transient regime
Influence of the operating plan	No influence	No influence	The coil must be readjusted	No influence
Triggering at first fault	Yes	Yes	No	No for phase-to-earth faults
Protective equipment	Fuses, reclosers	Amperometric protection with independent time. Increasing selectivity as we move away from the primary station	Permanent single-phase fault: homopolar wattmetric protection. Two- or three-phase fault: phase amperometric relay	Amperometric Earth voltage rise monitoring (European Standard EN 50522)
Countries	US, Canada	Belgium, France, the UK, Ireland, Japan, Sweden, Southern Europe	Germany, Finland, Norway, Northern Europe	Germany, Belgium, Italy, Japan, Norway
Disadvantages	Monitoring and over-insulation required, frequent triggers	Necessary shutdown upon detection of the fault	Expensive Double fault problem (5% of cases)	At the second fault, risk of two-phase fault Monitoring

Table 1.6. *The different neutral systems*

1.5. Opening of the energy markets: appearance of new players

1.5.1. *Market deregulation versus technical regulation*

The opening up of energy markets, also known as market deregulation, following the American debates of the 1970s (Bouttes and Trochet 2005), aims to improve the management of investments and the performance of electricity systems by putting an end to monopolies and introducing more competition. Historically, the management of electrical energy was monopolistic, that is to say, all activities related to the production, transmission, distribution and sale of electricity were regulated by the country. During the 1980s, countries such as the US, through the Energy Policy Act of 1992, as well as Chile and Great Britain, opted to deregulate their respective energy markets. The reasons for this evolution are numerous: a sufficiently mature system, a desire for greater competitiveness, manufacturers needing cheaper energy and significant differences in energy costs between different countries. Competition seems to offer many advantages to consumers, and governments need to privatize public companies to pay off their debts. Therefore, the term "deregulation" refers to the price of electricity, which is no longer imposed by the public authorities but which varies according to supply and demand. The break-up of monopolies implies the appearance of new players that complicate the management of the electricity system, whose primary objective is not an economic optimum, but the security of the system and continuity of service (access to electricity for all, at all times). The deregulation of the markets therefore generates more technical regulation to define the operating framework for all players in order to guarantee the proper functioning of the system.

1.5.2. *Historical players in the power system*

All of the players in the power system are constantly evolving. Some, whom we shall call historical players, see their role evolve over time through the stages of deregulation, while other players are born according to the emerging need to ensure the proper functioning of the system as a whole. In this section, we first define the historical players, who can either be grouped together (we speak of verticalization) or split up (de-verticalization) according to the market models:

– *Power plants*: these ensure the supply of energy to the power system from those producing small units to production plants generating several gigawatts. The types of energy have diversified over time: nuclear, thermal, hydraulic, wind, solar and biomass among others.

– *Transmission system operators (TSOs)*: these manage the transmission systems and are responsible for the security of the system, including the balance between production and consumption. To do this, they carry out the operations necessary for their functioning and manage the interconnections with other countries. They must provide fair access to all players and develop and plan the system from the short to the long term. In France, RTE is the sole operator of the transmission system, but there may be several TSOs in the same country, as in Germany (50 Hertz, Amprion, TenneT, TransnetBW). The US is a somewhat special example, given its division into states that function as mini-countries. Therefore, the TSO function can be performed either by electricity companies in a monopoly situation or by ISOs (independent system operators) or RTOs (regional transmission organizations). The main difference between the latter two entities is the geographical scope of action, which is greater for RTOs. The latter is discussed in greater detail in section 1.5.3.

– *Distribution system operators (DSOs)*: these are responsible for the maintenance and development of the distribution systems (for countries where the transmission and distribution sectors are separate) in order to provide access to consumers and producers all the while guaranteeing good quality and continuity of supply. In France, the distribution systems belong to the municipalities, which delegate their management to Enedis and the 150 LDCs via concession contracts. In Germany, for example, there are approximately 879 DSOs registered with the Federal Network Agency in Germany as of November 10, 2017 (Stock 2018), of which 78 are responsible for distributing 75% of the total energy consumed on the distribution system, each supplying over 100,000 distribution points. Note that 70% of German DSOs supply less than 30,000 delivery points each and about 20% between 30,000 and 100,000 delivery points. In Brazil, which is a federation of 26 states and one federal district, concessions are awarded to a total of 49 DSOs, of which the nine largest control 65.4% of the sector, according to ANEEL (Brazil's national electricity agency). Although there are more DSOs than states, the electricity systems of the states are usually operated by a single DSO, except in the most populous states (e.g. São Paulo and Rio).

– *Utilities* (Gilstrap et al. 2015): in the US production, transmission and distribution of electricity are generally (but not necessarily) bundled and carried out by electricity companies called utilities, which have been issued licenses to operate. They are charged with developing and maintaining the transmission and distribution facilities. In the US, there are more than 3,100 utilities supplying approximately 145 million consumers, which can be classified into three categories:

- private companies called investor-owned utilities (IOUs);
- public-owned utilities (POU) owned by cities or counties;
- cooperatives (Co-Op) whose shareholders are the consumers themselves.

In 2014, the U.S. Energy Information Administration listed 200 IOUs (about 72% of consumers), 2,000 POUs (about 15% of consumers) and 900 cooperatives (about 13% of consumers) (NRECA 2021). In California, there are more than 40 POUs, the largest being Los Angeles Department of Water and Power (LADWP) serving 3.9 million consumers and the smallest serving fewer than 400 consumers (CPUC 2021).

– *Retail electricity providers*: these sell electricity directly to end consumers and manage their contracts. They are either the producers themselves (such as EDF or Engie in France), utilities, or a third party that buys electricity from other producers. In 2018, 39 such electricity suppliers were listed in France (CRE 2018a). They can also provide curtailment services to consumers so that they participate in the system's production–consumption balance. Some ESCOs (Energy Service Companies) are providers of additional services such as gas and oil. Examples include Engie and Total in France and PG&E in the US.

– *Consumers*: these can have different requirements (rail systems, industry, residential, among others) with different operating constraints. They are often categorized according to these power requirements, the expected quality of supply, with some even having direct access to energy markets. This population is evolving considerably with new uses, as we will see in section 1.6.2.

– *Regulators*: an independent entity that ensures the proper functioning of energy markets (for deregulated systems), sets electricity tariffs (for regulated systems) and ensures non-discriminatory access to the power system for both consumers and producers:

- the French case: the regulator is the *Commission de régulation de l'énergie* (CRE 2021), which was created according to the laws set out on February 10, 2000 and January 3, 2003. It sets the TURPE applied by the DSOs and TSOs;

- the American case: there are two types of regulators:

– state regulator,

- State Public Service Commission or Public Utilities Commission or Corporation Commission (Gilstrap et al. 2015) for IOUs. In California, the California Public Utilities Commission (CPUC) regulates IOUs such as Pacific Gas and Electricity (PG&E), San Diego Gas and Electric (SDG&E) and Southern California Edison (SCE) supplying 75% of California's demand,

- cities and counties for the POUs,

- elected officials for Co-Ops;

– FERC (Federal Energy Regulatory Commission) within the Department of Energy, created in 1977, for interstate flow and financial transactions.

1.5.3. *Market models around the world*

Every country's energy strategy is evolving toward a deregulated energy market, or not, and can gradually change toward one model or the other. There are currently four models, ranging from the least deregulated to the most deregulated, whose distribution in the world is outlined in Table 1.7. In each market structure, the activities of production, transmission, distribution and supply can be carried out by a single entity or several.

Market models		Production	Transmission	Distribution	Supply
Model 1	Vertically integrated (10%)	Regulated			
Model 2	Single buyer (40%)	Deregulated (IPP)	Regulated		
Model 3	Wholesale market (30%)	Deregulated	Regulated		
Model 4	Wholesale and retail market (20%)	Deregulated	Regulated		Deregulated

Table 1.7. *Market models around the world*

1.5.3.1. *Model 1: vertically integrated model*

Certain countries have chosen not to deregulate their market and remain in a monopoly situation. In this case, we speak of a vertically integrated model since the entire electricity chain, from production to sale, is regulated by the State and generally owned by a single entity. This is the case for some parts of the US, such as the Southeast region shown in Figure 1.18, Hong Kong (CLP: China Light and Power) and in Macau (CEM: *Companhia de Electricidade de Macau*), and the Faroe Islands (SEV: Streymoy, Eysturoy and Vágar). However, some countries have chosen to separate the players involved in energy production and transmission, on the one hand, and distribution and sales, on the other hand. This is the case for the Tennessee Valley Authority in the Southeast region of the United States.

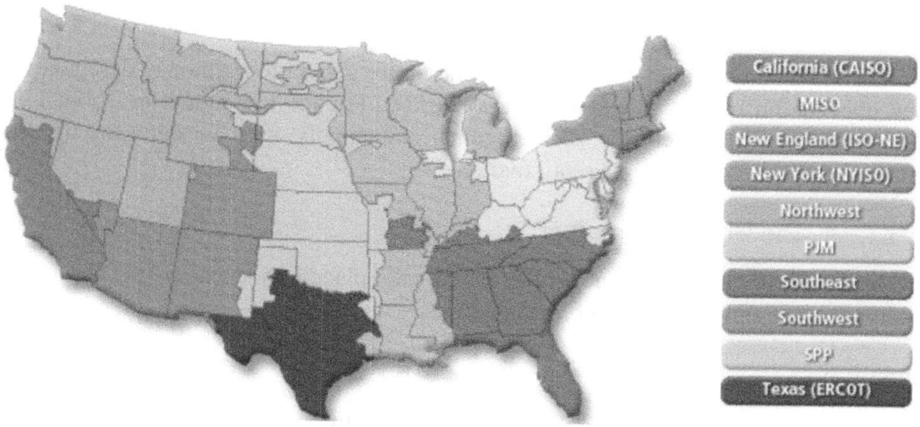

Figure 1.18. *Different market areas in the US (source: FERC (2018)). For a color version of this figure, see www.iste.co.uk/alvarezherault/distribution.zip*

1.5.3.2. *Model 2: single buyer model*

The most widespread model in the world is the single buyer model, in which private power generation companies called IPPs (independent power producers) emerge. The principle is that the government selects IPPs through calls for tender and sets up long-term power purchase contracts (Bouttes and Trochet 2005). This model allows a country, for example, to cater to a fast-growing rate of demand and also increase the share of renewable energy being fed into the grid. As with the previous model, the different areas (generation, transmission, distribution and supply) can be:

– integrated:

- in Egypt, we find the Egyptian Electricity Holding Company (EEHC) which manages generation, transmission, distribution and supply and IPPs;

– partially integrated:

- in China, generation is provided by five main producers, their subsidiaries and IPPs, and transmission-distribution-supply is provided by two entities (the State Grid Corporation of China (SGCC) and China Southern Power Grid (CSG)); it should be noted that the Chinese market model is evolving toward the wholesale and retail market model described below;

- in Thailand, there is one bulk power producer (EGAT), two distributors, Metropolitan Electricity Authority (MEA) and Provincial Electricity Authority (PEA), and IPPs.

1.5.3.3. *Models 3 and 4: wholesale and retail markets*

For the last two models in Table 1.7, the entire generation sector is deregulated and depends on market rules (wholesale market). For the last model in Table 1.7, some countries have chosen to deregulate the supply side in addition to the generation side and to set up a retail market.

1.5.3.3.1. The wholesale market

In the wholesale market, negotiations can take place either via an ISO or via an energy market and balance responsible parties.

Via an ISO

Producers submit bids to ISOs, which are entities in charge of operating the transmission system and energy markets. Depending on the demand, the ISOs will rank the producers' offers by increasing price until the cumulative power reaches the demand *(merit order*). This intersection sets the selling price of electricity, which will also be the price at which all energy producers (including those with lower bids) will be paid. Producers offering a price higher than the market price are not retained. Therefore, a notion of economic optimum appears, wherein the cheapest producers can be given priority. Large commercial and industrial consumers also have access to the wholesale market.

– *Examples of ISOs in countries that do not deregulate supply*: we can cite as examples, the American ISOs (CAISO, NYISO and ERCOT), the American RTOs (PJM, SPP) and the MISO ISO/RTO, as well as the Brazilian ISO ONS and the Ontario/Canadian ISO IESO, as shown in Figure 1.18. The generation, transmission, distribution and supply sectors can be fully integrated, as is the case with Brazil's CPFL, and in some parts of the US such as in California with PG&E, SDG&E, SCE and other smaller utilities. In other countries, the transmission, distribution and supply sectors may be separate from power generation, as is the case with Ontario, Canada, where Hydro One is the largest utility (38% of the Canadian population).

– *Examples of ISOs from countries deregulating supply*: this is the case with Australia's AEMO, New England's ISO-NE (which is also an RTO, see Figure 1.19) and Singapore's EMC.

Figure 1.19(a) shows the principle behind a wholesale market operating via an ISO. Load supply entities can be aggregators, suppliers, large consumers and others, and the supply market can be deregulated or not.

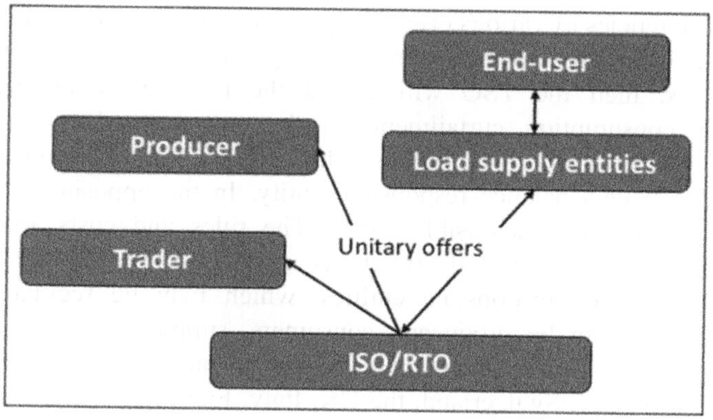

a)

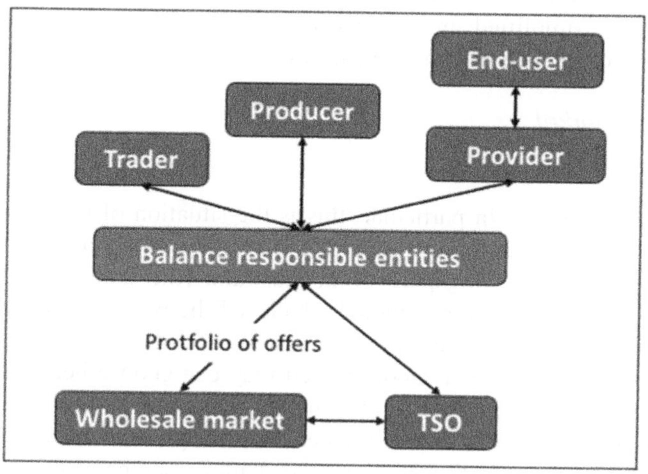

b)

Figure 1.19. *Wholesale market via: a) an ISO and b) via a market and balance responsible entities. For a color version of this figure, see www.iste.co.uk/alvarezherault/distribution.zip*

Via an energy market and balance responsible entities

Balance responsible entities carry out financial transactions on the energy market, according to the same principle as the ISOs defined in the previous section, in order to define a set of power injections and extractions for a given area (called the balance perimeter) which respects the injection/extraction balance. A posteriori, the TSOs carry out an injection/extraction balance on each balance perimeter. Due to operating contingencies of various origins (e.g. climatic), the perfect production-use-consumption balance may not be achieved for a given zone. If injections are lower than extractions, then the TSO will ensure the balance by buying electrical generation or consumption curtailment on the energy markets via balancing mechanisms. The cost of this difference that the TSO must compensate for is financed by the zone's balance responsible entity. In the opposite case, the TSO remunerates the balance responsible entity. The rules and costs applied to the differences between injections and extractions are defined in a contract between the TSOs and the balance responsible entities, which may be regulated. Balance responsible entities can be producers, consumers, suppliers or even banks. For example, there are four energy markets in Europe for buying and selling electricity: EPEX SPOT (Germany, Switzerland, the UK, Italy, France, Austria, Poland, Czech Republic, Romania, Denmark, the Netherlands, Spain, Slovenia, Belgium, Greece, Luxembourg, Norway, Slovakia), NORD-POOL (Finland, Norway, Sweden, Estonia, Great Britain, Germany), OMEL (Spain and Portugal), GME (Italy). Figure 1.19(b) shows the simplified interactions between key players in a model with an energy market and balance responsible entities.

1.5.3.4. *Retail market*

The supply sector can be:

– *completely separate*: in particular, this is the situation of the European Union (which has seen an increase in the number of electricity suppliers by 41% since 2011; CEER (2017)), in Singapore or in Australia (the latter's model is in the process of evolving toward a new model which will be presented shortly). In some countries, the producer/incumbent provider may continue to exist. If the consumer chooses to keep their historical supplier, then they can choose between a regulated tariff set by the public authorities and a market tariff proposed by the supplier. If the consumer chooses an alternative supplier, then the price of electricity will be set by the chosen supplier. Of the 28 European Union member states, 11 countries still offer regulated tariffs for residential consumers and seven countries for non-residential consumers (CEER 2017). In France, for example, since 2016, consumers with a subscribed power greater than 36 kVA (CRE 2018b) no longer have access to the regulated tariff;

– *partially separate* (alternative retail): as of 2017, in 13 US states (NREL 2017) consumers can choose an alternative supplier but are not required to do so (about 50% of large commercial and industrial consumers versus 10% of residential consumers).

Many innovations are expected in the retail market, with the coupling of electricity sales with other types of products giving rise to numerous commercial strategies. For example, supermarkets offer to accumulate the points obtained with the supply of electricity on the loyalty card.

1.5.4. *Additional players in deregulated systems*

The system of opening up the electrical energy market has also seen the appearance of other players:

– *Aggregators*: these act as intermediaries between small-scale production and consumers, on the one hand, and electricity system operators and the market, on the other hand, in order to provide flexibility in the electricity system, and can therefore only operate within a certain regulatory context. For this reason, they do not exist in all countries, for example, in Spain, Portugal, Italy and Estonia (SEDC 2017). Aggregators are also called FSPs (flexibility service providers) in Europe and CSPs (curtailement service providers) in North America. We distinguish two categories of aggregator:

- *demand side flexibility operators*: these contract for load curtailment with consumers who agree to be curtailed according to a predefined contract (a service that can also be offered by an electricity supplier). In this case, the demand side flexibility operators sell on the energy markets the block of curtailed energy within the supplier's balance perimeter, and in this way the injections then exceed the extractions. They then remunerate the suppliers for the electricity that they have not billed to consumers that has been sold by the demand-side flexibility operator. In France, RTE listed 22 demand-side flexibility operators in 2018 (RTE 2018);

- *direct sales operators*: in France and Germany, renewable energy producers now *sell* their electricity directly on the energy markets with additional remuneration and no longer go through the traditional purchase obligation mechanism at a fixed price. The objective is to encourage production when it is favorable for the system and to cut production when it is not in order to increase the penetration of renewable energy on the distribution system. Direct sales began in 2012 in Germany on an optional basis, then mandatory in 2014 for new generation units above 500 kW, and then for those above 100 kW in 2015. Therefore, in 2016, about 30% of the German generation fleet was doing direct sales (about 60 GW) consisting mainly of wind generation (about 44.7 GW) (SPOT 2017). In France, direct sales have been mandatory for all new installations with a capacity of more than 500 kW since January 1, 2016. Producers then receive the remuneration corresponding to the market price as well as a remuneration supplement corresponding to the difference between the reference tariff for each production and the average market price (SPOT 2017). In 2016, 90% of its wind generation (i.e. approximately 44.7 GW) was sold through direct sales, that is, approximately 40% of renewable generation

(SPOT 2017). Direct sales are more recent in France following the 2015 law on the energy transition for green growth.

The aggregator business is evolving, particularly because of the diversification of flexibilities services. We are talking more and more about load modulation rather than load curtailment, that is, not only load curtailment to avoid system constraints, but also increasing load consumption during production peaks for example. In addition, the possibilities offered by storage and the production curtailment multiply the possible regulatory models and the involvement of players going forward.

– *Private systems* (in Europe) or microgrid campuses (in the US): this type of system is private, that is, all the lines, equipment and production units belong to a private entity, which can entrust its management to a third party. In France, this type of system is quite rare (under exemption).

Deregulation therefore allows for an opening to competition and is thus favorable to the creation of new players and services. In a future context where small renewable energies, storage, electric vehicles and new services for optimizing system operation and investment will develop, transactions and exchanges of local electrical and non-electrical services will increase. Therefore, the electrical system will have to be more flexible and its profitability model more complex.

1.5.5. *Example of the European model*

1.5.5.1. *Key dates around the establishment of the European single market*

Figure 1.20 shows the different milestones pertaining to the opening of the energy market in Europe, detailed in this section.

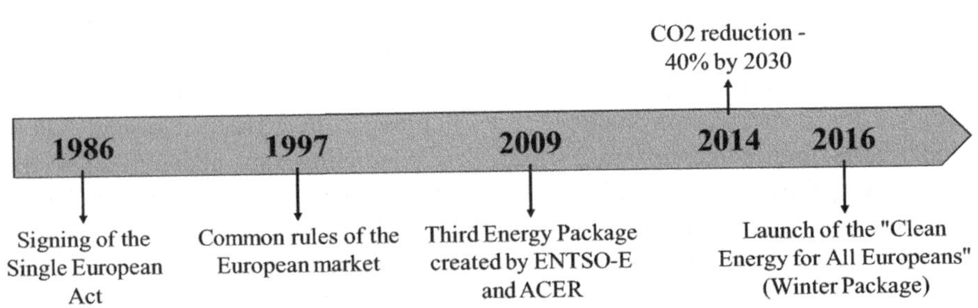

Figure 1.20. *Key dates around the creation of the European single market*

In 1986, the Single European Act was signed. It defined a 6-year program designed to remove barriers to trade within the EU in order to give rise to the "single market", also called the internal market (European Union 2018b). In 1997,

Directive 96/92/EC (European Parliament 1997) concerning common rules for the internal market in electricity was published. It sets out a timetable for the progressive opening of markets in various countries: the liberalization of production in stages, self-registration of ancillary players to access the system, separation of production/consumption and transmission/distribution activities, and gradual expansion of eligible customers, that is, those who can choose their supplier. In 2007, the European Commission issued two reports (European Commission 2007a, 2007b), noting the inadequacy of the means implemented to ensure interconnection and the objectives of an open energy market. In 2009, it published a set of laws called the Third Energy Package to put in place the regulatory framework to achieve Europe's 2020 strategic objectives (European Union 2018a) for electricity and gas. This package consists of two directives and three regulations detailed in Table 1.8.

Electricity	Directive 2009/72/EC concerning common rules for the internal market in electricity and repealing Directive 2003/54/EC
	Regulation 714/2009/EC on conditions for access to the system for cross-border exchanges in electricity and repealing Regulation 1228/2003/EC
Gas	Directive 2009/73/EC concerning common rules for the internal market in natural gas and repealing Directive 2003/55/EC
	Regulation 715/2009/EC on conditions for access to the natural gas transmission systems and repealing Regulation 1775/200/EC
Electricity and gas	Regulation 713/2009/EC of the European Parliament and of the Council of 13 July, 2009 establishing an Agency for the Cooperation of Energy Regulators

Table 1.8. *Third energy package*

In 2009, the European Commission created the European Networks of Transmission System Operators for Electricity (ENTSO-E), which today consists of 43 European transmission system operators from 36 countries (ENTSOE 2018a). ENTSO-E's scope has expanded since 2009 and covers:

– the drafting of the European power system regulations;

– long-term planning of the pan-European system (ENTSOE 2018b) for different time horizons (2025, 2030 and 2040), called the Ten-Year Network Development Plan (TYNDP);

– technical cooperation between TSOs including R&D plans, innovation activities and participation in research programs such as Horizon 2020 (FP7);

– ENTSO-E also regularly publishes its short-term generation and system balancing forecasts.

The pan-European system consists of 34 countries interconnected through 341 interconnection lines (RTE 2015) to optimize the security of the overall system, to support the creation of the single European market, and to integrate renewable energies. In 2013, the pan-European system supplied approximately 600 million consumers with 3,355 TWh of electricity (RTE 2015). RTE (2021b) estimates that the electricity produced and transiting on the pan-European system accounted for 35% of European greenhouse gas emissions. In order to ensure good coordination between the different TSOs, regional security coordinators (RSCs), in which the TSOs are shareholders, have been created to model electricity flows on the transmission systems of several countries. In Europe, there are five RSCs (Coreso, TSCNET Services, SCC, Nordic RSC, Baltic RSC) that advise the various TSOs on the actions to be taken in certain situations. The RSCs therefore do not have the role of controlling the system, but rather of providing information, thanks to the more global vision that they bring (ENSTOE 2021). In concrete terms, based on the information provided by the TSOs, up to a H-1 horizon, the RSCs will analyze the state of the system and relay these results to the TSOs, who in turn will then make the most appropriate decision. The RSCs predict the state of the system on time scales ranging from a few days to a few hours, while ENTSO-E focuses on the rules of operation and the development of the system on larger time scales (year).

At the same time, in 2009, in order to coordinate the activities of the National Regulation Authorities (NRAs) at the European Union (EU) level, and to work toward the establishment of a single European energy market for electricity and natural gas, the European Commission created the Agency for the Cooperation of Energy Regulators (ACER) through Regulation 713/2009/EC (European Parliament 2009) the Agency for the Cooperation of Energy Regulators, also known as ACER, of which CRE is a member. It therefore submits the code of systems to the European Commission and recommends its adoption. It supervises and can intervene in the cooperation of the transmission system operators represented by ENTSO-E to ensure the efficiency and transparency of the internal markets, in particular their system development plan at EU level. Finally, ACER monitors the functioning of the gas and electricity markets in general, and the wholesale energy market in particular (ACER 2018). ACER was given additional responsibilities in 2011 under Regulation 1227/2011/EU on Regulation on Wholesale Energy Market Integrity and Transparency (REMIT), and in 2013 under Regulation 347/2013/EU on guidelines for trans-European energy infrastructure.

In 2014, in order to become a world leader in the green energy transition, the European Commission revised CO_2 reduction targets to –40% by 2030. Energy efficiency, renewable energies and consumers, are the preferred targets of these actions.

In November 2016, it proposed to work on a new regulation, dubbed the "Winter Package", consisting of nine decrees calling for clean energy for all Europeans (European Commission 2016). Currently under discussion, it includes a revision of the directives on energy efficiency (Directive 2012/27/EU), renewable energy (Directive 2009/28/EC), energy performance of buildings (Directive 2010/31/EU), the framework and governance of the European electricity market (Directive 2009/72/EC and Regulation 714/2009/EC), eco-design, circular economy and energy labels.

1.5.5.2. *European power system codes*

Since 2009, following Regulation 714/2009/EU, a vast harmonization project of the European electrical codes has been undertaken by ENTSO-E, with guidance from ACER. There are eight codes, divided into three families: connection codes, operational system management codes and market codes, as detailed in Table 1.9. The connection codes for generators and consumers are discussed in further detail below.

Subject	**Name**	**Date**	**Enacted**
Connection codes			
HVDC Interconnection	2016/1447/EU High Voltage Direct Current (HVDC)	26/08/2016	07/09/2016
Consumer Connection	2016/1388/EU Demand Connection Code (DCC)	17/08/2016	07/09/2016
Producer Connection	2016/631/EU Request for Generators (RFG)	14/04/2016	17/05/2016
Power system management operational codes			
State of emergency and rebuilding of the system	2017/2196/EU Emergency and Restoration (ER)	24/11/2017	18/12/2017
System operation	2017/1485/EU, System Operation (SYS)	02/08/2017	14/09/2017
Market codes			
Forward capacity allocation	2016/1719/EU Forward Capacity Allocation (FCA)	26/09/2016	16/10/2016

Capacity allocation and congestion management	2015/1222/EU, Capacity Allocation and Congestion Management (CACM)	24/07/2015	14/08/2015
Balancing	Electricity Balancing Guideline (EB)	23/11/2017	18/12/2017

Table 1.9. *European power system codes*

Pan-European system legislation is done at the European Union level (European Union 2010, 2018b) following the primacy principle (EUR-Lex 2018). The typical process for electrical codes is:

– drafting of a proposal for a new code by ENTSO-E (timeframe: a few years);

– public consultation on the ENTSO-E website (timeframe: a few weeks);

– revision of the text by ENTSO-E (timeframe: a few months);

– validation of the text by ACER and the proposal of the text to the European Commission (timeframe: several months);

– validation of the text by the European Commission following the comitology process (European Commission 2018) (timeframe: several months);

– publication of the text in the European Official Journal (timeframe: several months);

– implementation according to European law (timeframe: approximately 3 years);

– member States must define national values according to the principle of proportionality (EUR-Lex 2018);

– the code must be transposed into the national code.

The eight codes are at the implementation stage. To have an accurate view of a code, it is strongly recommended to consult the text on the ENTSO-E website or on the European official journal, and to monitor the publication transposed into national regulations on the website of the official journal of the country in question. On the other hand, since the codes were drafted before harmonization, it is wise to consult them to ensure that they do not impose constraints that differ from those of international, European and national standards.

In France, in order to implement these regulations, consultation bodies have been set up for each regulation. For example, in the case of the Generator Code (RFG in Table 1.8), they are co-chaired by the public electricity transmission system operator (RTE) and the Association of Electricity Distributors in France (ADEeF), while the

CRE and France's energy and climate directorate (DGEC) play a consensus-building role.

1.5.5.2.1. Connection code for power generators

The 2016/631/EU code (see Table 1.9), which came into force on May 17, 2016, defines the rules for connecting all generators above 800 W to the European electricity system, regardless of the nature of the source (thermal, renewable), except for those classified as emerging technologies. The code classifies generators into four categories: A, B, C and D according to their observability, controllability and expected contribution to ancillary services. The transition from one category to another will depend on the voltage level, the peak active power of the generation and the location in one of the five European synchronous zones: Continental Europe, Northern Europe, Great Britain, Ireland and the Baltic region. It applies to all new installations, to existing installations with significant changes in installed power, or at the express request of the TSO should it require constructive capacity to ensure system security. For each synchronous area, each Member State may define its own connection values, which may be different from the ranges allowed by the code, but which must be included. An exemption procedure under conditions has also been defined. For France, for example, the list of installations under derogation known as emerging technologies is published on the website of the CRE, which is responsible for setting the conditions for derogation (CRE 2018c). This code will have to be implemented in spring 2019 throughout Europe.

1.5.5.2.2. Consumer connection code

Entering into force on September 7, 2016, the 2016/1388/EU code (see Table 1.9) concerns new facilities connected to the TSO, new DSO connections to the transmission system, and all new facilities providing demand-side flexibility services, that is, having the ability to reduce their consumption at certain times of the day, and which are independently connected to the TSO/DSO over 1 kV. The code requires functions for:

– *load shedding, frequency reloading*: disconnecting or reconnecting consumers in order to avoid frequency instability due to an unforeseen operating or climatic contingency on the system;

– *observability*: from a certain number of measurements on the system and its modeling, it is possible to reconstitute the power flows and the system voltage profile;

– *controllability*: because of control devices and embedded intelligence, we can control the system remotely from control centers called dispatch.

If there are significant changes to an existing facility then it will have to comply with the new code. As for the code for generator connections, exceptions are also defined in the code. In France, it will have to be implemented for the connection of new sites connected to the DSO, new DSO connections to the TSO, any new site offering load shedding services connected to the TSO or to the DSOs, including LDCs.

1.5.5.2.3. Impact of the new codes on the French market

This section describes the current functioning of the French market before the new European decrees transposed into French law come into force, expected late 2018 and 2019. Directive 2009/72/EC sets up the liberalization of the European electricity market, which is accompanied by the legal unbundling of vertical system players. These directives and regulations are transposed in France with the law 2000-108 and a series of decrees implementing the legislation and directives (Legifrance 1996). The legal dissociation resulted in a reorganization of EDF, which retained its producer activity, but its TSO and DSO activities were split into two new companies in 2008: *Réseau de transport d'électricité* (RTE) and *Électricité des réseaux de distribution français* (ERDF), which became Enedis in 2016. The TSO activity in France is exclusively reserved for RTE in order to ensure supply security. While the activity of DSO is ensured by both Enedis and the various LDCs still present, Enedis is divided into eight major geographical areas, covering 95% of the French territory. The national association of public utility companies and local government bodies, ANROC, reports that as of October 2012, there were 150 LDCs representing 2,800 municipalities and 1.8 million customers (ANROC 2012). LDCs serve 5% of the population and 5% of the electricity in the distribution system. At the European level, 2,000 LDCs supplying 75 million customers were listed.

1.6. Roles of consumers and producers

Historically, power systems were designed to transport electricity from large production centers to consumption areas. We can therefore assume that these two actors had separate roles and geographical locations. Figure 1.21 illustrates the future developments that are strengthening the links between these two actors, described in this section.

1.6.1. *Development of distributed energy resources based on renewable energies*

Global electricity generation, which stood at 25,082 TWh in 2016 (IEA 2018), is increasing year on year in non-OECD countries (56% of total generation) and stagnating or even decreasing depending on the type of generation for OECD countries. Table 1.10 shows the global distribution of production sources. This global distribution differs from country to country depending on its development policy but also according to its resources. For example, in 2016, the share of nuclear generation was 72.3% for France as compared with only 3.6% for China, while the share of fossil fuels was 8.6% for France and 65% for China (RTE 2016; China Energy Portal 2017). Between 2016 and 2017, the share of fossil fuel-based production declined for the fifth consecutive year, while renewable energy-based production continued to grow with good momentum (e.g. an increase of 21.9% between 2016 and 2017 for wind power).

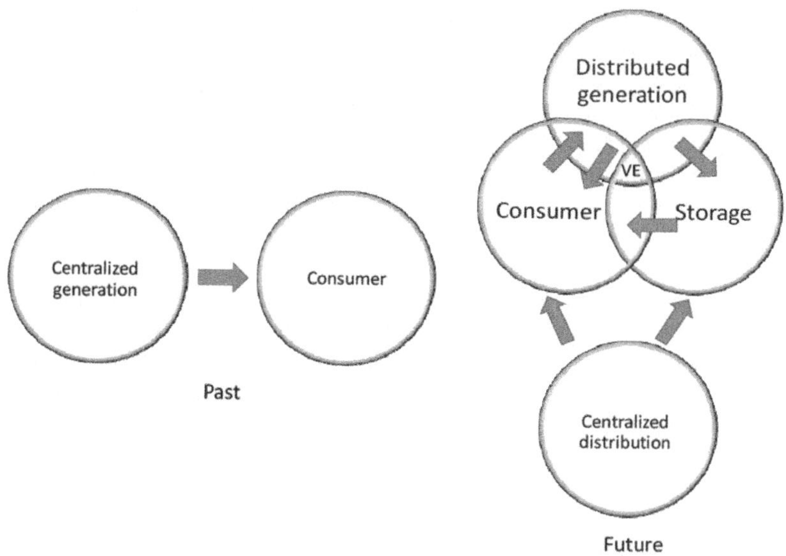

Figure 1.21. *Evolution of the definitions of production and consumption. For a color version of this figure, see www.iste.co.uk/alvarezherault/distribution.zip*

Fossil fuels (65.1%)	**Coal**	38.3%
	Fuel	3.7%
	Natural gas	23.1%
Biofuels (biomass, animal products) and waste (renewable or not)		2.3%
Hydraulic (including PHES)		16.6%
Nuclear		10.4%
Geothermal, solar, wind, tidal and others		5.6%

Table 1.10. *Global electricity production (IEA 2018)*

Distributed generation is generally close to the consumer and small in size, and thus connected to the distribution system. Depending on their power, their proximity to the power system, costs and connection constraints (contribution to system services), these have different connection voltages.

1.6.1.1. *Development of distributed generation*

Until the end of the 1990s, distributed generation was not very widespread because it was not profitable. A preliminary development started in 1982, after the creation of the critical peak pricing (CPP) tariff was introduced by EDF. The principle of this tariff is that electricity is more expensive during periods of high consumption (22 days a year; CRE (2010b)) so as to encourage people to consume less. Indeed, the means of production used during peak periods are coal or gas power plants, which are extremely polluting and expensive. Some manufacturers therefore decide to install their own means of production, which they use during periods when electricity is expensive. In 1995, they were even authorized to sell their surplus production at the same rate as the CPP. Along the same lines, the decree of July 25, 1996 authorized the resale of electricity by cogeneration units based on the same price as fossil fuels.

A second vector for the evolution of distributed generation is the opening of the energy markets, detailed above. In particular, it allows anyone to become eligible to produce electricity and inject it into the grid. In the US, for example, following the opening of the markets in 1992, the number of private generators doubled between the years 1990 and 1996 and came to represent about 7% of total American production (Height 2021). However, what has really enabled the development of distributed generation is the rise of renewable energy. In fact, not only have significant technological advances in renewable energy production systems made it possible to improve their cost and efficiency, but the problems of global warming and environmental pollution have also led to various laws encouraging the development of distributed, renewable energy generation. For example, we cite the law of February 10, 2000 in which the DSO cannot refuse a connection request from a producer and is required to make a technical and financial proposal for connection that is as economical as possible. European directives have also set targets in terms of renewable energy to be achieved, such as 22.1% of electricity from renewable sources by 2010 (Directive 2001/77/EC of 27 September 2001), which France has not yet achieved (16.4% in 2012; RTE (2012)). The December 2008 Climate Plan sets another target for 2020 called the "3 × 20" goal reflecting a 20% renewable electricity target in Europe, a 20% reduction in European CO_2 emissions and a 20% increase in energy efficiency compared to that of 1990 (European Commission 2018). In the law of August 17, 2015 on the energy transition for green growth, the previous targets were raised to 32% renewable energy by 2030, 40% reduction of greenhouse gases compared to 1990, 20% reduction of final energy consumption and 30% reduction in primary energy consumption of fossil fuels as compared to 2012. French events have also taken place, such as the *Grenelle de l'Environnement* between September and December 2007 leading to Grenelle I (August 3, 2009) and Grenelle II (July 12, 2010). In California, a set of laws have also been enacted for the development of renewable energies: the SB 1078 of September 12, 2002 (20% renewable energy by 2017), the SB X1 2 of March 29, 2011 (33% renewable energy by 2020), the SB 350 of October 7, 2015 (50% renewable energy by 2030) and recently the SB 100 of September 10, 2018 (50% renewable energy by 2026, 60% by 2030 and 100% by 2045). The latter legislation is also under discussion in other states, notably Hawaii (Proctor 2018).

1.6.1.2. *France's situation*

The order of April 23, 2008 (Abadie 2008) gives the connection voltages of the generation units according to their maximum power, summarized in Table 1.11.

<table>
<tr><th>Connection voltage</th><th>Maximum power</th><th>Conditions</th><th>Administrator</th></tr>
<tr><td rowspan="2">≤1 kV (400 V in France)</td><td>250 kVA</td><td>Three-phase</td><td rowspan="4">Enedis, LDC and EDF SEI for the ZNI</td></tr>
<tr><td>18 kVA</td><td>Single-phase</td></tr>
<tr><td rowspan="2">≤50 kV (usually 20 kV in France)</td><td>12 MW</td><td>Producer located in a zone not interconnected with the continental metropolitan system</td></tr>
<tr><td>17 MW</td><td>Producer located in a zone interconnected to the continental metropolitan grid, if the connection is technically possible</td></tr>
<tr><td>≥50 kV (usually 63, 90, 225 or 400 kV in France)</td><td>≥12 MW</td><td></td><td>RTE (Réseau de transport d'électricité) and EDF SEI for the ZNI</td></tr>
</table>

Table 1.11. *Connection voltage level of generation units in France according to their maximum power*

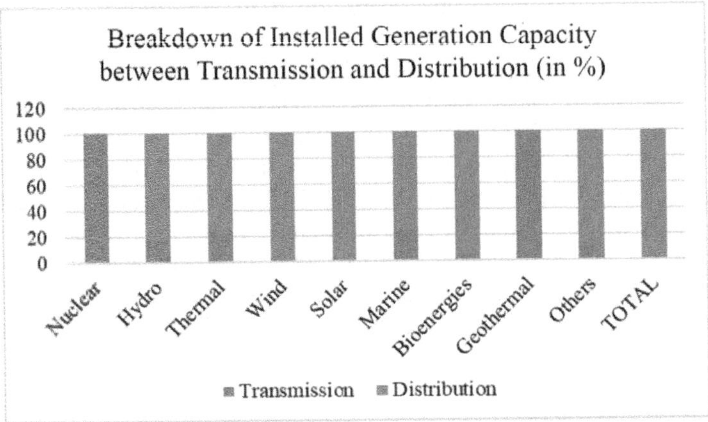
Breakdown of Installed Generation Capacity between Transmission and Distribution (in %)
120
100
80
60
40
20
0
Nuclear
Hydro
Thermal
Wind
Solar
Marine
Bioenergies
Geothermal
Others
TOTAL
Transmission
Distribution

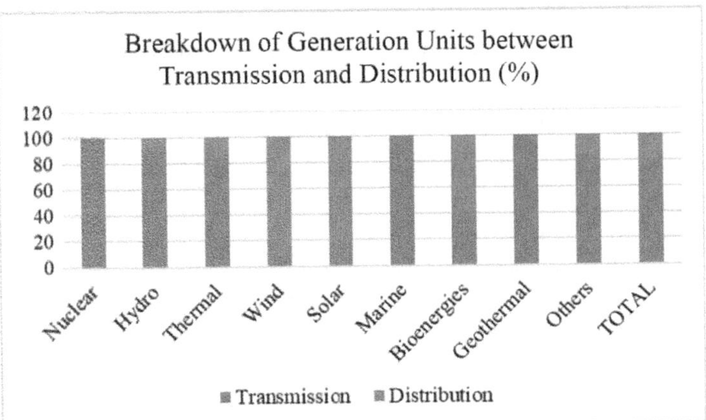
Breakdown of Generation Units between Transmission and Distribution (%)
120
100
80
60
40
20
0
Nuclear
Hydro
Thermal
Wind
Solar
Marine
Bioenergies
Geothermal
Others
TOTAL
Transmission
Distribution

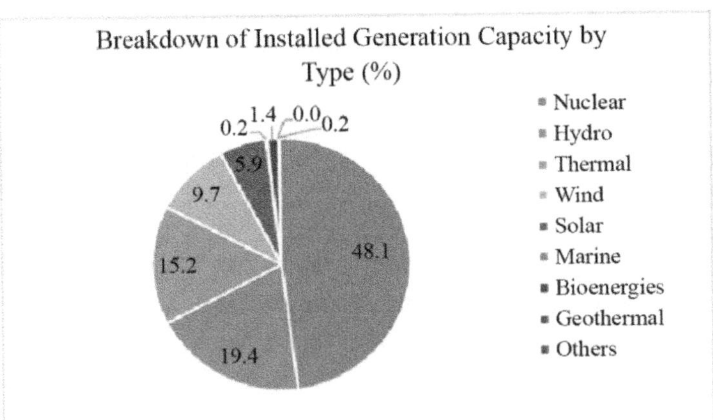
Breakdown of Installed Generation Capacity by Type (%)
0.2
1.4
0.0
0.2
5.9
9.7
15.2
48.1
19.4
Nuclear
Hydro
Thermal
Wind
Solar
Marine
Bioenergies
Geothermal
Others

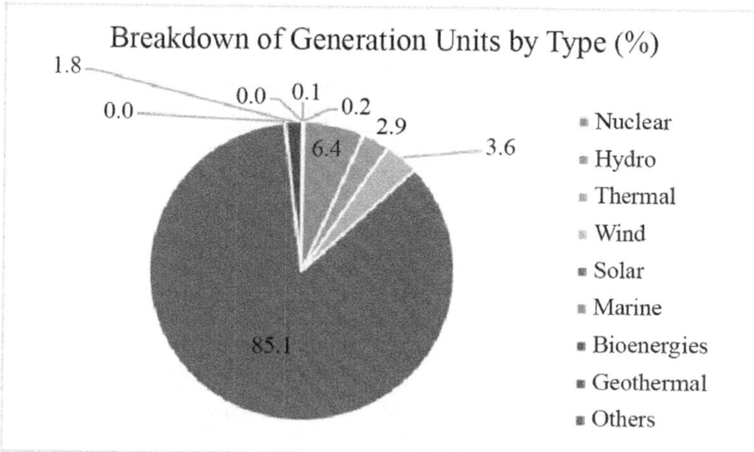

Figure 1.22. *Generation facilities in France and in non-interconnected zones (excluding LDC)*

In the national register of electricity generation and storage facilities, as of December 31, 2017 (Government of the French Republic 2017), we find that all the generation connected to RTE grids (transmission system operator in mainland France), Enedis (operator of 95% of the French distribution system) and EDF SEI (transmission and distribution system operator in the zones not interconnected to the mainland grid, known as ZNI). In this database, generators with a power output of less than 36 kW are not listed individually but aggregated into groups of at least 10 generators. We can see that in France, there are more than 37,253 generating units representing a total installed capacity of 131.3 GW (excluding generation connected to the distribution grid of the LDCs). Figure 1.22 shows the distribution of the number of generation units as well as their total capacity according to the sector and the voltage level. Therefore, we can see that 81% of the installed capacity injects 2.4% of the generation units onto the transmission system managed by RTE and EDF SEI, and 19% of this installed capacity injects 97.6% of the generation units onto the system of Enedis and EDF SEI. We therefore understand the distinction between centralized generation (generation units connected to the transmission system) and distributed generation (generation units connected to the distribution system). Figure 1.22 also shows that centralized generation comes mostly from nuclear, hydro and thermal sources, and distributed generation from renewable sources (wind, solar, bioenergy), but not exclusively, given that about 16% of the total capacity of thermal generation is connected to the distribution grid.

Type	Connection voltage	Maximum power	Remarks
Type 1		25 kW	Inverter-based power generation (photovoltaic, PV or storage)
Type 2	U < 4 kV	1 MW	The maximum connection power is increased by 1 MW if the production devices are connected: – to a line with a rated current greater than 600 A – or within 2.5 miles of the source station
	5 kV < U < 14 kV	2 MW	
	15 kV < U < 30 kV	3 MW	
	31 kV < U < 60 kV	4 MW	
Type 3		10 MW	– does not export power after the Point of Common Coupling (delivery point)
Type 4		10 MW	– that are not type 1, 2 or 3

Table 1.12. *Connection of distributed power generators in the US*

1.6.1.3. *The American situation*

The case of the US is more complex than that of France because it is subject to two types of rules: Federal laws and State laws. As in France, there are no laws regulating connection voltage levels according to the power range. However, depending on the type of generation, there are a number of procedures for validating the connection and calculating its cost. The Interstate Renewable Energy Council (IREC) offers a methodology that synthesizes best practices for connecting generation across several US states (IREC 2013). The maximum capacity of the considered generation is 10 MW; however, this maximum value may vary by state. Generators are thus classified into four types, summarized in Table 1.12, which are subject to different connection rules and costs.

1.6.2. *Change in the status of the consumer: the "prosumer"*

Historically, the electrical system was built to supply consumers while trying to minimize costs and ensure the continuity of supply. The customer had no role in the electrical landscape other than to consume. With the deregulation of markets, and the desire of society to reduce greenhouse gas emissions and energy consumption, the customer–consumer is becoming more and more active. From being a consumer, they are becoming *prosumers* (an amalgamation of *producer* and *consumer*). They can be producers, or consumers offering curtailment services, or both, and join Local Energy Communities:

– *physical communities*: these are non-profit organizations that make community investments (renewable energies, fleets of electric vehicles, storage facilities, etc.). The principle is that the consumer is a shareholder in the renewable resources. For example, in Europe, RESCoops (Renewable Energy Sources Cooperatives) is a cooperative grouping together 1,250 communities, that is, a total of 1 million customers. In Europe, the largest community is in Belgium, ECO-POWER, with over 50,000 members;

– *virtual communities*: in Germany, the company Sonnen, which markets batteries for storing photovoltaic generation for private individuals, has created a virtual community based on sharing the surplus energy produced among its members instead of being sold back to suppliers (Sonnen 2018). In some US states (e.g. California, Illinois, Ohio, Massachusetts, and New York), community choice aggregation (CCA) allows the end consumer to opt to be powered by renewable generation (Lean Energy 2018).

The consumer impacts the investment strategies of the power systems because they can no longer be regarded as a passive variable represented by simplistic historical models.

The consumer can be the owner of production and storage facilities, and thus resell all the electricity being generated to the electricity system. In France, Law no 2017-227 of February 24, 2017 followed by its implementation in April 2017 also allows the consumer to self-consume what they produce and to resell only the surplus. The consumer can therefore have three different statuses in the same day: consumer, producer, or invisible to the system (they produce what they consume). Buildings (residential, tertiary or industrial) can be equipped with generation and storage facilities, sized to be completely autonomous or even producers (energy-plus-building).

Second, consumers can also provide services to the power system, such as curtailing their consumption to avoid system constraints, particularly during peak periods (demand-side flexibility). These actions can be automatic via suppliers or curtailment operators who determine the electrical uses and the total power that can be curtailed. Consumers can also take action themselves by monitoring their consumption via the measurements communicated by electricity meters (smart meters) that are available through specific companies via an online interface (hourly energy consumption, associated costs, temperature, etc.).

Finally, with the development of battery electric vehicles (BEVs) and plug-in hybrid and electrical vehicles (PHEVs), consumers will also be able to choose to recharge their vehicles when the grid is less charged, and eventually feedback from their vehicles if grid consumption is too high. In 2016, about 750,000 electric vehicles (just under 2 million in 2019) were sold (40% of which were sold by

China), bringing the total number of electric vehicles in use to about 2 million (OECD/IEA 2017). In Europe in 2016, there were 460,000 charging stations, of which 70,000 were public and 390,000 privately owned (Meuus and Nouicer 2018). As policies aim to reduce greenhouse gas and nitrogen oxide emissions, this number will continue to increase with a dynamic dependent on the country. For example, the governor of California has set a goal of 5 million electric vehicles by 2030 (Brown 2021). Significant penetration of electric vehicles thus increases the overall electric demand and thus the action of the consumer will be increasingly essential to the proper functioning of the system.

1.6.3. *Distributed energy resources*

The terms producers, consumers and prosumers have been evolving over the last few years toward a more generic and broad term: distributed energy resources (DERs). The definition of DERs varies from one country to the next, and between one actor and another. The common point between all of these definitions is the small size of DERs and their location downstream of distribution systems. Drawing from FERC (2018), which compares several of them, we can define DERs as small-scale resources located near to end-users that either produce electricity that is consumed locally or reduce consumption. Thus, DERs include:

– electrical energy production such as photovoltaic panels, gas turbines, small wind and water power plants, heat pumps, fuel cells, etc.;

– electrical or thermal energy storage systems (batteries);

– electric vehicles and their charging/discharging systems;

– all devices and associated controls for demand-side flexibility and energy efficiency;

– microgrids defined as small-scale systems that can be disconnected from the main grid and operate autonomously when needed.

Some actors, like Naruc (2016), go as far as to include ancillary services in the role of DERs. Finally, considering that historically, distribution systems had only consumers as actors, then we can say that DERs include all new devices and associated services not present in historical distribution systems. However, there is currently no common regulation in Europe concerning DERs, specifically:

– in relation to the placement of storage facilities in the distribution systems. For example, in France, the DSO and TSO are not allowed to own storage, while in the UK and Spain, DSOs can own storage as long as it is less than 50 MW (or in some cases less than 100 MW, e.g. in the UK) (Sonnen 2018). In Italy, Germany and

Finland, the use of storage facilities by DSOs is accepted by the regulator in certain justified cases;

– on the deployment of electric vehicle charging stations, Sonnen (2018) lists three different models:

- in Spain, the DSO deploys, installs and connects charging stations, and management is performed by the providers,

- in Ireland, the DSO deployed the charging stations and recovered the cost of deployment through the distribution tariff. However, database management of these vehicles is with another company,

- in the Czech Republic, third parties install and manage the charging stations and the DSO only provides the connection to the electrical grid.

Significant development of DERs is expected in the coming years. For example, according to FERC (2018), the installed capacity of DERs by 2024 is expected to multiply by about 2.5 in comparison to 2015. CSIRO (2017) states that by 2050, between 30% and 45% of electricity needs will be met by DERs.

1.7. Conclusion

In this chapter, we have provided historical, technical and regulatory elements to understand how power systems are structured and managed today in Europe and around the world. The legacy of the past and the state of the different laws in the countries influence the direction for the development of the actors and the electricity systems. The massive development of distributed energy resources (renewable generation, storage, electric vehicles, flexible loads, etc.) in the future suggests significant changes in the power systems and, in particular, distribution systems. Indeed, these resources are of small power generation capability and thus connected to the distribution system, which historically has been sized by only considering traditional consumption. In Chapter 2, we thus focus on distribution systems, and in particular, on the way in which they have been planned around the world, before looking at how they must evolve in order to accommodate these distributed energy resources.

1.8. References

Abadie, P. (2008). Arrêté du 23 avril 2008 relatif aux prescriptions techniques de conception et de fonctionnement pour le raccordement à un réseau public de distribution d'électricité en basse tension ou en moyenne tension d'une installation de production d'énergie électrique [Online]. Available at: https://www.legifrance.gouv.fr/affichTexte.do?cidTexte=JORFTEXT000018698 004 [Accessed 2018].

ABB (2021). The Gotland HVDC link [Online]. Available at: https://new.abb.com/systems/hvdc/references/the-gotland-hvdc-link [Accessed 2021].

ACER (2018). Électricité [Online]. Available at: https://acer.europa.eu/fr/Electricity/Pages/default.aspx [Accessed 2018].

Allerhand, A. (2017). A contrarian history of early electric power distribution. *Proceedings of the IEEE*, 105(4), 768–778.

Anderson, P.L. and Geckil, I.K. (2003). Northeast blackout likely to reduce US earnings by $6.4 billion. Economic Impact of 2003 Blackout. Working document, AEG, East Lansing/Chicago.

ANROC (2012). Les entreprises locales de distribution d'énergie [Online]. Available at: http://www.anroc.com/index.php [Accessed 2018].

Bakke, G. (2016). *The Grid: The Fraying Wires between Americans and Our Energy Future.* Bloomsbury Publishing, New York.

BBC (2018). Volta battery [Online]. Available at: http://www.bbc.co.uk/ahistoryoftheworld/objects/-ZClk5X_Sj-ytwQi64xE8A [Accessed 2018].

Berseneff, B. (2010). Réglage de la tension dans les réseaux de distribution du futur. PhD Thesis, Université de Grenoble, Grenoble.

Blondel, C. and Wolff, B. (2021). L'expérience de Hans-Christian Oersted (1820) [Online]. Available at: http://www.ampere.cnrs.fr/histoire/parcours-historique/oersted/decouverte-hasard [Accessed 2021].

Bouttes, J.-P. and Trochet, J.-M. (2005). La conception des règles des marchés de l'électricité ouverts à la concurrence. *Économie publique/Public Economics,* 14, 1–37.

Brown, E. (2021). Governor Brown takes action to increase zero-emission vehicles, fund new climate investments [Online]. Available at: https://www.gov.ca.gov/2018/01/26/governor-brown-takes-action-to-increase-zero-emission-vehicles-fund-new-climate-investments/ [Accessed 2021].

Carrive, P. (1991). Réseaux de distribution : structure et planification. *Techniques de l'ingénieur Génie électrique*, 8, 1–32.

CEER (2017). Retail markets monitoring report [Online]. Available at: https://www.ceer.eu/documents/104400/-/-/56216063-66c8-0469-7aa0-9f321b196f9f. [Accessed 2017].

Centre, A.P. (2016). Impact of power quality on lighting technology. Working note.

China Energy Portal (2017). 2016 detailed electricity statistics [Online]. Available at: https://chinaenergyportal.org/en/2016-detailed-electricity-statistics/.

Coiffier, A. (2013). Schémas directeurs de développement des réseaux électricité de distribution MV (D4211). *Techniques de l'ingénieur Génie électrique*, 1–28.

CPUC (2021). Differences between publicly and investor-owned utilities [Online]. Available at: https://www.energy.ca.gov/pou_reporting/background/difference_pou_iou.html [Accessed 2021].

CRE (2010a). Les dispositifs de couverture des pertes d'énergie des réseaux publics d'électricité [Online]. Available at: http://gtpe.cre.fr/media/documents/100323Rapport Pertes.pdf.

CRE (2010b). Tarif Bleu [Online]. Available at: file:///C:/Users/alvarezm/AppData/Local/Temp/100815_Grille_TRV2010.pdf [Accessed 2018].

CRE (2018a). Liste des fournisseurs nationaux d'électricité enregistrés sur le site energie-info1 et actifs du 2 au 10 janvier 2018 [Online]. Available at: https://www.cre.fr/L-energie-et-vous/Mon-fournisseur [Accessed 2018].

CRE (2018b). Marché de détail de l'électricité [Online]. Available at: https://www.cre.fr/Electricite/Marche-de-detail-de-l-electricite.

CRE (2018c). Les codes de réseau relatifs aux conditions de raccordement [Online]. Available at: https://www.cre.fr/Electricite/Reseaux-d-electricite/Codes-de-reseau-europeens [Accessed 2018].

CRE (2021). Glossaire : foisonnement [Online]. Available at: https://www.cre.fr/Pages-annexes/Glossaire/FOISONNEMENT [Accessed 2021].

CSIRO (2017). Electricity network transformation roadmap [Online]. Available at: https://www.energynetworks.com.au/assets/uploads/entr_final_report_april_2017.pdf.

DOE (2016). Maintaining reliability in the modern power system [Online]. Available at: https://www.energy.gov/sites/prod/files/2017/01/f34/Maintaining%20Reliability%20in%20the%20Modern%20Power%20System.pdf.

Enedis (2018). Principes d'étude et de développement du réseau pour le raccordement des clients consommateurs et producteurs LV. Study document, Niort.

ENTSOE (2018a). [Online]. Available at: https://www.entsoe.eu/ [Accessed 2018].

ENTSOE (2018b). TYNDP 2018 executive report – Connecting Europe: Electricity 2025-2030-2040 [Online]. Available at: https://tyndp.entsoe.eu/tyndp2018/ [Accessed 2018].

ENTSOE (2021). Regional security coordinators [Online]. Available at: https://docstore.entsoe.eu/Documents/SOC%20documents/RSC%20Factsheet.pdf [Accessed 2021].

Escane, J.-M. and Bastard, P. (2004). Réseaux électriques linéaires – systèmes triphasés. *Techniques de l'ingénieur*, 1–11.

EUR-Lex (2018). Glossaire des synthèses [Online]. Available at: https://eur-lex.europa.eu/summary/glossary/proportionality.html?locale=fr [Accessed 2018].

European Commission (2007a). Inquiry pursuant to Article 17 of Regulation (EC) No 1/2003 into the European gas and electricity sectors [Online]. Available at: https://eur-lex.europa.eu/legal-content/EN/TXT/PDF/?uri=CELEX:52006DC0851andfrom=en [Accessed 2018].

European Commission (2007b). Prospects for the internal gas and electricity market [Online]. Available at: https://ec.europa.eu/energy/sites/ener/files/documents/2006_internal_market_country_reviews_0.pdf [Accessed 2018].

European Commission (2016). Commission proposes new rules for consumer centred clean energy transition [Online]. Available at: https://ec.europa.eu/energy/en/news/commission-proposes-new-rules-consumer-centred-clean-energy-transition [Accessed 2018].

European Commission (2018). 2020 climate and energy package [Online]. Available at: https://ec.europa.eu/clima/policies/strategies/2020_en [Accessed 2018].

European Commission (2018). Comitology [Online]. Available at: https://ec.europa.eu/info/implementing-and-delegated-acts/comitology_fr [Accessed 2018].

European Parliament (1997). Directive 96/92/CE du parlement européen et du conseil du 19 décembre 1996 concernant les règles communes pour le marché intérieur de l'électricité [Online]. Available at: https://eur-lex.europa.eu/legal-content/FR/TXT/PDF/?uri=CELEX:31996L0092andfrom=FR [Accessed 2018].

European Parliament (2009). Règlement (CE) no. 713/2009 du parlement européen et du conseil du 13 juillet 2009 instituant une agence de coopération des régulateurs de l'énergie. *Journal Officiel de l'Union européenne* [Online]. Available at: https://eur-lex.europa.eu/legal-content/FR/TXT/PDF/?uri=CELEX:32009R0713andfrom=EN [Accessed 2018].

European Union (2010). Versions consolidées du traité sur l'Union européenne et du traité sur le fonctionnement de l'Union européenne (2010/C 83/01).

European Union (2018a). Energy [Online]. Available at: https://europa.eu/european-union/topics/energy [Accessed 28 August 2018].

European Union (2018b). L'histoire de l'Union européenne [Online]. Available at: https://europa.eu/european-union/about-eu/history_fr#1990-1999 [Accessed 28 August, 2018].

European Union (2018c). Versions consolidées du traité sur l'Union européenne et du traité sur le fonctionnement de l'Union européenne – *Journal Officiel de l'Union européenne* [Online]. Available at: https://eur-lex.europa.eu/legal-content/FR/TXT/HTML/?uri=CELEX:12012E/TXTandfrom=EN [Accessed 5 September 2018].

Expositions universelles (2021). Exposition universelle de 1878, l'exposition des nouvelles technologies [Online]. Available at: http://www.expositions-universelles.fr/1878-exposition-universelle-paris.html [Accessed 2021].

FERC (2018a). Distributed energy resources: Technical considerations for the bulk power system [Online]. Available at: https://cms.ferc.gov/sites/default/files/2020-09/staff-report.pdf.

FERC (2018b). Electric power markets: National overview [Online]. Available at: https://www.ferc.gov/industries-data/market-assessments/electric-power-markets [Accessed 2018].

FERC (2019). 2010 ISO/RTO metrics report [Online]. Available at: https://www.ferc.gov/industries/electric/indus-act/rto/metrics/caiso-rto-metrics.pdf.

Fortescue, C.L. (1918). Method of symmetrical co-ordinates applied to the solution of polyphase networks. *Proceedings of the 34th annual convention of the American Institute of Electrical Engineers,* Atlantic City.

Gilstrap, M., Amin, S., De Corla Souza, K. (2015). Unites States electricity industy primer. (Department of Energy) [Online]. Available at: https://www.energy.gov/sites/prod/files/2015/12/f28/united-states-electricity-industry-primer.pdf.

Global Transmission (2017). TSO focus. Report [Online]. Available at: https://www.globaltransmission.info/archive.php?id=28836.

Government of the Republic of France (2017). Registre national des installations de production d'électricité et de stockage au 31 décembre 2017 [Online]. Available at: https://www.data.gouv.fr/fr/datasets/registre-national-des-installations-de-production-delectricite-et-de-stockage-au-31-decembre-2017-1/ [Accessed 2018].

Government of Canada (2006). Ontario-U.S. power outage – Impact on critical infrastructure. Report, Ottawa.

Groupe Cahors (2018). [Online]. Available at: https://www.groupe-cahors.com [Accessed 2018].

Guinet, J.N. (2015). Bellegarde, première ville de France éclairée à l'électricité grâce à Dumont. *Le Progrès* [Online]. Available at: https://www.leprogres.fr/ain/2015/08/07/bellegarde-premiere-ville-de-france-eclairee-a-l-electricite-grace-a-dumont [Accessed 2018].

Gupta, A. (2014). The world's longest power transmission lines [Online]. Available at: https://www.power-technology.com/features/featurethe-worlds-longest-power-transmission-lines-4167964/.

Harvey, A., Larson, A., Patel, S. (2017). History of power: The evolution of the electric generation industry [Online]. Available at: https://www.powermag.com/history-of-power-the-evolution-of-the-electric-generation-industry/?pagenum=1 [Accessed 2018].

Hässermann, H., Geneviève, M.Z., Pierre, Z. (1992). Les infrastructures urbaines en Allemagne avant 1945. *Flux*, 10, 25–31.

Height, M. J. (2000). Distributed electricity generation. *Sustainable Energy*, 617, 1–19 [Online]. Available at: http://web.mit.edu/10.391J/www/proceedings/DistributedGeneeration_Height2000.pdf [Accessed 2021].

IEA (2018). Electricity information overview [Online]. Available at: https://www.iea.org/events/statistics-electricity-information-2018-overview.

IEA (2020). Key world energy statistics 2020 [Online]. Available at: https://www.iea.org/reports/key-world-energy-statistics-2020

IREC (2013). Model interconnection procedures [Online]. Available at: https://ppp.worldbank.org/public-private-partnership/library/irec-model-interconnection-standards-and-procedures-small-generator-facilities.

JEPIC (2018). Operating and financial data [Online]. Available at: https://www.jepic.or.jp/en/.

Lacroix, B. and Calvas, R. (1998). Les schémas des liaisons à la terre dans le monde et les évolutions [Online]. Available at: https://download.schneider-electric.com/files?p_enDocType=Cahier+Technique&p_File_Name=CT173.pdf&p_Doc_Ref=CT173&_ga=2.186680365.462099605.1666021843-369737043.1666021843.

Lean Energy (2018). CCA by state [Online]. Available at: http://leanenergyus.org/cca-by-state/ [Accessed 2018].

Legifrance (1996). Directive 96/92/CE du parlement européen et du conseil du 19 décembre 1996 [Online]. Available at: https://www.legifrance.gouv.fr/affichTexte.do?cidTexte=JORFTEXT000000889133 [Accessed 2018].

Massé, R. (2002). Histoire de l'électrification rurale en France [Online]. Available at: http://www.reseau-cicle.org/wp-content/uploads/riaed/pdf/Histoire-electrification-rurale-France.pdf [Accessed 2018].

Meuus, L. and Nouicer, A. (2018). The EU clean energy package. Technical report, European University Institure, Italy.

Multon, B. (2013). Conversion d'énergie – comment produire de l'électricité à partir des mouvements d'un solide ou d'un fluide en 3 questions à Bernard Multon (partie 1/2) [Online]. Available at: http://culturesciencesphysique.ens-lyon.fr/ressource/conversion-energie-Multon1.xml.

NARUC (2016). Distributed energy resources rate design and compensation [Online]. Available at: https://pubs.naruc.org/pub/19FDF48B-AA57-5160-DBA1-BE2E9C2F7EA0.

Neidhofer, G. (2011). 50-Hz frequency [history]. *IEEE Power and Energy Magazine*, 9(4), 66–81.

NEXANS (2018). Câbles de réseau basse tension [Online]. Available at: https://www.nexans.ch/Switzerland/group/doc/en/NX-%20BT%20(F).pdf [Accessed 2018].

NRECA (2021). America's electric cooperatives: 2017 fact sheet [Online]. Available at: https://www.electric.coop/electric-cooperative-fact-sheet/ [Accessed 2021].

NREL (2017). An introduction to retail electricity choice in the United States [Online]. Available at: https://www.nrel.gov/docs/fy18osti/68993.pdf.

Oddi, M. (2011). Plan de protection des réseaux de distribution publique à moyenne tension. *Techniques de l'Ingénieur*, 1–26.

OECD/IEA (2017). Global EV outlook 2017 [Online]. Available at: https://www.iea.org/publications/freepublications/publication/GlobalEVOutlook2017.pdf [Accessed 2018].

Ormazabal (2018). [Online]. Available at: https://www.ormazabal.com/ [Accessed 2018].

Owen, E.L. (1997). The origins of 60-Hz as a power frequency. *IEEE Industry Applications Magazine*, 3(6), 8–14.

Persoz, H. (1984). *Planification des réseaux électriques*. Eyrolles, Paris.

PG&E (2020). Local RA substations are determined pursuant to the CPUC's annual designation of local RA areas and CAISO's local capacity technical analysis [Online]. Available at: https://www.pge.com/pge_global/common/pdfs/for-our-business-partners/energy-supply/standard-contracts-for-multiple-facilities-pursuant-to-ab-1613/lcr-substation-list.pdf.

Proctor, D. (2018). California mandates 100% renewable energy [Online]. Available at: https://www.powermag.com/california-mandates-100-renewable-energy/ [Accessed 2018].

RTE (2004). Memento de la sûreté du système électrique [Online]. Available at: http://clients.rte-france.com/htm/fr/mediatheque/telecharge/memento_surete_2004/memento_surete_2004_complet__.pdf [Accessed 2018].

RTE (2009). Performances du réseau public de transport – plages de tension et de fréquence normales et exceptionnelles. Technical reference document [Online]. Available at: https://clients.rte-france.com/htm/fr/mediatheque/telecharge/reftech/24-04-09_article_3-1__v2.pdf [Accessed 2018].

RTE (2012). Bilan électrique 2012 [Online]. Available at: https://www.rte-france.com/sites/default/files/rte_bilan_electrique_2012.pdf [Accessed 2018].

RTE (2015). Les échanges électriques européens, vecteur de la transition énergétique [Online]. Available at: https://www.rte-france.com/fr/actualite/les-echanges-electriques-europeens-vecteur-de-la-transition-energetique [Accessed 2018].

RTE (2016). Bilan électrique français 2016 [Online]. Available at: https://www.rte-france.com/sites/default/files/2016_bilan_electrique_synthese.pdf [Accessed 2018].

RTE (2017). Bilan électrique 2017 [Online]. Available at: http://bilan-electrique-2017.rte-france.com [Accessed 2018].

RTE (2018). Liste des Opérateurs d'Effacement [Online]. Available at: https://clients.rte-france.com/lang/fr/visiteurs/vie/nebef_operateurs.jsp [Accessed 2018].

RTE (2021a). Capacités de Transfert Nettes (NTC). Available at: https://www.services-rte.com/fr/visualisez-les-donnees-publiees-par-rte/ntc-previsionnelles-export-import.html. [Accessed 2021].

RTE (2021b). Les réseaux électriques européens à l'heure de la transition énergétique [Online]. Available at: https://www.rte-france.com/sites/default/files/carnet_de_presse_europe_reduit_02.pdf [Accessed 2021].

RTE (2021c). Savoie-Piémont : 190 km de solidarité européenne entre Chambéry et Turin [Online]. Available at: https://www.rte-france.com/projets/nos-projets/savoie-piemont-190-km-de-solidarite-europeenne-entre-chambery-et-turin#Leprojet [Accessed 2021].

SDE35 (2011). Origines et évolutions du système de distribution électrique français [Online]. Available at: http://www.sde35.fr/iso_album/les_lois_fondatrices_mises_en_formeoct2011.pdf [Accessed 2018].

SEDC (2017). Explicit demand response in Europe [Online]. Available at: https://www.smarten.eu/wp-content/uploads/2017/04/SEDC-Explicit-Demand-Response-in-Europe-Mapping-the-Markets-2017.pdf [Accessed 2018].

Smithsonian Institution (2019). Documenting history in the making [Online]. Available at: http://americanhistory.si.edu/powering/ [Accessed 2021].

SONNEN (2018). What is the sonnenCommunity? [Online]. Available at: https://sonnengroup.com/sonnencommunity/ [Accessed 2018].

Spectrum (2012). San Francisco's secret DC Gridsur [Online]. Available at: https://spectrum.ieee.org/tech-history/dawn-of-electronics/san-franciscos-secret-dc-grid [Accessed 2018].

SPOT (2017). Vente directe des énergies renouvelables sur la bourse de l'électricité – rctour d'expérience Franco-Allemand sur l'intégration au marché des énergies renouvelables [Online]. Available at: http://static.epexspot.com/document/38036 [Accessed 2018].

State of New York Public Service Comission (2016). Order adopting a ratemaking and utility revenue model policy framework [Online]. Available at: https://documents.dps.ny.gov/public/Common/ViewDoc.aspx?DocRefId=%7BD6EC8F0B-6141-4A82-A857-B79CF0A71BF0%7D.

Stock, R. (2018). DNOS as system managers of the energy transition [Online]. Available at: https://www.renac.de/fileadmin/renac/media/Projects/US_German_Energy_Dialogue/US_Power_System_Operator_2018/07_VKU_Stock_web.pdf [Accessed 2018].

Sulzberger, C. (2021). Milestones: Pearl Street Station, 1882 [Online]. Available at: https://ethw.org/Milestones:Pearl_Street_Station,_1882 [Accessed 2021].

Travin, L. and Khudyakov, V. (2013). HVDC Advances in the Russian Federation [Online]. Available at: https://www.tdworld.com/substations/hvdc-advances-russian-federation [Accessed 2021].

UCTE (2018). The 50 year success story – Evolution of a European interconnected grid [Online]. Available at: https://docstore.entsoe.eu/fileadmin/user_upload/_library/ publications/ce/110422_UCPTE-UCTE_The50yearSuccessStory.pdf [Accessed 20 July 2018].

UNDP (2015). Human development. Report. United Nations, New York.

Usborne, T., Eastman, J., Freeman, A. (directors) (2011). Shock and Awe: The Story of Electricity: Episode 2 – The Age of Invention. Film. Open University/BBC, UK.

Valentin, G., Fondeur, R., Joyeux-Bouillon, B., Turpain, J.-C. (1996). Postes à moyenne tension. *Techniques de l'ingénieur*, 1–28.

Wikitravel (2021). Electrical systems [Online]. Available at: https://wikitravel.org/en/Electrical_systems.

Wolf, G. and Adapa, R. (2018). 2017 Was a fantastic year for HVDC technologies [Online]. Available at: https://www.tdworld.com/hvdc/2017-was-fantastic-year-hvdc-technolo-gies.

World Bank (2018). Indicators [Online]. Available at: https://data.worldbank.org/indicator/EG.USE.ELEC.KH.PC [Accessed 19 July 2018].

Yang, X.-J. (2016). Pumped hydroelectric storage. *Storing Energy*, 25–38.

2

Principles of Power Distribution System Planning

2.1. Methods of power distribution system planning

2.1.1. *Definition*

The power distribution system has historically been built in three main phases, as explained in Carrive (1991). The first phase corresponds to the electrification of a given area in order to satisfy the needs of consumers. In the second phase, the growth in demand for electrical energy, linked to economic development, leads to a necessary evolution of the power system in order to satisfy this increase in consumption. In the last phase, the planner must respect the constraints of supply reliability and power quality. From these three phases, the notion of power system planning has emerged, which we can define as the set of operational decisions and investments required by the power system in order to meet the three fundamental objectives mentioned above: to serve end-users (consumers and producers), to forecast their electrical needs (power extraction and injection) and to ensure the quality and continuity of the supply, with a view to profitability over the lifetime of these investments. Accordingly, planning time scales range from near real-time (less than an hour) to the long term (several decades).

Power system planning is a complex and strategic issue. First, it is conditioned by the country's energy policy, which in France is reflected in the publication of the Multi-Year Investment Plans (MYIP) (DGEC 2016) defining the context for the development of the power system, and at the European level by the European Fund

---

This chapter was co-authored by Oana IONESCU RIFFAUD.

for Strategic Investments (EFSI). Similarly, the regulatory context of countries influences the objectives of DSOs. For example, in France, they have a public service obligation to serve all customers, as well as strategic orientations and major technical choices defined in their technical reference documentation (Enedis 2008). The DSO must therefore invest in the power system and each decision can have very important socioeconomic repercussions, particularly due to the size of the system and the economic-electricity dependence. For example, in 2019, investments made by the main French distribution system operator, Enedis, amounted to €3.9 billion (Enedis 2021b), which represents about 7% of industrial investments in France (National Institute of Statistics and Economic Studies 2019). In addition to these investments, there are operating costs related to technical losses (joule losses, iron losses in transformers, phase imbalance) and non-technical losses (metering device malfunctions, discrepancies between metering information and the commercial billing base, electricity theft, among others). The Council of European Energy Regulators (CEER) Report on Power Losses (2017) provides the results from a survey on losses, expressed as a percentage of total power injected into the grid, by regulators and grid operators in 27 countries, between 2010 and 2015, although estimation methods may vary from country to country.

The smallest observed losses were in Finland (around 2%), while the countries with the highest losses vary year by year. In 2015, Romania had the highest losses (around 10%). The California Public Utilities Commission estimates that losses in California ranged between 6.58% and 7.95% between 2009 and 2014 (Mutialu and Zafar 2017). In France, losses averaged 6.5% for the total energy consumed from 2008 to 2016 (RTE 2018). The cost of failure can also have heavy repercussions on users (such as the major demonstrations at the port of Harcourt in Nigeria, where people protested the repetitive power cuts; Energy Mix (2018)) and can lead to industrial production losses or human losses (e.g. hospitals). Therefore, the different development strategies decided upon during planning studies depend on a large number of parameters that vary according to the characteristics of the system under study, its location and the country's regulations. These are carried out in the context of an uncertain future in which consistent assumptions must be made in order to simplify the planning problem. The planning of distribution systems must thus be the result of a thorough analysis in order to decide the appropriate technical-economic solution that respects the different predefined objectives and constraints. For this reason, planning studies are generally divided into four time horizons, for which the methodology is quite different, mainly due to the increasing uncertainties and the responsiveness of planners.

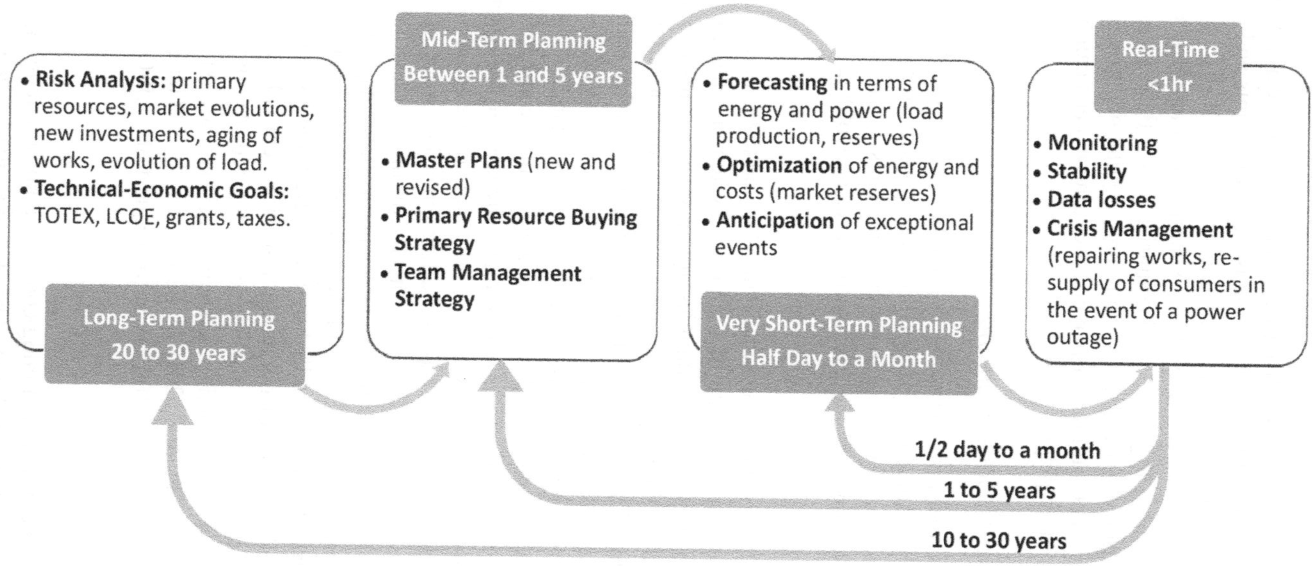

Figure 2.1. *Different types of planning according to the time horizon. For a color version of this figure and all subsequent figures in this chapter, see www.iste.co.uk/alvarezherault/distribution.zip*

2.1.2. *The different time scales in planning*

Depending on the time horizon (from less than 1 h to 30 years), the planning of distribution systems is not subject to the same issues, constraints and objectives. For this reason, we distinguish between real-time (less than 1 h), very short-term (half-day to 1 month), medium-term (1–5 years) and long-term (10–30 years) planning, as summarized in Figure 2.1. All countries follow these planning steps but, depending on the structure of their system and regulations (monopoly or not, presence of energy market, number of players, etc.), these steps can be grouped and carried out by one or more players.

2.1.2.1. *Real-time operational planning*

The objective of real-time planning is to monitor the stability of the power system and to quickly resolve critical situations, as opposed to investment planning, the objective of which is to analyze these past incidents, study several investment strategies in order to improve the quality of service (QoS) and select the most profitable one.

In the first case, the time horizon is less than 1 h and the uncertainties associated with loads, generation and the state of the power system are therefore very small, whereas in the second case, the uncertainties are very large and some are difficult to predict. Decisions taken in long-term planning can be questioned later on, unlike in real-time planning. This is why real-time planning is often called operational planning. It is essentially a matter of monitoring the system to ensure the generation–consumption balance (also called frequency stability), maintaining the voltage plan (also called voltage stability) and diagnosing inconsistent data feedback from the system's metering devices. Depending on the state of the system, actions are taken with the priority objective of system security and not the economic optimum. In the event of an incident, such as a fault on a line that takes some of the consumers offline, the fault must be located and the cut-off consumers must be re-supplied as quickly as possible. To do this, staff must be available 24 h a day and incidents must be recorded for subsequent analysis. In France, for example, system stability and the handling of inconsistent data is handled by the single transmission system operator RTE, while the handling of incidents on the power system is handled by both RTE and the DSOs depending on where the incident is located. Another example of real-time scheduling at the distribution system level is the load shedding (Base Interruptible Program (PG&E 2019)) performed by PG&E but also other aggregators for consumers above 100 kW at the request of CAISO (California ISO).

2.1.2.2. *Very short-term operational planning*

Very short-term operational planning concerns a horizon ranging from half a day to one month. The objective is to forecast production and storage reserves in order to cover the system's energy and peak power needs, to optimize the system's overall costs (purchase of losses, energy exchange) and to anticipate exceptional events which could create dangerous situations for the system (cyclones and storms in island power systems, events that could create a significant consumption peak, such as a sporting event).

2.1.2.3. *Medium- and long-term planning*

Medium- and long-term planning concerns the power system equipment defined in Chapter 1 (generation facilities, transformer substations, power lines and remote-controlled switches) for the connection of new users, the replacement of equipment that is too old or has become obsolete, the relocation of facilities, or to reduce a failure rate that would otherwise be economically detrimental (Enedis 2008). These investment decisions concern facilities with a lifespan of up to several decades, with equally long lead times, particularly for administrative procedures. The weight of the past is therefore very influential and the impact of each decision on the future must be carefully evaluated.

2.1.2.3.1. Medium-term planning

Medium-term planning refers to studies ranging from 1 to 5 years. We can cite as examples the connection request of a consumer or a producer; master plans for the development of the system, in order to strengthen it to overcome a technical constraint (the construction of an MV line takes between 1 and 3 years and in rare complex cases such as a railroad or highway crossing, it can take up to 6 years); maintenance work (e.g. the pruning of trees which is planned over 3 years); strategies for the purchase of primary resources; and team management strategies.

2.1.2.3.2. Long-term planning

In long-term planning studies, the main challenge is to anticipate the long-term (20–30 years) energy context (primary resources, market developments, consumption and production trends) while minimizing the economic risk. The main obstacle is therefore the high degree of uncertainty and the quantification of the risks associated with a decision. Long-term planning studies therefore rely on discounted technical and economic calculations (TOTEX for TOTal EXpenditures) to take into account the value of money. These calculations depend on the structure of the system and the regulations of the country, and can therefore include financial aid and taxes. Finally, for monopoly systems or if the DSO also owns generation systems, for example, the LCOE (levelized cost of energy) calculation makes it possible to integrate an additional energy cost. Long-term planning is an important step because it makes it possible to anticipate future evolutions and to establish economic strategies. It also allows managers to bring in new investors and to have discussions with regulators. As an example of long-term planning studies, we can mention the construction and/or dismantling of lines (between 10 and 15 years for the construction of HV lines), the modification of architectures and/or their back-up scheme, through the addition of new feeders, merging existing feeders, or the creation of primary substations (on average between 5 and 10 years, the time between the moment when the decision to build was taken and the delivery of the primary substation).

The various scientific challenges related to Smart Grids, such as the integration of flexibility in planning, the modification of the load behaviors that can be mixed (load/production) in the case of the electric vehicle, for example, or even curtailment, will disrupt the current planning methods. Not only is the number of variables increasing, but the economic models of these new players must be defined and integrated into the calculation of planning costs. In addition, there is a high degree of uncertainty, especially regarding user behaviors. Therefore, planning is a complicated exercise because it is multi-temporal and has a degree of uncertainty that can be significant. It is also an opportunity to consider innovative solutions that break with current solutions (particularly with regard to long-term planning), calling into question historical ways of thinking.

2.1.3. *France's power distribution system planning*

In France, the four planning levels defined above are not all managed by the DSO. At the level of real-time and very short-term operational planning, the role of the DSO is limited as it is mainly the RTE TSO that is in charge of system security and planning of energy requirements. The DSO's only real-time task is to plan the stages of fault detection and localization, and then the resupply to customers in the event of incidents on its system. In the very short-term, the DSO forecasts the consumption of the power system in order to adjust the forecast of its losses and their purchase on the energy market. The objective is to minimize the additional quantity of losses to be purchased at the last moment, and which are due to consumption forecasting errors, given that electricity is very expensive at this time.

In terms of medium- and long-term planning, the TSO is responsible for developing the transmission system, while the DSO is responsible for developing the distribution system, which is the subject of this book. Long-term planning consists of imagining what the power system would look like in a few decades (usually 30 years in France since, beyond that, the uncertainty about load growth is considered too great), without this necessarily becoming the target to be reached, but rather giving a direction for development. We can distinguish three types of long-term planning, according to Temraz and Quintana (1993); Willis et al. (1995, 1996); Khator and Leung (2007): "greenfield" planning (if the system is entirely new), power system extension (if there is an existing system), and maintenance or improvement of the power system's efficiency (if a set of system performance criteria are deemed unsatisfactory). Given the changing energy and societal context presented in Chapter 1, it is legitimate to ask whether the current power system architecture will still be the most appropriate in 30 years' time. Therefore, the distribution system imagined at this horizon, the "long-term target", may be radically different from the initial system. The French DSO defines an intermediate target at 10 years, the "medium-term target", which will not be questioned a priori, except for a modification during the course of the project following additional information brought to the distributor. In order to find this medium-term target, the planner considers all possible strategies to move from the initial system to the long-term target. The best strategy is selected because of a techno-economic study and a 10-year projection is used to define the medium-term target. All of the operations to be carried out each year (from year 0 to year 10) in order to reach this target must be described. In the case of medium- and long-term planning studies, an in-depth study is then necessary to establish the targets. In the case of Enedis, this study is called *Schéma d'orientation des réseaux électriques* (Score). It consists of five main steps (Carrive 1991; Enedis 2008; Coiffier 2013), relatively common to the methodologies of foreign DSOs, described below.

2.1.3.1. *Step 1: MV system assessment*

First, a preliminary analysis of the system in its initial state is necessary to highlight its strengths and weaknesses. It consists of making an inventory of the current system elements such as transformer stations (HV/MV and MV/LV), customers (consumers and producers), lines, protection devices and switchgear. All new customers to be connected and their characteristics are also listed and an external analysis is performed to assess the sensitivity of customers to environmental factors, such as weather. Finally, the problems currently encountered by the system are analyzed and integrated into the planning objectives. To do this, the planner calculates the following criteria, which will be explained in detail later in this chapter:

– the maximum power of each feeder (kVA) (called P_{tmb} in France; Garry (2016));

– the developed length of each feeder (km);

– the linear load density (kVA/km);

– the number of remote-controlled points per feeder;

– statistical average number of annual incidents;

– average annual outage time (min);

– the probability of reaching more than three incidents per year, per century (e.g. a probability of 2 years per century means that there are on average 2 years per century for which the number of incidents will be higher than 3);

– the voltage drops in normal operation (%);

– the total voltage drops in backup (%);

– short circuit power (MVA).

This analysis is carried out every 5 years in order to be able to readjust the long-term target if the system does not evolve as expected. In the future, the development of smart meters on the power system, associated with control and optimization software platforms, will allow for better knowledge of the system and should be integrated into the planning. These platforms, known as ADMS (advanced distribution management systems) internationally, will be able to provide services such as load and generation control according to the needs of the power system (DOE 2015). In France, advanced control functions are already deployed on the system, for example, with the automation of service recovery functions in the event of system incidents. They are based on the information provided by fault indicators via private or public communication systems (e.g. radio, telephone and Internet systems). Another historical French example is the control of the load because of the full/off-peak signal sent by the TCFM (*Télécommande à fréquence musicale*: Ripple control communication system). Two tariffs programmed at certain times make it possible to encourage consumers not to consume during peak periods.

2.1.3.2. *Step 2: definition of load growth and eventual production scenarios*

French and international regions are not homogeneous and load evolution scenarios must be developed based on national and local studies. Therefore, in order to define the rate of evolution of electricity consumption, the planner relies on different criteria such as MV and LV consumption history, demographic evolution, rail and road infrastructure development, but also activity zones. Coiffier (2013) details the current method used to define its load evolution rates. This step is crucial since it has a strong influence on investments and is subject to significant uncertainties.

The load growth rate may not be constant over the entire study period under consideration, denoted as N_{years}. If $N_{periods}$ is the number of periods with different load growth rates and if we denote τ_p^{peak} the peak load growth rate and $N_{years}(p)$ the number of years in period no. p, with $1 \leq p \leq N_{periods}$, then the maximum load power at the last year N_{years}, denoted $P_{max}(N_{years})$, is given by equation [2.1], with $P_{max}(0)$ being the maximum load power at year 0.

$$P_{max}(N_{years}) = P_{max}(0) \times \prod_{p=1}^{N_{periods}} \left(1 + \tau_p^{peak}\right)^{N_{years}(p)} \qquad [2.1]$$

For example, for a long-term planning study with $N_{years} = 30$ years with two evolution rates ($N_{periods} = 2$): 2% in the first 20 years and 1% in the last 10 years, the maximum load power at the end of the study period is given by equation [2.2].

$$\begin{aligned} P_{max}(30) &= P_{max}(0) \times (1 + 0.02)^{20} \times (1 + 0.01)^{10} \\ &\approx 1.64 \times P_{max}(0) \end{aligned} \quad [2.2]$$

Two types of growth rates are usually used: the growth rate of the annual peak and the growth rate of the average consumption. It is possible, for example, that the peak increases significantly without any change in average consumption. In a historical context of conventional loads only, the uncertainties remain limited and a forecast error will only impact the date of appearance of the constraints, while the long-term target remains unchanged (Coiffier 2013). Historically, the distribution system was not designed to have generation connected to it. Until now, the penetration rate of new producers has been negligible, so that the system is sized solely by consumption (e.g. the winter peak in France). In the context of the development of renewable energies, it is possible that in certain case studies, the production becomes a determinant factor and that it is necessary to integrate it in the long-term planning studies. This is likely to introduce additional complexity to the planning studies as the generation connected to the distribution systems is mainly renewable, which is particularly intermittent. The hypotheses for the evolution of generation are thus quite complex, with formulas that can include daily, annual, average target peaks and aging coefficients. In addition, there is the emergence of the electric vehicle and the flexibility of sources and loads, as presented in Chapter 1. The planning of distribution systems in this new context will be further explored in Chapter 3.

2.1.3.3. *Step 3: definition of 30-year targets*

This step consists of imagining the architecture of the possible long-term target to satisfy the desired quality and continuity of the electrical power supply. It depends on the main technical choices, such as the voltage level, the neutral system (distributed neutral or not), the maximum permissible currents and short-circuit currents in normal and backup mode (modification of the system topology in the event of a fault in order to accommodate the resupply of consumers), the distributor's policy (burying of conductors for example). At the end of the study, the long-term target must contain all the transformer stations and lines to be added and/or removed. The long-term target does not necessarily represent the final system, which may change as the study hypotheses evolve, but defines the main structural orientations. This is why the economic criterion is not taken into account, the idea being to find an ideal system that allows the optimization of the various technical objectives. The planner's approach is therefore to calculate the criteria listed in Step 1 and to check whether the set objectives are achieved. These objectives vary from one country to the next since they depend on the structure of the system and its history, as well as on the policies of the DSOs. The planner will therefore not perform an in-depth techno-economic calculation of their long-term target at this stage, but will modify the architecture of their system in order to meet the targets they have set.

Therefore, the general methodology is to identify lines for which the technical criteria defined in Step 1 are not met and to remove these from the system. Similarly, the planner removes lines that are more than 20/25 years old, and those that are unsuitable for climatic hazards (Coiffier 2013). The planner's objective is then to devise the best possible structure to reconnect all customers which allows the planning criteria to be met or even improved, as schematized in Figure 2.2.

The major modifications of the target can be a topological modification (creation/deletion of lines, creation/modification of remote-controlled switches (RCSs), a creation of primary substations or a coupling of the two. A specific case of long-term planning study that we can cite is the replacement of a system with a different voltage. Even if the main MV voltage in France is 20 kV, there are still some historical voltages, such as 15 kV (11% of the MV system in 2004 (Enedis 2008)) or 5.5 kV. In case of technical constraints on these systems, increasing the voltage level can be more economical than reinforcement. We cite here the example of Grenoble and its historical 5.5 kV system. The arrival of the tramway triggered the progressive transition of the 5.5 kV system into a 20 kV because of the considerable increase in the power needing to be drawn. In the case where the existing equipment is adapted to the new voltage, we can replace it gradually (the majority of cases) and in the opposite case, it is necessary to start by duplicating parts of the existing system, followed by a progressive re-commissioning (Coiffier 2013).

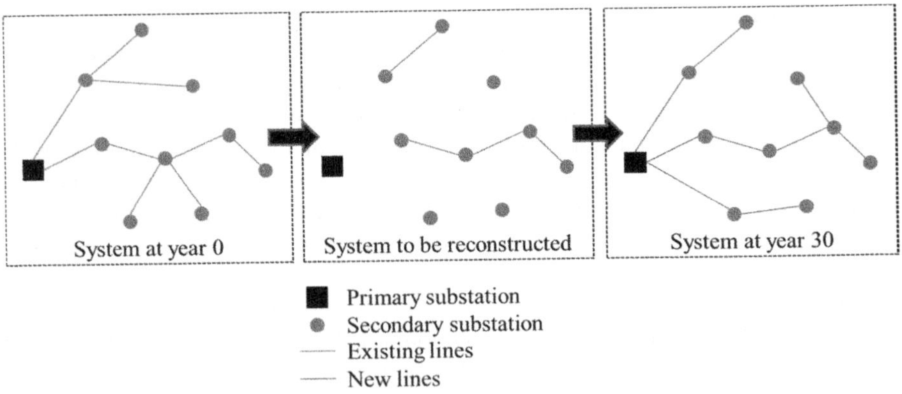

Figure 2.2. *Definition of the 30-year target*

In this stage, the long-term target is therefore established mainly through field expertise. Even if decision support tools for planners exist, such as calculation modules that allow for the evaluation of different technical and economic criteria, there are no tools to propose structural solutions to the planner in an automated way.

2.1.3.4. *Step 4: establish intermediate systems*

In Step 4, a set of coherent strategies for moving from the initial target to a long-term target are determined. Each strategy defines intermediate systems between the initial target and the long-term target (usually between two and eight depending on the difference between the two systems), as well as the list of works to be done to transition from one system to the other. For each intermediate system, economic calculations are performed to verify the overall consistency. The selection of the final strategy is done by qualifying each project according to the economic means implemented: the need to finance a high CAPEX, or not, the minimization of OPEX, amount of working capital and cash flow, financing costs, return on investment and customer benefits. Some environmental elements are also taken into account, such as the ability to obtain land use permits (public or private land use rules) and to respect system implementation requirements (rural, urban and ecological niches). It should be noted that in most cases it is in our interest to carry out the work early on, as it has a significant impact on operating costs (which will be defined later in this chapter). A cost–benefit analysis explained below allows us to find the optimal strategy and the dates on which the various works should be carried out, that is, the dates of the intermediate systems. In the particular case where some works are not considered economical, further reflection on their interest will be carried out before deciding whether they should be eliminated or continued. At the end of this step, the 10-year target is obtained with all the works and their execution date over the period [0–10 years] guaranteeing the absence of technical constraints on the system (current and voltage).

2.1.3.5. *Step 5: integration of practical constraints into the theoretical target*

In reality, distribution companies have a certain budget allocated per year for planning. A cost–benefit ratio (CBR) calculation allows for the allocation of system investments over the period considered. A second aspect that is being studied at this stage is the possibility of mutualizing works with road works. The option to save on the cost of asphalt, for the opening of a trench or to bury cables, can justify advancing or postponing works, even if the CBR recommends doing it in a given year. Finally, the above study is conducted with certain assumptions, such as average line fault rates. However, it is possible that for some system equipment, the failure rates are higher than the average rates considered and require priority improvement work.

Scores should be reviewed every 5 years to take into account any changes in load growth assumptions, significant deviations between the theoretical and actual target or the emergence of new technology, among others.

2.1.3.6. *International planning methods*

In California, PG&E's planning methodology, known as the distribution capacity program (DCP) (PG&E 2015), involves the same steps as in France but with different time scales. Stress occurrence studies are performed for the current year, for the next 2 years, and for the next 5 years in the case of a newly installed transformer or the development of power distribution lines. Updates to these studies may be made during the year as new information becomes available.

2.1.4. *Indicators used in planning and the solutions commonly employed to meet them*

The objectives of planning are to develop a more reliable, economical and environmentally friendly electricity system. It is therefore necessary to define indicators that can be used to distinguish between the planning solutions under consideration.

2.1.4.1. *Electrotechnical indicators*

– *Voltage*: to guarantee the proper functioning of customers connected to the distribution system, the mandate of December 24, 2007 relating to "quality levels and technical requirements for the quality of public electricity distribution and transmission systems" in accordance with the European standard EN 50160 indicates that MV and LV voltages must be within ±10% of the nominal voltage (230 V phase-to-neutral and 400 V phase-to-phase in LV and 20 kV or 15 kV phase-to-phase in MV). However, the French DSO is contractually committed to ensuring that the voltage at the MV delivery point remains within ±5% of a contractual voltage in normal operation (this contractual voltage itself within ±5% of the nominal voltage) (Enedis 2008).

– *Current*: the current must not exceed the maximum permissible current of the system equipment (lines, transformers). There is therefore a maximum constraint on the power that can flow through the lines to prevent them from being damaged (degradation of the insulation, even fire).

– $3I_0$: in the case where the system has capacitive lines (e.g. in the case of long underground lines), the capacitive current (noted I_0) may no longer be negligible compared to the rated current, or even of the same order of magnitude if the system customers consume little energy. The adjustment of the circuit breaker at the feeder head can therefore be complicated: it must be able to switch on in case of a fault without switching on unexpectedly due to a capacitive current flowing on the feeder and feeding a fault located on a part of the system not concerning it (Oddi 2011).

– I_{ccbi}: in the case of an LV system with high impedances (e.g. in the case of short sections of overhead lines), the fault current following a two-phase short-circuit (denoted I_{ccbi}) can be low and thus of the same order of magnitude as the rated current.

2.1.4.2. *Reliability indicators: SAIDI, SAIFI and ENS*

The CEER (Council of European Energy Regulators) is a non-profit association allowing a group of European regulators to exchange best practices. In particular, it conducts comparative reports on reliability. There are three main reliability metrics that are considered in planning studies. The first is *SAIDI* (System Average Interruption Duration Index), which is defined in CEER (2008). In France, grid operators use the terminology of *B criterion* (or respectively, *M criterion*) instead (FTICL Energy 2016), which is the average outage time per year and per LV (or MV, respectively) customer and is expressed in minutes per consumer per year. The second reliability indicator is the *SAIFI* (System Average Frequency Duration Index) (CEER 2008). In France, we instead talk about *F-LV criterion* (respectively, *F-HVA criterion*), which is the average outage frequency per year and per LV (respectively, MV) customer (FTICL Energy 2016). Finally, the *ENS* (Energy Not Supplied), known as END (*Énergie non distribuée*) in France, represents the additional quantity of energy that could have been consumed if there had been no outages and is expressed in kWh per year (Enedis 2008).

The general formulas for SAIDI, SAIFI and ENS are given by equations [2.3]–[2.5]. In the context of a planning study on a given area, these formulas are refined and depend on the architecture of the system and its operation mode. We detail these expressions for the French situation later on in this chapter.

$$SAIDI = \frac{\sum_{d=1}^{N_d} T_{co}(d) \times N_{CCO}(d)}{N_{cust}} \quad [2.3]$$

$$SAIFI = \frac{\sum_{d=1}^{N_d} N_{CCO}(d)}{N_{cust}} \quad [2.4]$$

$$END = \sum_{d=1}^{N_d} T_{co}(d) \times P_{avg}^{cut}(d)$$ [2.5]

with:

– $T_{co}(d)$: the outage cut-off time due to the fault d;

– $N_{CCO}(d)$: the number of customers cut off during the fault d;

– N_{cust}: the total number of customers;

– $P_{avg}^{cut}(d)$: the average power cut during the fault d;

– N_d: the number of faults in a year.

France's LV and MV SAIDI and SAIFI for the year 2017 are given in Table 2.1. To encourage DSOs to improve the service quality of their systems, bonus–malus systems have been adopted by several countries. In France's case, for example, the CRE sets an annual target in terms of SAIDI to be achieved by distribution systems. For example, in 2019, the SAIDI to be met was 63 min in LV and 45.1 min in MV (Enedis 2018). If the SAIDI is higher than the target (or lower), penalties (or bonuses, respectively) are included in the calculation of system costs (CEER 2016) called TURPE (*Tarifs d'utilisation des réseaux publics d'électricité*: Tariffs for the use of public electricity systems) in France (FTICL Energy 2016).

SAIDI (min)			SAIFI (by customer)		
Types of cuts	BT	MT	**Types of cuts**	BT	MT
Planned	14.14	11.35	**Brief**	1.32	1.49
Not planned	50.13	31.41	**Unplanned long cuts**	0.63	0.63
			Planned long cuts	0.12	0.12
Total	64.27	42.76	**Total long cuts**	0.75	0.75

Table 2.1. *France's SAIDI and SAIFI in 2019 (Enedis 2018)*

Table 2.2 gives some distribution system characteristics for Germany, France, Italy, the Netherlands, a California DSO for the year 2016, and Japan for the year 2015, as well as for the country with the best reliability (Singapore in 2016), and one with the worst (Nigeria). The notations used are as follows: L_{tot} (total distribution system length), τ_{bur} (buried system rate), $SAIDI_{EPXE}$ and $SAIFI_{EPXE}$ (SAIDI and SAIFI excluding planned and exceptional events). The total system lengths and buried rates are given for the year 2015 for Japan. The sources from which the various values are derived are also specified in the table. Burying lines significantly reduces the probability of a fault. We can cite, for example, the case of the Netherlands where almost 100% of the distribution system is buried (compared to 46% in France). Their SAIDI for 2016 was 17.7 min as compared to 48.7 min for France. Finally, the way the systems are operated has a direct impact on reliability. We can cite the case of Japan, which had a SAIDI in 2015 of 18 min where only 5.8% of the system is underground and the length of the distribution system is similar to France. We revisit the case of Japan later on in this chapter.

We distinguish two types of QoS based on the frequency of occurrence and their impacts:

– *daily quality*: these are minor incidents that can occur several times a year, such as a line short-circuit due to a tree branch, a bird and even snow;

– *quality in the face of large-scale events*: this concerns rare and exceptional events of very high intensity that occur every 20 years, such as storms. Following the French storm of 1999, electrical zones were classified into four types of Climate Change Plan (CCP) zones: "snow", "wind", "frost" and "wooded", which experience increased failure rates due to these climatic factors.

The solutions used to improve these two types of quality can be:

– *replacement or renovation of lines* that have increased failure rates and do not meet the criteria;

– *replacement oil-impregnated paper-insulation cables* by cables with cross-linked polyethylene insulation. These were installed up until 1977 in France, but are still present in the power system today and are more prone to failure;

– *burial of lines underground;*

– *change of overhead technology* such as the use of twisted cables rather than bare conductors;

– *the use of additional devices on the system* (anti-rotation weights to prevent snow or ice mantles, controlled elongation devices, etc.).

	Germany (CEER, Energy Supply Quality Working Group (EQ-WS)) - CEER 6.1 Benchmarking Report on the Continuity of Electricity and Gas Supply - (Data update 2015/2016, 2018)	Japan (OCCTO, 2016) and (TEPCO, 2016)	France (CEER, Energy Supply Quality Working Group (EQ-WS)) CEER 6.1 Benchmarking Report on the Continuity of Electricity and Gas Supply - (Data update 2015/2016, 2018)	Italy (CEER, Energy Supply Quality Working Group (EQ-WS)) - CEER 6.1 Benchmarking Report on the Continuity of Electricity and Gas Supply - (Data update 2015/2016, 2018)	Netherlands (CEER, Energy Supply Quality Working Group (EQ-WS) - CEER 6.1 Benchmarking Report on the Continuity of Electricity and Gas Supply - (Data update 2015/2016, 2018)	California (PG&E) (PG&E, PG&E Overview, 2014) and (PG&E, 2016 Annual Electric Reliability Report, 2017)	Singapore (Energy Market Authority, 2017) and (EMA, 2015)	Nigeria (Treasury Department, 2016)
L_{tot} (km)	1,711,030	1,368,709	1,348,876	1,257,130	256,026	228,526.8*	26,458**	
T_{bur} (%)	86.5%	5.8%	46%	49.3%	99.9%	20%	100%	
$SAIDI_{EPXE}$ (min)	12.8	18[2]	48.7	37.11	17.7	83	0.23	30% of the year or 2,628 h
$SAIFI_{EPXE}$	0.51	0.10**	0.08	1.5	0.21	0.818	0.006	

Table 2.2. *Parameters influencing SAIDI for different countries in 2016 (*PG&E data from 141,000 miles. Conversion rate: 1 mile = 1.60934 km; **data from 2015)*

Additional solutions can be used to improve the service quality during large-scale events:

– *mechanical reinforcement*: the use of components in the system that have better mechanical strength and can withstand stronger gusts of wind or large amounts of snow, such as metal poles rather than wooden poles, cables with a larger cross-section, and through technologies that have better mechanical strength, such as self-supporting insulated twisted cables;

– *moving the lines* to less wooded areas, for example;

– *raising the height of the lines* to prevent nearby trees from falling on them and causing a short circuit, or even, disconnecting the line and pushing poles over.

The technical solution chosen will be the one with the best techno-economic balance, or the lowest cost of implementation, and whose technical feasibility is proven. In most cases, the preferred solution is to bury the lines since the administrative procedures (obtaining the right-of-way agreements) are facilitated by the discreet nature of the project. However, if this solution is too complex or too expensive to implement due to land constraints or an increase in the total length of the lines (in the case where detours are required), then other solutions are considered. Finally, we will see in section 2.2.1 that the quality of the daily service can also be improved by acting on the system architecture and the use of switching devices to resupply consumers outside the faulty area.

New solutions are emerging where the distributor will not systematically seek to increase the reliability of the lines but will seek to increase the indicators as close as possible to the users. We can cite, for example, the installation of storage in distribution stations or the creation of microgrids. These solutions are presented in the following chapters.

2.1.4.3. *French indicator: the PL criterion*

The *PL* product is a purely French indicator. For a given area of the electrical distribution system, it is the product of the total maximum power injected in this area by the total length of the conductors that supply the customers. The French planner develops the system according to two objectives in terms of PL product:

– each zone must have a PL product that does not exceed a certain limit value;

– the PL products of each zone must be close (ideally, they should be identical).

Limiting the PL product allows the planner to ensure a good quality of supply and not to develop systems that are too long (with the risks of undervoltage) or too loaded (with the risks of violating the maximum allowable current). Second, the definition of equivalent PL product zones ensures an equipartition of the fault risks, and thus, an equality amongst consumers with respect to the continuity of supply. Indeed, a group of customers who consume little electricity will be supplied by a portion of the system with a greater length than a group of customers who consume a lot of electricity. The probability of a fault being proportional to the length of the conductor, thereby minimizing the duration of a power cut, thus improving the QoS.

In Coiffier (2013), the values used by the DSO Enedis in 2007 (ERDF at the time) are indicated: in rural areas, the maximum PL product varies from 55 to 100 MVA·km for maximum power of feeders varying from 3.75 to 5 MVA, while in urban areas, it is not necessary to set a PL product limit because of the high power density.

Possible planning solutions to reduce excessive PL and/or balance areas can be:

– *the creation of new zones*: the PL product of the old zone decreases because the total maximum power decreases;

– *the creation of a new primary substation*: if the loads are too far away from the primary substation, for example, then the PL product will be large because the length of the conductors will increase.

2.1.4.4. *Environmental indicators*

When developing strategies, the planner takes into account a set of constraints related to the environment, such as not deploying overhead systems in wooded areas and raising a transformer in a flood zone. On a more subjective level, the investor may also consider the visual impact by using discreet techniques such as burying or fronting cables when justified.

2.1.4.5. *Economic indicators*

2.1.4.5.1. The discounting principle

In a planning study, the DSO wanted to compare investment and expenditure options that occur at different times, and which as such are not comparable. To remedy this, they used the discounting principle to take into account the "value of time" and the "value of risk". The value of time is the preference for the present, that is, the fact that a euro today is worth more than a euro tomorrow when invested, for example. The value of risk is the uncertainty of events leading to the loss of money after a financial or economic operation, for example. The value of an investment C_{inv} made in year n compared to year 0 is given by equation [2.6].

$$C_{inv}' = \frac{C_{inv}}{(1+i)^n} \qquad [2.6]$$

with:

– i: the discount rate (in %);

– C_{inv}: the investment made in year n;

– C_{inv}': the value of the investment C_{inv} made in year n and brought back to year 0.

The simple example in Sim (2021) illustrates the principle of discounting. A person offers to gift me €1,000 today or in 5 years. It is intuitive to answer that I want to receive the €1,000 today because:

– I can invest this money in the bank and after 5 years have a sum higher than €1,000 or I can use this money for an immediate need (notion of "time value");

– in 5 years, the person may no longer want to give me the €1,000 because, for example, their personal economic situation has changed (the "risk value" concept).

The value of the discount rate is complex to define and depends on many country-specific parameters. We can see that a low discount rate will drive investment from the beginning while a high discount rate will discourage investment. For example, if the discount rate is 10%, then, if I get €620 today, I will have €1,000 in 5 years' time; however, if the discount rate is 50%, then, if I get €132 today, I will have €1,000 in 5 years. In the case of planning studies, the discount rate set by the public authorities in France has been at 8% since 1985 (Coiffier 2013), but in 2020 this was revised to 4.5% (Enedis 2020).

2.1.4.5.2. Calculation of discounted cost

The international terminology for the overall cost spent by DSOs on electricity systems is TOTEX (TOTal EXpenditure). It is composed of investment costs called CAPEX (CAPital EXpenditures) and operating costs called OPEX (OPerational EXpenditures). Equation [2.7] gives the value of TOTEX[1] back to year 0 based on the TOTEX of each year of the study period.

$$C_{TOTEX}' = \sum_{n=0}^{N_{years}} \frac{C_{TOTEX}(n)}{(1+i)^n} \qquad [2.7]$$

with:

– C_{TOTEX}': all C_{TOTEX} back to year 0;

– n: the year index;

– $C_{TOTEX}(n)$: the TOTEX for year n;

– N_{years}: the number of years in the study period.

1 The expression of TOTEX may vary slightly depending on when CAPEX and OPEX are paid, that is, either at the beginning of the year or at the end of the year. In this formulation and for simplicity, the initial year is year 0 and the planning study is performed up to and including N_{year}. We assume that all costs are paid at the beginning of the year and therefore there are N + 1 payments. In reality, we can have work sites that span several years.

CAPEX

The CAPEX is the investment made by the DSO to create future benefits (e.g. in terms of losses and reliability) by purchasing system equipment (conductors, trenches, and remote-controlled switches). Equation [2.8] gives the CAPEX expression.

$$C_{CAPEX}(n) = \left[\sum_{j=0}^{N_{lines}} L(j) \times \left(C_c(j) + C_t(j)\right) \right] + N_{NOS} \times C_{NOS} - V_u(n) \qquad [2.8]$$

with:

- N_{lines}: the number of lines and cables in the system;
- $L(j)$: the length of the line or cable no. j (in km);
- $C_c(j)$: the cost of the line or cable no. j (in k€/km);
- $C_t(j)$: the cost of trench no. j (in the case of a cable) (in k€/km);
- N_{RCS}: the number of remote-controlled operating elements;
- C_{RCS}: the cost of a remote-controlled shunting device (in k€);
- V_u: the use value (k€).

The use value is the residual value of the structure at the end of the study period (Coiffier 2013). It is therefore zero if the study period corresponds to the life of the structure, which only happens if a new system is built. However, for short- and medium-term studies, such as conductor reinforcement or transformer replacement, the use value is no longer negligible. Equation [2.9] gives the expression for the use value (Coiffier 2013) in year *n*.

$$V_u(n) = V_F + (V_N - V_F) \times \frac{(1+i)^{T_{DV}} - (1+i)^n}{(1+i)^{T_{DV}} - 1} \qquad [2.9]$$

with:

- n: the index of the year;
- V_N: the replacement value of the work (in k€);
- V_F: the value of structural reinforcement (in k€);
- T_{DV}: the structural lifetime (in years).

OPEX

OPEX includes all the costs of operating the power system. It currently consists of maintenance costs, including spare parts, labor, technical losses and reliability. These costs must take into account the growth rates of the loads each year.

In the future, if advanced flexibility solutions are developed, their costs will be included in the OPEX calculation. For example, in the case of load management, the consumers' remuneration for this service will be introduced by calculating the cost of the total energy curtailed in a given year for example.

Equation [2.10] gives the cost of OPEX at a given year *t* (undiscounted), and equation [2.11] gives the cost of OPEX discounted to year 0. To encourage strategies to improve reliability, the number of outages and the total amount of undelivered energy are valued. Depending on the countries' market model, it may be interesting to include the cost for energy production in the economic calculations. The TOTEX is then divided by the production cost of the energy, commonly referred to as LCOE.

$$C_{OPEX}(n) = \sum_{j=1}^{N_{lines}} C_{peak} \times P_{lmax}(j,n) + \sum_{z=1}^{N_Z} C_{ENS} \times ENS(z,n) + \sum_{z=1}^{N_Z} C_{PC} \times P_{cut}(z,n) + \sum_{h=1}^{N_M} C_m(h,n) \quad [2.10]$$

$$C_{OPEX}' = \sum_{n=0}^{N_{years}} \frac{C_{OPEX}(n)}{(1+i)^n} \quad [2.11]$$

with:

– C_{peak}: the cost of joule losses at peak (€/kW);

– $P_{lmax}(j,n)$: the peak losses for line or cable j in year n (kW);

– C_{ENS}: the cost of energy not supplied (fixed at 9.15 €/kWh in France (Doulet 2002; Marignac and Legrand 2003));

– $ENS(z,n)$: the energy not supplied for zone *z* in year n (kWh);

– N_Z: the number of system zones for which the different reliability indices are calculated (depends on the system architecture);

– z: the index of the zone under study;

– C_{PC}: the cost of interruptions due to failures (0.75 €/kW in France (Dfeeder 2002; Marignac and Legrand 2003));

– $P_{cut}(k,t)$: power cut in zone k in year t (kW) due to faults;

– N_M: the number of system components for which maintenance work is required;

– h: the index of the component requiring maintenance work;

– $C_m(h,t)$: the maintenance cost of component h in year n (k€).

Estimated cost of joule losses

The general expression for peak joule losses is given by equation [2.12]. Since the voltage varies little around the nominal voltage, we can apply the load growth rates defined in equation [2.1] to express the current in a line or cable no. j at year *t* as a function of that at year 0.

Therefore, the maximum loss equation at year *t* becomes equation [2.13] and then equation [2.14].

$$P_{lmax}(j,n) = 3 \times r(j) \times L(j) \times \left(I_{max}(j,n)\right)^2 \quad [2.12]$$

$$P_{lmax}(j,n) = 3 \times r(j) \times L(j) \times \left(I_{max}(j,0) \times \sum_{p=1}^{N_{periods}} \left(1+\tau_p^{peak}\right)^{T_p} \times \left(1+\tau_{N_{periods}+1}^{peak}\right)^{\left(n-\sum_{p=1}^{N_{periods}} T_p\right)} \right)^2 \quad [2.13]$$

$$P_{lmax}(j,n) = P_{lmax}(j,0) \times \left(\sum_{p=1}^{N_{periods}} \left(1+\tau_p^{peak}\right)^{T_p} \times \left(1+\tau_{N_{periods}+1}^{peak}\right)^{\left(n-\sum_{p=1}^{N_{periods}} T_p\right)} \right)^2 \quad [2.14]$$

with:

– $r(j)$: the linear resistance of the line or cable no. j (in Ω/km);

– $I_{max}(j, 0)$: the maximum peak current in line or cable no. j at year 0 (in A);

– $N_{periods}$: the number of whole periods with different load growth rates in the interval [0;t] years;

– τ_p^{peak}: annual growth rate in maximum power consumption (in %) for the period p;

– T_p: the number of years of the period no. p;

– $\tau_{N_{periods}+1}^{peak}$: annual growth rate in maximum power consumption for the last period (in %).

As shown in equation [2.11], the annual joule loss cost is obtained by multiplying the peak loss value by the empirically constructed peak joule loss cost, C_P, from the DSOs.

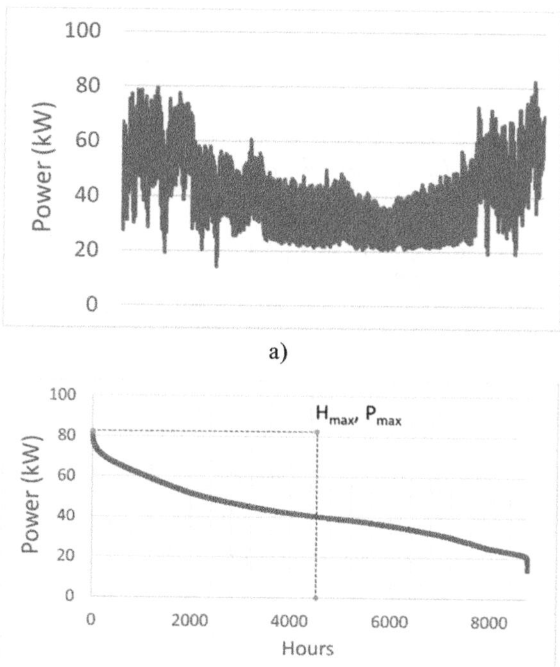

Figure 2.3. *Annual load curve of: a) a distribution substation at hourly intervals and b) its equivalent monotone load (source: Enedis 2021a)*

To calculate the peak losses, we evaluate the peak consumption using the parameter H_{max}, which is expressed in hours. H_{max} corresponds to the equivalent time for which a customer operating at maximum power will consume the same energy as when operating at average power throughout the year. H_{max} is therefore such that the product of the maximum annual power by H_{max} is equal to the annual energy consumed. Figure 2.3(a) shows an example of a real load curve of a substation in the Greenlys Project measured in hourly time intervals (Enedis 2021a). Figure 2.3(b) shows the corresponding load monotone, that is, the ranking of the power in decreasing order as well as the parameter H_{max}. The cost of peak losses, C_P, takes into account not only the cost of the energy lost for 1 kW of power injected at peak, but also the annual capital cost of delivering a kilowatt to the consumption areas (i.e. the cost of the system infrastructure) and eventually the cost of subscribed power.

Energy not supplied and power loss estimation

In general, the distribution system is partitioned into feeders with protection devices. The energy not supplied and power interruption in year t of a zone z of a given feeder is given by equation [2.15] through equation [2.18].

$$ENS(z,n) = \tau_{fail} \times L_{zone}(z) \times \left(T_{iso} \times P_{avg}^{feeder}(n) + T_{rep} \times P_{avg}^{z}(n)\right) \quad [2.15]$$

$$\begin{aligned} ENS(z,n) = \tau_{fail} \times L_{zone}(z) &\times \left(T_{iso} \times P_{avg}^{feeder}(0) + T_{rep} \times P_{avg}^{z}(0)\right) \\ &\times \sum_{p=1}^{N_{periods}} \left(1+\tau_p^{avg}\right)^{T_p} \\ &\times \left(1+\tau_{N_{periods}+1}^{avg}\right)^{\left(n-\sum_{p=1}^{N_{periods}} T_p\right)} \end{aligned} \quad [2.16]$$

$$P_{cut}(z,n) = \tau_{fail} \times L_{zone}(z) \times P_{fail}^{feeder}(n) \quad [2.17]$$

$$\begin{aligned} P_{cut}(z,n) = \tau_{fail} \times L_{zone}(z) \times P_{avg}^{feeder}(0) &\times \sum_{p=1}^{N_{periods}} \left(1+\tau_p^{avg}\right)^{T_p} \\ &\times \left(1+\tau_{N_{periods}+1}^{avg}\right)^{\left(n-\sum_{p=1}^{N_{periods}} T_p\right)} \end{aligned} \quad [2.18]$$

with:

– τ_{fail}: the failure rate (in number of incidents/year/km);

– $L_{zone}(z)$: the length of the zone no. z (in km);

– T_{iso}: the fault isolation time (in hours);

– T_{rep}: the time to repair the fault (in hours);

– P_{avg}^{feeder}: average power of the studied outlet (in kW);

– P_{avg}^{z}: average power of the zone no. z (in kW);

– τ_{p}^{avg}: the growth rate of average power consumption (in %);

– $\tau_{N_{periods}+1}^{avg}$: annual rate of change in average power consumption for the last period (in %).

In He et al. (2011), the Swedish DSO, Vattenfall Distribution, conducted a statistical study of the causes of 10,710 failures that resulted in a power outage of more than 6 h on its distribution system between 2004 and 2008. The conclusions are that between 77.9% and 96.7% of LV and MV overhead failures are due to natural outdoor conditions (vegetation, climate and animals), depending on the type of equipment (busbars, isolated lines, bare lines). In contrast, the main causes of underground failures are due to equipment defects: due to age (55.8% in LV and 63.1% in MV); sectioning during construction (15.4% in LV and 16.3% in MV); and tree falls at junctions between overhead and underground parts of the system in the case of mixed systems (12.4% in LV and 7.3% in MV). Table 2.3 shows the average, minimum, and maximum failure rates as a function of voltage level in the US (Brown 2008) and on about 30 French concessions in 2012 (Syndicat départemental d'énergie électrique de la Gironde 2015). The average rates indicated vary from one area to another as they depend on the conditions of use (voltage levels, climate among others).

	Number of incidents/year/100 km			
	US		**France**	
	Average rate	**Minimum and maximum rate**	**Average rate**	**Minimum and maximum rate**
Underground MV	4.35	[0.19; 36.47]	4.8	[2.4; 9.6]
Underground LV	6.21	[0.31; 9.32]		
Overhead MV	6.21	[1.24; 18.64]	6.67	[3.7; 9]
Overhead LV	5.47	[1.24; 18.64]		

Table 2.3. *MV line and cable failure rate*

2.1.4.5.3. Investment decision criteria and selection of the optimal date

The strategies identified in the planning studies are compared using the overall discounted cost calculation presented earlier (Enedis and ADEeF 2017). The strategy selected is the one with the lowest discounted cost. In economics, this comparison is done by calculating the net present value of a given strategy (NPV). This is the difference between the TOTEX of the "Do nothing" strategy and the TOTEX of the "investing" strategy. For example, suppose the planner has three possible planning strategies: "Build a new line", "Reinforce the system", or "Do nothing", that is, do not invest in the system for the duration of the project.

It will then calculate $NPV_{new\ line}$ and $NPV_{reinforcement}$ according to equation [2.19] and choose the maximum NPV.

$$\begin{cases} NPV_{new\ line} = TOTEX_{DN} - TOTEX_{new\ line} \\ NPV_{reinforcement} = TOTEX_{DN} - TOTEX_{reinforcement} \end{cases} \quad [2.19]$$

with:

– $NPV_{new\ line}$: the net present value of the "Build a new line" alternative;

– $NPV_{reinforcement}$: the net present value of the "Reinforce the system" alternative;

– $TOTEX_{DN}$: the total discounted cost of the "Do nothing" alternative;

– $TOTEX_{new\ line}$: total discounted cost of the "Build a new line" alternative;

– $TOTEX_{reinforcement}$: total discounted cost of the "Reinforce the system" alternative.

The NPV calculation allows us to determine the economic return of each alternative ("Build a new line" or "Reinforce the system") assuming they are implemented immediately or in year 0. However, in a context of increasing electrical load over time, the decision-maker may prefer to defer the investment for a number of years until the benefits of each alternative are sufficient to cover the costs[2]. Finding a point in time when the electrical load increases sufficiently to justify the implementation of a solution is the optimal time to make the investment. Specifically, once the optimal strategy has been selected, the timing of the work to be done to achieve the intermediate systems can be obtained either by calculating the

2 For further detail, the reader can refer to the pioneering work of Jorgenson (1963) who emphasizes that waiting can delay the payment of the investment cost and thus its discounting. The investment becomes optimal if the marginal benefit of an additional unit of capital is equal to the cost of using capital (discount rate).

CBR or by maximizing the NPV[3] between the selected strategy and the "Do nothing" strategy.

Let C_{inv} be the investment necessary to carry out the work making it possible to obtain the intermediate system 1 expressed in euros of year 0. If this investment is made in year *n*, then its value in year 0 will be $\frac{C_{inv}}{(1+i)^n}$. The fact of anticipating this investment in year n-1 generates for the firm the additional cost given by equation [2.20].

$$\frac{C_{inv}}{(1+i)^{n-1}}-\frac{C_{inv}}{(1+i)^n}=\frac{i\times C_{inv}}{(1+i)^n} \quad [2.20]$$

It will be in its interest to invest in year n-1 if this additional cost is compensated by the gain in operating cost, that is, the difference between the operating costs of the solution "I do not invest in year n-1" and that of "I invest in year n-1". This amounts to satisfying the conditions in equation [2.21], which we can simplify by equation [2.22] and then equation [2.23].

$$\frac{i\times C_{inv}}{(1+i)^n}\leq\frac{C_{OPEX}^{DN}(n-1)-C_{OPEX}^{STR}(n-1)}{(1+i)^{n-1}} \quad [2.21]$$

$$\frac{i}{1+i}C_{inv}\leq C_{OPEX}^{DN}(-1)-C_{OPEX}^{STR}(n-1) \quad [2.22]$$

$$\frac{C_{OPEX}^{DN}(n-1)-C_{OPEX}^{DN}(n-1)}{C_{inv}}\geq\frac{i}{1+i} \quad [2.23]$$

with:

– $C_{OPEX}^{STR}(n-1)$: the operating cost of the "I invest from year n-1" strategy (in k€);

– $C_{OPEX}^{DN}(n-1)$: the operating cost of the "I do not invest in year n-1" strategy (in k€).

We define in equation [2.24] the cost–benefit ratio in year n $(CBR_r(n))$ corresponding to the work to move from intermediate system *r* to intermediate system *r* + 1.

3 In the economic literature, Marglin (1963) was among the first to show that when an investment is profitable but irreversible, it may not be optimal to act immediately but to wait for a more advantageous time to make it. This corresponds to maximizing the NPV of the investment opportunity by choosing the optimal time.

$$CBR_r(n) = \frac{Gain_{OPEX}(n)}{C_{inv}^r} = \frac{C_{OPEX}^{DN}(n) - C_{OPEX}^{STR}(n)}{C_{inv}^r} \quad [2.24]$$

with:

– $Gain_{OPEX}(n)$: the difference between the operating costs of the "Do nothing in year n" alternative and the "I invest in year n" alternative;

– C_{inv}^r: the investment required to move from the r-1 system to the r system (in € of the year 0).

The date n on which the investment should be made will be the first date for which $CBR_r(n)$ is greater than $\frac{i}{1+i}$.

2.1.5. *Planning options*

In order to respect these technical constraints, several planning options, combined or not, can be considered. We can list:

– *reinforcement of conductors*: replace an existing line with a line of larger cross-section, and/or made of a less resistive material (copper rather than aluminum, for example) in order to increase the maximum current carrying capacity or split a line to spread the load if the cross-section is already at maximum. As the resistance of the lines is inversely proportional to their cross-section, the reinforcement allows for a decrease in the resistance, and consequently the voltage drop (see equation [1.16] of Chapter 1);

– *transformer reinforcement*: the adaptation of an existing transformer using a transformer of higher apparent power, or the creation of an additional transformer, the location of which depends on the barycenter of the load, taking into account topology and spatial constraints;

– *reducing the length of the feeders* in order to reduce voltage drop: by creating new feeders or by transferring a load to a shorter feeder, the length of the lines is thereby reduced;

– *increasing the voltage level*: increasing the voltage level decreases voltage drops but also the losses (see equation [1.25] in Chapter 1), and thus the power flows.

Other solutions not currently used in distribution are being studied:

– *taking into account the real technical constraint of the line*: the design data gives a maximum admissible current of a line under given operating conditions. We speak here of *static line rating*, since these operating conditions do not change and generally correspond to the worst-case scenario, that is, low wind speed, high ambient temperatures and high solar radiation (Wang and Pinter 2014). New technologies consisting of sensors and a communication system with a control center for the power grid operator, called *dynamic line rating*, can determine the maximum power that the line can support in real time. In Wang and Pinter (2014), the New York Power Authority (NYPA) and the Oncor Electric Delivery Company (Oncor) tested dynamic line rating technologies on their transmission system as part of a Smart Grids Demonstrator Program (SGDP) in partnership with the U.S. Department of Energy (DOE). In particular, it was shown that it was possible to gain 25% more capacity on the lines by considering actual conditions rather than the constructor's conditions. Therefore, having access to the technical data of the line in real time can defer unnecessary investments by considering the real maximum power of the line in question;

– *MVDC reinforcement*: in the UK, the increase in consumption and renewable generation on the Isle of Anglesey required additional reinforcement from the AC grid. In the Angle-DC project, over the 2016–2020 period, an AC reinforcement was implemented. This involved the construction of a ±27 kV bi-directional medium voltage DC line connecting the island's 33 kV AC system to the 33 kV AC system in North Wales, by means of two AC/DC converter stations. This reinforcement allowed, among other things, for the better control of power flows and voltage.

2.1.6. *Application of techno-economic formulas on simple examples*

2.1.6.1. *Example 1: MV customer connection (short-term planning)*

A large industrial customer with a consumption of up to 5 MW and a power factor of 0.92 wishes to be connected to the distribution system. Considering their power consumption, the DSO decides to connect them with a 3 km underground direct feeder to the nearest HV/MV primary substation. The objective of this case study is to consider two possible solutions:

– solution no. 1: selection of the minimum cable cross-section (i.e. the least expensive cable) to meet the current and voltage constraints;

– solution no. 2: selection of the economic cable section making it possible to minimize the global cost discounted over the cable life (investment cost and joule losses cost) and to respect the current and voltage constraints. The reliability criteria will not be calculated in this example since the solution studied has no impact on the reliability of the system (same conductor length in underground technology).

The technical and economic information needed to solve this case study is given in Table 2.4.

Typical cables used in MV distribution					
Type	**Cross-section (mm^2)**	**r (Ω/km)**	**x (Ω/km)**	**I_{max} (A)**	**Cost (C_c, k€/km)**
Aluminum	95	0.32	0.1	230	9
Aluminum	150	0.206	0.1	300	10
Aluminum	240	0.125	0.1	405	12
Copper	240	0.075	0.1	505	50
Supply voltage (U)		20 kV	**Discount rate (i)**		4.5%
P_{max}		5 MW	**Load growth rate (τ_p)**		0.5%/year
H_{max}		4,000 h	**Power factor (cosφ)**		0.92
Cable length (L)		3 km	**Average cost of an urban trench (C_t)**		110 k€/km
Cost of peak losses (C_P, €/kW)		204	**Duration of the study (N)**		40 years

Table 2.4. *Technical and economic data*

2.1.6.1.1. Minimum cable section

The goal is to select the least expensive cable that satisfies the maximum allowable current and voltage drop constraints. According to Table 2.4, the load will increase by 0.5% per year for the duration of the study (30 years). The maximum current consumed by the industrial customer in year 30 is given by equation [2.25]. Using the relationship in equation [2.1] between the maximum power at year 30 and the load growth rate, we obtain equation [2.26].

$$I_{max}(N_{years}) = \frac{P_{max}(N_{years})}{\sqrt{3} \times U \times \cos\varphi} \qquad [2.25]$$

$$I_{max}(N_{years}) = \frac{P_{max}(0) \times \left(1 + \tau_p^{peak}\right)^{N_{years}}}{\sqrt{3} \times U \times \cos\varphi} = 191.5\ A \qquad [2.26]$$

Table 2.4 shows that the 95 mm² cable is sufficient because its maximum current rating is 230 A. We check that the maximum voltage drop given by equation [2.27] is respected. As the maximum voltage drop is 1.58% (equation [2.28]), we can conclude that a 95 mm² aluminum cross-section is sufficient to respect the voltage and current constraints.

$$\frac{\Delta U}{U}\left(N_{years}\right)=\frac{R \times P_{max}\left(N_{years}\right)+X \times P_{max}\left(N_{years}\right) \times \tan \varphi}{U^{2}} \quad [2.27]$$

$$\frac{\Delta U}{U}\left(N_{years}\right) = \frac{(r \times L + x \times L \times \tan \varphi) \times P_{max}(0) \times \left(1+\tau_{p}^{peak}\right)^{N_{years}}}{U^{2}} = 1.66\% \quad [2.28]$$

2.1.6.1.2. Economic section of the cable

To determine the economic section of the cable, we need to calculate the overall actuated cost (CAPEX + OPEX – value in use) over the life of the cable for the different cross-sections and then select the cross-section with the minimum cost. Taking the approach from equation [2.7] through equation [2.11], we find that the overall discounted cost of laying a cable to connect the industrial customer can be given by equation [2.29].

$$C_{TOTEX}' = L \times (C_c + C_t) + \sum_{n=0}^{N_{years}} \frac{C_{peak} \times P_{lmax}(n)}{(1+i)^{n}} \quad [2.29]$$

Adapting equation [2.13] to our case study and feeding it back into equation [2.29], we obtain equations [2.30] through [2.33].

$$P_{lmax}(n) = 3 \times r \times L \times \left(I_{max}(0) \times \left(1+\tau_{p}^{peak}\right)^{n}\right)^{2} \quad [2.30]$$

$$P_{lmax}(n) = 3 \times r \times L \times \left(\frac{P_{max}(0) \times \left(1+\tau_{p}^{peak}\right)^{n}}{\sqrt{3} \times U \times \cos \varphi}\right)^{2} \quad [2.31]$$

$$P_{lmax}(0) = 3 \times r \times L \times \left(\frac{P_{max}(0)}{\sqrt{3} \times U \times \cos \varphi}\right)^{2} \quad [2.32]$$

$$C_{TOTEX}' = L \times (C_c + C_t) + P_{ermax}(0) \times C_{peak} \times \sum_{n=0}^{N_{years}} \left(\frac{\left(1+\tau_p^{peak}\right)^2}{(1+i)} \right)^n \quad [2.33]$$

Or:

$$\frac{\left(1+\tau_p^{peak}\right)^2}{(1+i)} = \frac{\left(1+\frac{0.5}{100}\right)^2}{\left(1+\frac{8}{100}\right)} = 0.94 < 1 \quad [2.34]$$

Hence $\sum_{n=0}^{N_{years}} \left(\frac{\left(1+\tau_p^{peak}\right)^2}{(1+i)} \right)^n$ is the sum of the first n+1 terms of a geometric sequence and common ratio $\frac{\left(1+\tau_p^{peak}\right)^2}{(1+i)}$ less than 1.

Therefore, we know that:

$$\sum_{n=0}^{N_{years}} \left(\frac{\left(1+\tau_p^{peak}\right)^2}{(1+i)} \right)^n = \frac{1 - \left(\frac{\left(1+\tau_p^{peak}\right)^2}{(1+i)} \right)^{N_{years}+1}}{1 - \left(\frac{\left(1+\tau_p^{peak}\right)^2}{(1+i)} \right)} \quad [2.35]$$

Equation [2.33] thus becomes equation [2.36].

$$C_{TOTEX}' = L \times (C_c + C_t) + P_{lmax}(0) \times C_{peak} \times \frac{1 - \left(\frac{\left(1+\tau_p^{peak}\right)^2}{(1+i)} \right)^{N_{years}+1}}{1 - \left(\frac{\left(1+\tau_p^{peak}\right)^2}{(1+i)} \right)} \quad [2.36]$$

The TOTEX calculation for the four cable types gives the results in Table 2.5. We can see that 240 mm² aluminum is the cross-section that respects the current and voltage constraints while minimizing the overall discounted cost over its lifetime.

This simple example illustrates the interest for the planner to take a long-term view by considering all of the system costs and not only the investment costs.

Type of cable	CAPEX (k€)	OPEX discounted to year 0 (k€)	TOTEX (k€)
95 mm² aluminum	354	325.1	679.1
150 mm² aluminum	357	209.3	566.3
240 mm² aluminum	366	127	493
240 mm² copper	450	76.2	526.2

Table 2.5. *TOTEX calculation for the four cable types*

2.1.6.2. *Example 2: topological modification of the system (long-term planning)*

Let us consider the initial system shown on the left in Figure 2.4, which consists of two tree-like feeders. In order to improve the service quality of the system, the planner proposes the target system on the right, which consists of three feeders.

Table 2.6 gives the general technical and economic characteristics of the study, and Tables 2.7 and 2.8 give the technical characteristics of the two systems. The objective is then:

– to calculate the total discounted cost of the target system over the next 30 years, as well as the cost of the "do nothing" alternative, which is to maintain the original system;

– to determine the optimal time period for the work to be carried out.

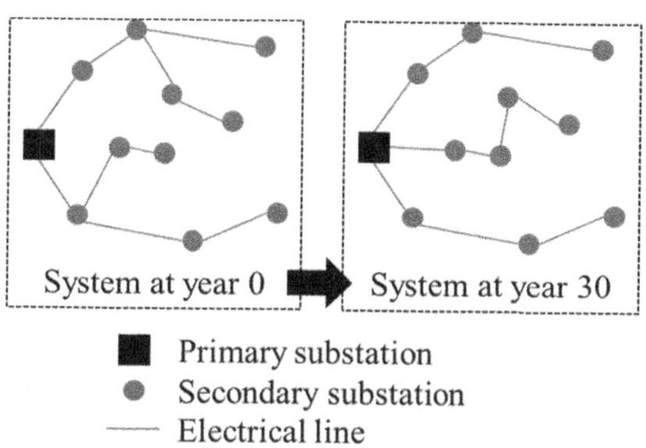

Figure 2.4. *Example no. 2*

<table>
<tr><td>Supply voltage (U)</td><td>20 kV</td><td>Discount rate (i)</td><td colspan="2">4.5%</td></tr>
<tr><td>Resistance of the cable used (150 mm² alu, Ω/km)</td><td>0.206</td><td>Load growth rate (τ^{avg})</td><td colspan="2">1.5%/year</td></tr>
<tr><td>H_{max}</td><td>4,000 h</td><td>Power factor ($\cos \varphi$)</td><td colspan="2">0.92</td></tr>
<tr><td rowspan="2">Period of study (years)</td><td rowspan="2">40</td><td rowspan="2">Failure rate (number of interruptions per year per 100 km)</td><td>Overhead</td><td>Underground</td></tr>
<tr><td>6</td><td>4</td></tr>
<tr><td>Cost of peak losses (C_{peak}, €/kW)</td><td>204</td><td>Cost of energy not supplied (C_{END}, €/kWh)</td><td colspan="2">10</td></tr>
<tr><td></td><td></td><td>Cost of interruptions (C_{PC}, €/kW)</td><td colspan="2">1</td></tr>
</table>

Table 2.6. *General technical and economic data*

<table>
<tr><td colspan="5">Characteristics of the initial system</td></tr>
<tr><td></td><td colspan="2">Feeder 1</td><td colspan="2">Feeder 2</td></tr>
<tr><td>P_{max} (MW)</td><td colspan="2">3</td><td colspan="2">3</td></tr>
<tr><td>Length (km)</td><td>Underground</td><td>Overhead</td><td>Underground</td><td>Overhead</td></tr>
<tr><td></td><td>10</td><td>7</td><td>18</td><td>12</td></tr>
<tr><td>Average outage duration (hours)</td><td colspan="2">3</td><td colspan="2">3</td></tr>
<tr><td>Maximum joule losses (P_{lmax}, kW)</td><td colspan="2">228.2</td><td colspan="2">243.4</td></tr>
</table>

Table 2.7. *Technical data of the initial system*

<table>
<tr><td colspan="7">Characteristics of the final system</td></tr>
<tr><td></td><td colspan="2">Feeder 1</td><td colspan="2">Feeder 2</td><td colspan="2">Feeder 3</td></tr>
<tr><td>P_{max} (MW)</td><td colspan="2">2</td><td colspan="2">2</td><td colspan="2">2</td></tr>
<tr><td rowspan="2">Length (km)</td><td>Underground</td><td>Overhead</td><td>Underground</td><td>Overhead</td><td>Underground</td><td>Overhead</td></tr>
<tr><td>10</td><td>0</td><td>18</td><td>0</td><td>19</td><td>0</td></tr>
<tr><td>Average outage duration (hours)</td><td colspan="2">2</td><td colspan="2">2</td><td colspan="2">2</td></tr>
</table>

Maximum joule losses (P_{lmax}, kW)	54.8	98.6	76.7
Total investment needed: 2,280 k€			

Table 2.8. *Technical data of the final system (with year 0 loads)*

2.1.6.2.1. Calculation of the total discounted cost

In order to calculate the overall discounted cost of the two systems, it is first necessary to calculate the operating costs, that is, the cost of losses, energy not supplied and the cost of outages as outlined in equation [2.10]. Since we have the global information for entire feeders but not at the level of lines, the cost calculations are done for each feeder. Moreover, we do not consider in this example any maintenance cost. Therefore equation [2.10] becomes equation [2.37].

$$C_{OPEX}' = \sum_{n=0}^{N_{years}} \frac{C_{losses}(n) + C_{ENS}(n) + C_{PC}(n)}{(1+i)^n} \quad [2.37]$$

$$C_{losses}(n) = \sum_{f=1}^{N_{feeders\,MV}} C_{peak} \times P_{lmax}(f,n) \quad [2.38]$$

$$C_{ENS}(n) = \sum_{f=1}^{N_{feeders\,MV}} C_{ENS} \times ENS(f,n) \quad [2.39]$$

$$C_{PC}(n) = \sum_{f=1}^{N_{feeders\,MV}} C_{PC} \times P_{cut}(f,n) \quad [2.40]$$

with: $N_{feeders\,MV}$: number of feeders of the studied system.

The losses cost is obtained using the formula in equation [2.14].

$$C_{losses}(n) = \sum_{f=1}^{N_{feeder\,MV}} C_{peak} \times P_{lmax}(f,0) \times (1+\tau^{avg})^{2\times n} \quad [2.41]$$

The energy not supplied cost can be estimated using equation [2.16] in order to obtain equation [2.42].

$$C_{ENS}(n) = \sum_{f=1}^{N_{feeders\,MV}} C_{ENS} \times \left(\tau_{fail}(overhead) \times L_{overhead}(f) + \tau_{fail}(underground) \times L_{underground}(f)\right) \times T_{co}(f) \times P_{avg}^{f}(0) \times (1+\tau^{avg})^{n} \quad [2.42]$$

As defined in section 2.1.4.5.2, the average power can be estimated using equation [2.43], resulting in equation [2.44].

$$P_{avg} = \frac{P_{max} \times H_{max}}{8{,}760} \quad [2.43]$$

$$C_{ENS}(n) = \sum_{f=1}^{N_{feeders\,MV}} C_{ENS} \times \left(\tau_{fail}(overhead) \times L_{overhead}(f) + \tau_{fail}(underground) \times L_{underground}(f)\right) \times T_{co}(f) \times \frac{P_{max}^{f}(0) \times H_{max}}{8{,}760} \times (1+\tau^{avg})^{n} \quad [2.44]$$

Similarly, the cost of the power outage is described by equation [2.45].

$$C_{PC}(n) = \sum_{f=1}^{N_{feeders\,MV}} C_{PC} \times \left(\tau_{fail}(overhead) \times L_{overhead}(f) + \tau_{fail}(underground) \times L_{underground}(f)\right) \times \frac{P_{max}^{f}(0) \times H_{max}}{8{,}760} \times (1+\tau^{avg})^{n} \quad [2.45]$$

Table 2.9 gives the numerical values of the average power per feeder for the two systems as well as the operating costs (cost of joule losses, energy not supplied, and interruptions) per feeder and per system for year 0. The total discounted cost is also calculated. By taking equation [2.36] and adapting it to the second case study, we obtain equation [2.46].

$$
\begin{aligned}
C_{OPEX}' = C_{losses}(0) \times \frac{1 - \left(\frac{(1+\tau^{peak})^2}{(1+i)}\right)^{N_{years}+1}}{1 - \left(\frac{(1+\tau^{peak})^2}{(1+i)}\right)} \\
+ \left(C_{ENS}(0) + C_{PC}(0)\right) \\
\times \frac{1 - \left(\frac{(1+\tau^{avg})}{(1+i)}\right)^{N_{years}+1}}{1 - \left(\frac{(1+\tau^{avg})}{(1+i)}\right)}
\end{aligned} \qquad [2.46]
$$

We find that the optimal overall discounted cost is for the construction of the target system (see Table 2.9). This strategy reduces the cost of energy not supplied by about 2.7, the cost of outages by 1.8 and the cost of losses by 2.25. We can also see that in the target system, the investment represents about 59% of the overall discounted cost; however, this diminishes after reducing the OPEX costs by a factor of 2.5.

2.1.6.2.2. Optimum date for completion of works

The NPV of the project is positive, so the planner will choose to carry out the adaptation of their system. In order to determine when the works will be carried out, in economic terms, a CBR calculation must be performed. We recall that the date for the execution of the works (if it exists) is the first one that satisfies the conditions described by equation [2.23], namely:

$$
\frac{C_{OPEX}^{DN}(n_a) - C_{OPEX}^{STR}(n_a)}{C_{inv}} \geq \frac{i}{1+i} \qquad [2.47]
$$

with: n_a the year of the beginning of the work.

The construction date obtained with the CBR criterion is year 5. As mentioned in section 2.1.4.5.3, this value can also be found by looking for the investment year that maximizes the NPV of the project. In Figure 2.5, showing the NPV as a function of the year of investment, we can observe that date n = 0 is the break-even point for the system reinforcement project, when the NPV is positive. From this point on, the implementation of a new solution (adding a new start date) is justified. Beyond this point, deferring the investment project for 1 year will produce a net gain as the benefit increases and the NPV increases until it reaches a maximum value at year 5. After this optimal point, the gain by deferring another year becomes negative and the NPV begins to decrease.

<table>
<tr><th></th><th colspan="2">Initial system
"Do nothing" strategy</th><th colspan="3">Final system
(data from year 0)
"Invest" strategy</th></tr>
<tr><th></th><th>Feeder 1</th><th>Feeder 2</th><th>Feeder 1</th><th>Feeder 2</th><th>Feeder 3</th></tr>
<tr><td>Average capacity (MW)</td><td>1.4</td><td>1.4</td><td>0.9</td><td>0.9</td><td>0.9</td></tr>
<tr><td>Cost of losses at year 0 ($C_{losses}(0)$, k€)</td><td>19</td><td>33.5</td><td>5</td><td>8.9</td><td>9.4</td></tr>
<tr><td>Cost of ENS at year 0 ($C_{ENS}(0)$, k€)</td><td>33.7</td><td>59.2</td><td>7.3</td><td>13.2</td><td>13.9</td></tr>
<tr><td>Cost of interruptions at year 0 ($C_{PC}(0)$, k€)</td><td>1.1</td><td>2</td><td>0.4</td><td>0.7</td><td>0.7</td></tr>
<tr><td>OPEX per feeder for the year 0</td><td>53.8</td><td>94.7</td><td>12.7</td><td>22.8</td><td>24</td></tr>
<tr><td>System OPEX for year 0</td><td colspan="2">148.5</td><td colspan="3">59.5</td></tr>
<tr><td>OPEX discounted over 30 years</td><td colspan="2">3,972.6</td><td colspan="3">1,605.4</td></tr>
<tr><td>TOTEX updated over 30 years</td><td colspan="2">3,972.6</td><td colspan="3">3,885.4</td></tr>
<tr><td rowspan="2">Details of TOTEX</td><td colspan="2">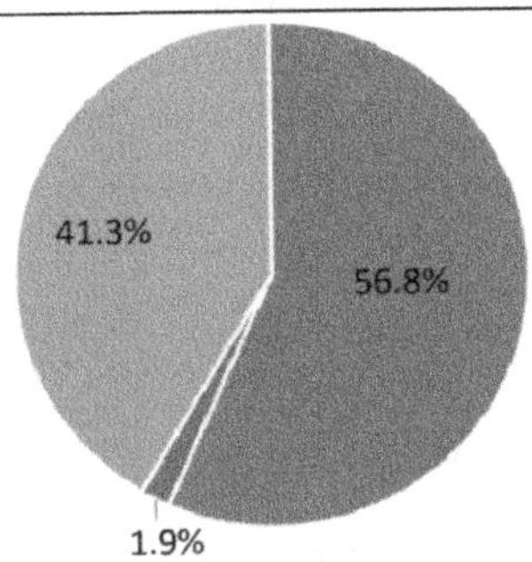</td><td colspan="3">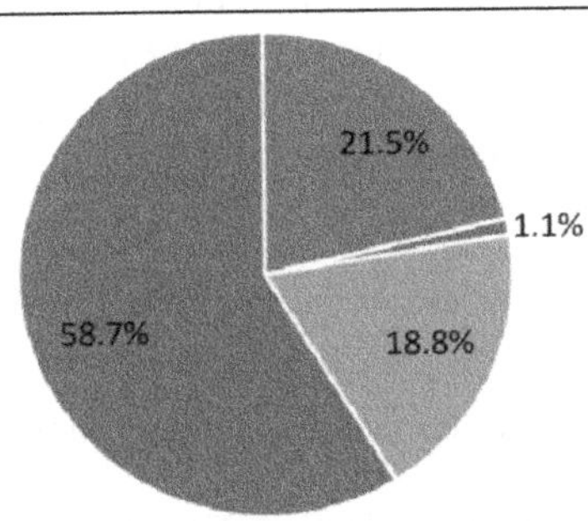</td></tr>
<tr><td colspan="5">■ ENS ■ Interruptions ■ Losses ■ Investment</td></tr>
<tr><td>NPV of project</td><td colspan="5">87.2 k€</td></tr>
</table>

Table 2.9. *Economic comparison of "Do nothing" and "Invest" strategies*

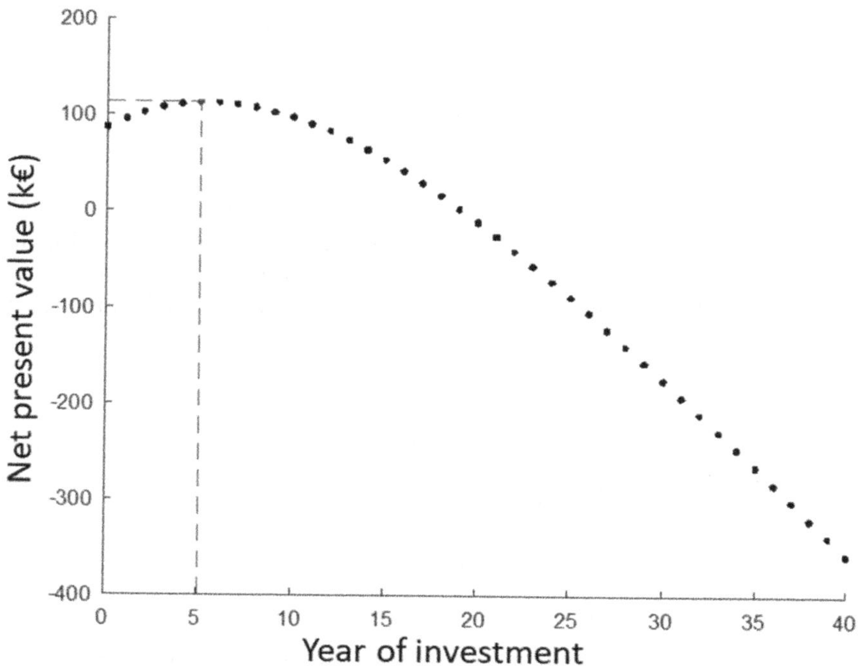

Figure 2.5. *Net actual value according to the year of investment*

2.2. Typical architectures of non-distributed neutral distribution systems (European system)

Historically, the development of distribution systems has been done on an ad hoc basis, without any overall optimization of the system. The first objective was to provide access to electricity, the growing dependence of societies on energy then led to the emergence of the notion of service quality. It is in this context that distribution systems have been developed with different architectures around the world, depending on several factors such as load density (urban, semi-urban or rural), the desired QoS and the investment we are willing to commit. In Chapter 1, we have seen that there are two main distribution systems, the European and North American systems, supplying consumers with three-phase, two-phase or single-phase power. The objective of this is to describe in more detail the different architectures encountered in the so-called European system where the neutral is not distributed in MV but only in LV.

2.2.1. *MV system architectures*

The most traditional distribution system architectures are classified into three categories (Carrive 1991; Enedis 2008; Alvarez-Hérault 2009; Coiffier 2013; Gouin 2015): two-connection-point and double-branch architectures reserved for the urban environment and single-connection-point architectures reserved for the rural environment.

2.2.1.1. *Architectures with two connection points (urban areas)*

Secured feeder architectures use distribution substations with two connection points. Main MV lines start from an HV/MV primary substation then successively supply several MV/LV substations before joining:

– either another primary substation (source-to-source secured feeder structure);

– or the same primary substation from which they originated (loop structure);

– or one or more reflection points (spindle and herringbone structures).

These structures are described and compared in this section, as well as other less used structures, such as the mesh, daisy chain and grid structures.

2.2.1.1.1. Classic secured feeder structure (urban and dense urban)

General description

Figure 2.6 illustrates the source-to-source secured feeder structure. This is the most widely used system structure in the world and is the benchmark for good service quality. When designing master plans, if the system is not already a secured feeder architecture, the strategies implemented will tend to develop it toward this structure. Circuit breakers located at the top of the feeder in the primary substations allow the short-circuit current to be cut off in the event of a fault. Each MV/LV transformer is connected to the main line via two switches, which are all normally closed except for one, thus allowing radial operation. This so-called loopable, radially operated structure allows the system to be reconfigured in the event of a fault so as to resupply the maximum number of consumers. The fault can be isolated by opening the two switches that surround it. Closing the normally open switch allows the rest of the loads not affected by the fault to be re-supplied. We term the *main feeder* as the set of conductors connecting consumers from one source to another, and the *feeder* as the set of conductors connecting the consumers of a source to a normally open switch. Therefore, a main feeder is composed of two outgoing feeders.

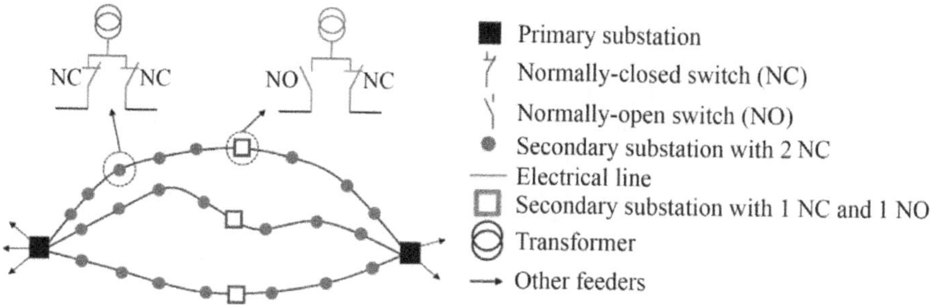

Figure 2.6. *Source-to-source switch break structure and the connection of distribution substations*

Equivalent PL product

Figure 2.7 shows the equivalent PL product areas as defined in section 2.1.4.3. All main feeders must have an equivalent PL and the two feeders that form the main feeder must have an equivalent PL, thus defining the position of the normally open remote-controlled switch (RCS) devices for each feeder. A switching device is composed of two switches that can either be remote-controlled or not. The automation of all switching devices is not economically justified, and Coiffier (2013) indicates that the optimized number of RCSs is 3.5 per feeder, arranged to create equivalent PL areas. A given feeder is thus made up of three RCSs dividing it into four zones of equivalent PL product, with the half RCS forming the boundary between the two feeders of a main feeder.

2.2.1.1.2. Power-up procedure in the event of a fault

Figure 2.8 illustrates the procedure for detecting, locating and isolating the fault. In normal operation, the system is in stage 0. The primary substation S_1 supplies the consumers located on the left of the RCS_3 and the primary substation S_2 supplies the consumers located on the right of the RCS_3. Suppose a fault occurs on a portion of the system located between RCS_1 and RCS_2 (Step 1). The circuit breaker of the S_1 feeder cell trips and all customers on the feeder are disconnected. Based on the fault passage indicators, the system operators can locate the fault within a few minutes and operate the RCSs to isolate the fault and restore power to some of the customers (Step 2). In our example, the right switch of RCS_1 and the left switch of RCS_2 are open and the normally open switch of RCS_3 is closed. The primary substation S_1 supplies the loads up to RCS_1 and the primary substation S_2 supplies the loads up to RCS_2. Once Step 2 is completed, a team goes into the field to operate the manual disconnectors of the MV/LV substations in order to supply additional customers (Step 3). They can then carry out the repair on the faulty conductor. This last step for resupply takes between 40 and 60 min, corresponding to the time needed to get to the site, depending on the length of the systems. In order to make this re-feeding process possible in the event of a fault, the cables must be oversized above normal operation requirements. For example, the cable located just downstream of the primary substation S_1 must be able to pass all the power for the entire main feeder in the event of a fault on the cable located just downstream of the primary substation S_2. The maximum cross-section commonly used in French distribution systems is 240 mm² aluminum and corresponds to a maximum apparent power of 14 MVA at 20 kV. The power of each feeder of the secured feeder structure must not exceed 7 MVA.

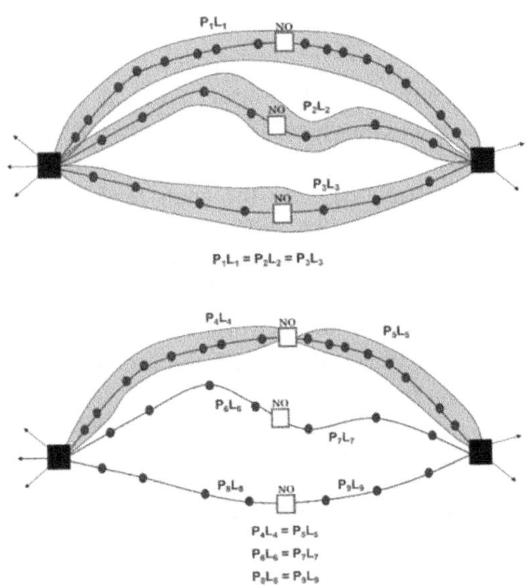

Figure 2.7. *Equivalent PL areas for secured feeders*

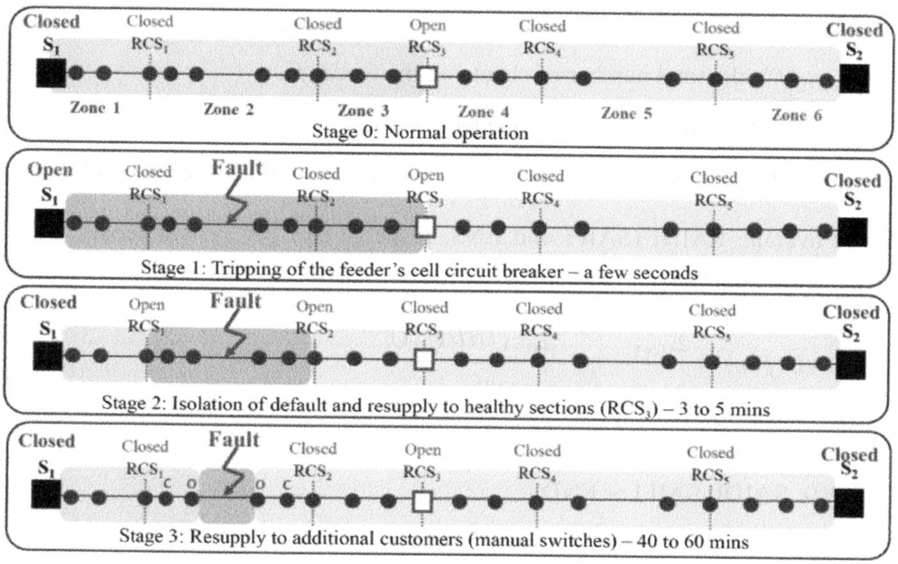

Figure 2.8. *Procedure for locating and isolating a fault*

Expression of SAIDI, SAIFI and END

Each feeder i in a secured feeder system is theoretically divided into four zones. For each zone Z_k of each feeder f, we define:

$$SAIDI(f,z) = \tau_{fail}(f,z) \times L(f,z) \times \left(T_{iso} + T_{rep} \times \frac{N_{cust}(f,z)}{N_{cust}(f)}\right) \qquad [2.48]$$

$$SAIFI(f,z) = \tau_{fail}(f,z) \times L(f,z) \times \left(\frac{N_{cust}(f,z)}{N_{cust}(f)}\right) \qquad [2.49]$$

$$ENS(f,z) = \tau_{fail}(f,z) \times L(f,z) \times \left(T_{iso} \times P_{max}^{feeder}(f) + T_{rep} \times P_{max}^{zone}(f,z)\right) \qquad [2.50]$$

with:

– $\tau_{fail}(f,z)$: the conductor failure rate in zone z of feeder f (in number/year/km);

– $L(f,z)$: total length of zone z of the feeder f (in km);

– T_{iso}: time of Step 1 and 2 (fault isolation using RCS);

– T_{rep}: time of Step 3 (isolation of the fault using the manual operating devices);

– $N_{cust}(f,z)$: the number of clients in zone z of the feeder f;

– $N_{cust}(f)$: the total number of clients on the feeder f;

– $P_{max}^{feeder}(f)$: total power of the feeder f (kW);

– $P_{max}^{zone}(f,z)$: the maximum total power of the zone z of the feeder f (in kW).

The average SAIDI (SAIFI and ENS, respectively) per system feeder is thus described by equation [2.51].

$$IND(f,z) = \frac{\sum_{f=1}^{N_{feeders\ MV}} \sum_{i=1}^{4} IND(f,i)}{N_{feeders\ MV}} \qquad [2.51]$$

with:

– IND: SAIDI, SAIFI or END;

– $N_{feeders\ MV}$: the number of system feeders.

The source-to-source secured feeder structure is flexible because the open points can be moved in order to move the load to other feeders in the event of faults, temporary congestion, and maintenance works on the system. In the case of the latter, it is necessary for the two MV feeders to be temporarily looped so that the consumers do not have a short interruption. HV/MV transformers must therefore have the same hourly index to avoid the circulation of loop currents that can be important (Coiffier 2013). In France, HV/MV transformers are therefore mostly coupled in Δyn11, that is, the high-voltage winding of the transformer is delta-coupled and the medium-voltage winding of the transformer is star-coupled with neutral out and connected to earth via an impedance, so that the single voltage at the secondary is +30° out of phase with the single high-voltage.

We also encounter this structure in Germany with some slight differences (Schlabbach and Rofalski 2014). First, the position of the opening point can be at the ends of the lines (see Figure 2.9), so that in normal operation, each feeder is loaded to 100% (instead of 50% in France) while the substations remain loaded to 50% each. This structure is not optimal in terms of losses because in this configuration these are proportional to the square of the current, and increase as the lines become more loaded. In addition, the length of the feeders supplied by the same substation in normal operation becomes longer, and thus the voltage drops are greater. The position of these normally open switching devices can be modified to reduce losses and improve the voltage plan. However, there are no planning rules, such as the PL product, to place these devices in the initial state. Second, MV/LV transformers can be connected by switches but also by fuses. Finally, the cable cross-sections are standardized and generally a single (or even two) cross-section is used. This structure can also be operated in a looped manner, that is, in areas with very high load densities and industrial sites all the switches normally open are kept closed. In this case, it is necessary to adapt the protection system by using distance or differential protection devices. If MV/LV transformers are connected via circuit breakers then continuity of service is guaranteed in the event of a fault.

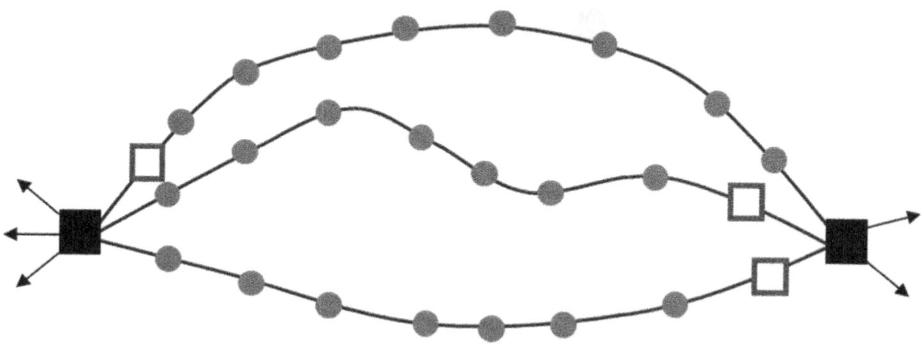

Figure 2.9. *German version of the secured feeder*

2.2.1.1.3. Looped structure

When the primary feeder is looped back to the same primary substation, we speak of a loop structure (Figure 2.10). It is easy to operate with good service quality and low joule losses. The looped feeders must belong to two different transformers fed by two different HV feeders to allow for backup. On the other hand, the safety of the HV/MV substations is not ensured, and as a consequence the size of the system is relative to the source-to-source feeder. This structure is employed particularly in small towns that are supplied by only one primary substation. In the case of Germany, we also find loop structures. In addition, the normally open position is determined on a case-by-case basis according to the set objectives (losses, voltage, etc.), and only one standardized cable cross-section is used (Schlabbach and Rofalski 2014). In anticipation of future reinforcement, we normally find open-loop connections in Germany in order to provide alternative paths in the event of a fault. However, this is not widespread due to the high cost and complexity of the protection scheme to be implemented (Schlabbach and Rofalski 2014). In the case of large loads, such as those needed by industrial applications, in Germany this structure can be operated as a closed loop.

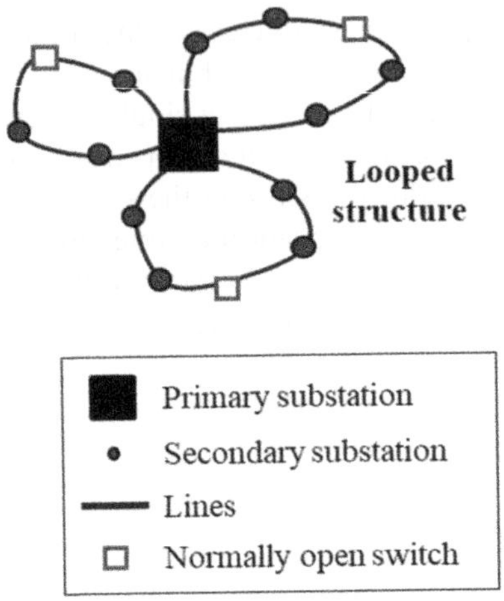

Figure 2.10. *Looped structure*

2.2.1.1.4. Spindle, herringbone and tuple structure

In the case of the spindle structure (Figure 2.11(a)), all the cables coming from the same substation converge at the same point, called the reflection point (RP in Figure 2.11). This is actually a station containing remote-controlled switches (one switch per cable that converges on the reflection station). This point is strategically chosen because it could become a future HV/MV primary substation. As the reflection point does not allow for the supply of loads, it is best to add a backup cable connecting it to an HV/MV primary substation. This cable under normal operations will be kept under voltage but not empty. We thus speak of "specialized" backup. Loads can also be connected to this backup cable, in which case, we talk about "integrated" backup. In general, we attribute one backup cable for a maximum of six "working" cables, that is, those intended to supply the loads under normal operation. The switches connected to the working cable in the reflection station are normally open and the one connected to the backup cable is normally closed. This structure is simple, easy to operate and has good supply quality. On the other hand, it is not flexible in the long term because it is difficult to evolve if the future location of the primary substation changes. When the different working cables do not meet at a single reflection point but at different normally open points on the backup cable, we speak of a herringbone structure, as shown in Figure 2.11(b). This variant is more flexible than the spindle structure and takes into account the actual load distribution; it is therefore more economical.

As with the secured feeder structure, in the German case, in the case of an integrated backup, the normally open switches are not necessarily all at the reflection point, as shown in Figure 2.12, and only one standardized cross-section is used. The architecture is the same in the case of dedicated backup, and the cross-sectional area of the backup cable can be increased if the backup cable is required to supply two faulty feeders at the same time (Schlabbach and Rofalski 2014). Similar to the secured feeder structure, in the case of large loads, such as industrial applications, in Germany this structure can be operated as a closed loop.

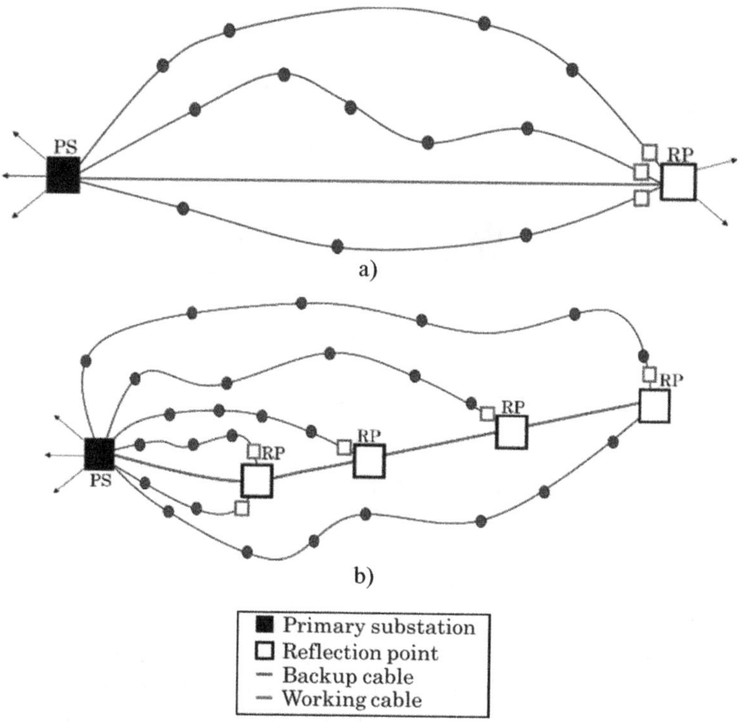

Figure 2.11. *a) Spindle structure and b) herringbone structure*

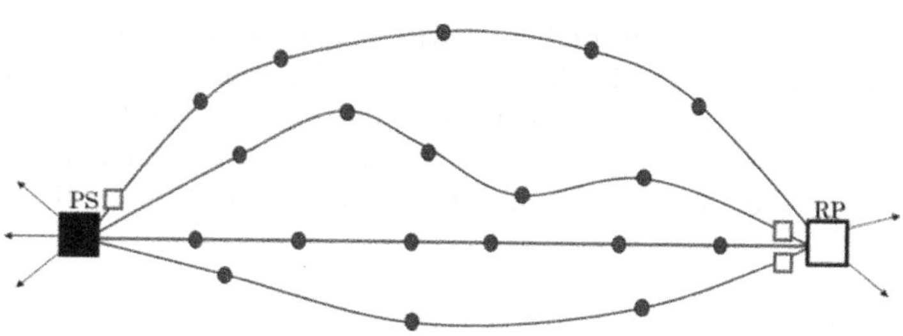

Figure 2.12. *German spindle structure*

In Germany, we find another rarer variant of the spindle and spur: the tuple, as shown in Figure 2.13. This structure is usually used temporarily while waiting for the construction of primary substations. Here, the lines can be loaded to 67% instead of 50% because in the event of a fault on a line, the backup is provided by two lines (a backup cable and a feeder) instead of one feeder. As in the case of the secured feeder, in the case of large loads such as industrial plants, this structure can be operated in a closed loop.

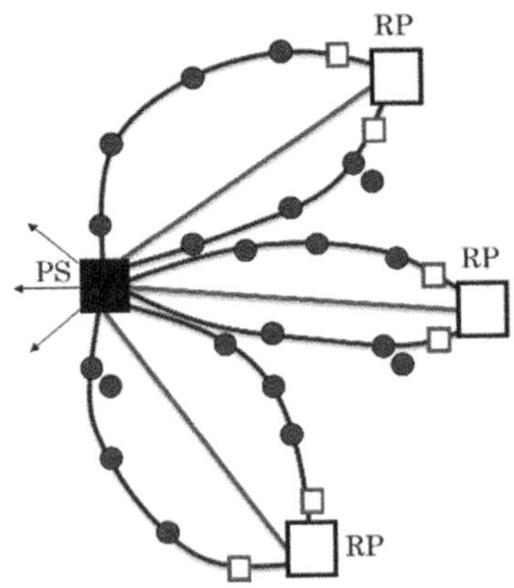

Figure 2.13. *Tuple structure*

2.2.1.1.5. Mesh and daisy chain structure

The mesh structure (Figure 2.14(a)) is composed of loops fed directly by the primary substations or via "loop-head" substations connected to the primary substations by large cross-section conductors (240 mm²) called structure cables. The loop-head substations have the same structure as the primary substations except there is no HV/MV transformer. Three inter-loop links allow the load to be transferred from one loop to the other in the event of the loss of a structural cable. Although this structure has very good power quality, it is expensive and difficult to operate. We find it in the city of Lyon and its' suburbs; however, it is gradually being phased out and replaced by the classic secured feeder structure. Indeed, due to the evolution of the loads, the Lyon structure has moved away from the structure represented in Figure 2.14, in particular the secured feeder structure which has developed at the level of the structural cables that therefore impacts the robustness of the architecture.

The daisy chain structure is convenient when the center of gravity of the loads is off-center of the primary substation. In this case, we supply a substation, named the *petal head* substation, by one or two feeders for a large section (typically 240 mm²) coming from the HV/MV primary substation and which supplies two or three loops. If loads also exist around the HV/MV substation, loops can be added around it. Moreover, the power of the cables is limited to the maximum power divided by half, so that they can withstand a fault. Unlike the mesh structure, there are no additional links between the loops. The petal structure is easy to operate with good power quality and low joule losses. On the other hand, safety of the primary substations is not ensured and, as a result, the size of the system is important compared to the source-to-source feeder. It is therefore used for specific cases (localized and dense loads).

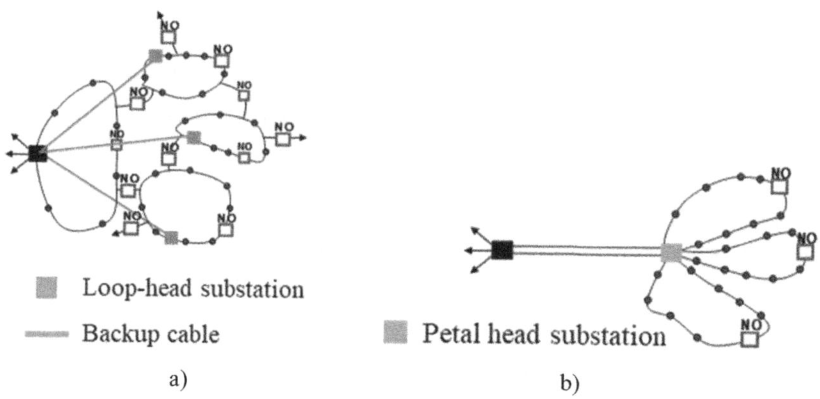

Figure 2.14. *a) mesh and b) daisy chain structures*

2.2.1.1.6. Grid structure

The grid structure (Figure 2.15) is composed of *intra-station* conductors (coming from the same primary substation) and *inter-station* conductors (connecting the intra-station conductors together). Normally open switching devices are distributed in the mesh structure to allow radial operation. The operation of this structure is very complex. In fact, if the system has to be reconfigured as a result of a fault, the number of switching operations to be performed is important and must be done in the right order. In addition, knowledge of the flows circulating within the system is necessary in order not to cause constraints on line sections during load transfers. Circuit breakers distributed across the system could allow for a meshed operation, but this is rarely used. This structure allows for excellent service quality but is very expensive and the protection plan is complex to implement.

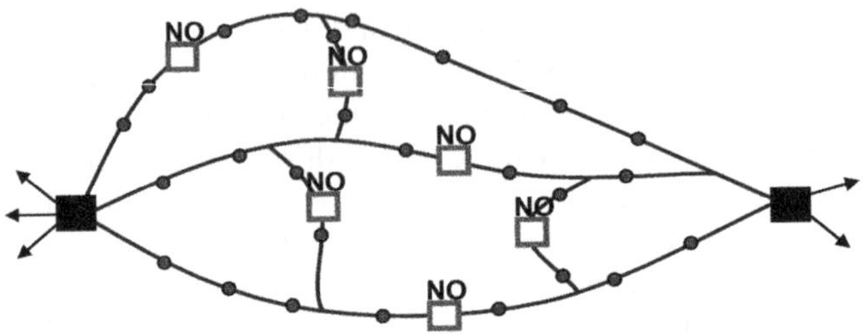

Figure 2.15. *Grid structure*

2.2.1.1.7. Single and multiple branching architectures (dense urban)

The simple double branch structures (Figure 2.16(a)) are of the secured feeder or tree type structure (depending on the load density), and have double the cables coming from the primary substations. Each MV/LV transformer is connected to two cables through inverter devices. In most cases, the loads are evenly distributed over the two cables in order to reduce the joule losses. This structure is difficult to operate manually but can easily be automated despite the costs.

A good technical and economic compromise is to install remote-controlled switches every 10th to 15th MV/LV substations in order to limit the number of operations required to clear a fault. In France, we find double branches in Nice, Bourgoin-Jallieu and in L'Isle-d'Abeau in Isère. We also find double branching in some German cities with very high load density. However, there are two major differences (Schlabbach and Rofalski 2014): each MV/LV transformer is connected to the two main feeders, via switches or without, and two operating modes are possible, open loop as in France (NB: the normally open position can vary) and closed loop.

In the case of high load densities or when a higher service quality is required (certain industries, hospitals, etc.), the multiple double branch structure (Figure 2.16(b)) can be considered. Each MV/LV primary substation has only two supply paths, but these two paths are alternately connected to between three and six cables leaving the MV/LV primary substation. In the event of a fault, the load can be switched either to a backup cable (dedicated backup) or onto the other cables (integrated backup). These structures perform well if they are well maintained, requiring significant OPEX. In regions where this structure is in the minority (opposite to Paris), operators are less accustomed to using it and it therefore tends to disappear if alternative opportunities, such as reinforcements, arise.

The Paris distribution system is a rare example of the application of multiple branching (like Moscow). The Île-de-France region is made up of a primary 400 kV loop and a secondary 225 kV loop connected to the 400 kV loop by nine transformer stations. From this 225 kV loop, radial lines connect three 20 kV rings as shown in Figure 2.17. Some 158 225 kV/20 kV primary substations are connected two by two, in double multiple branching (five working cables and one backup cable). The distribution system represents 78,750 km of lines and cables and the average outage time for Greater Paris was 28 min in 2011, less than half the national average (Canepa 2012).

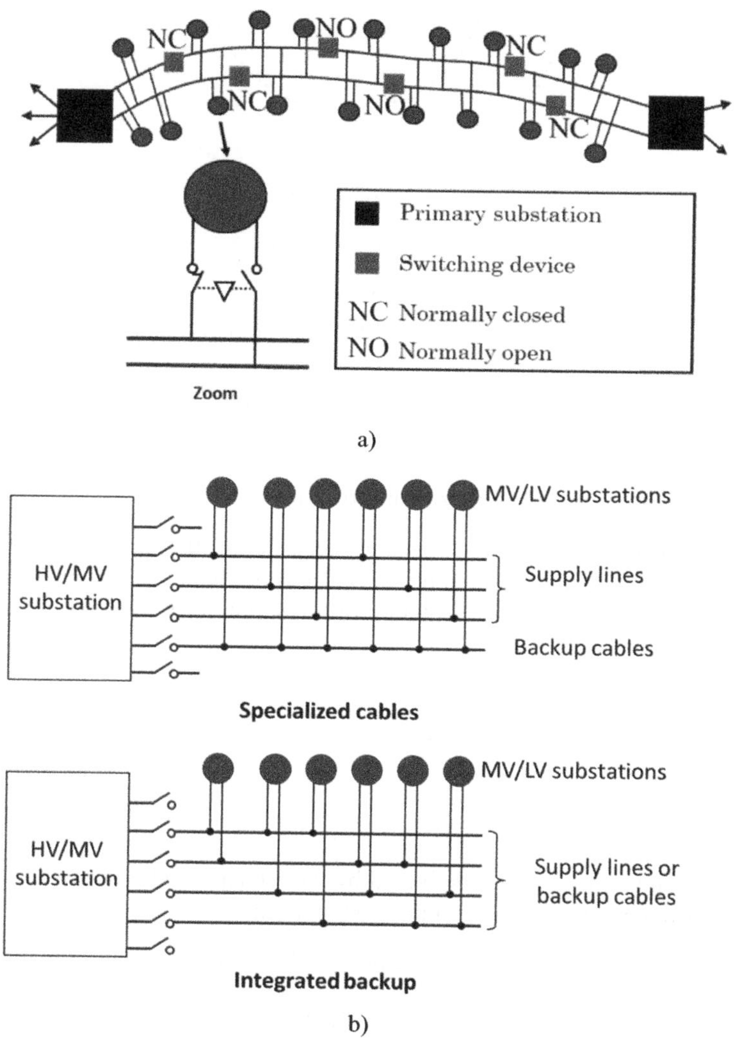

Figure 2.16. *Double branching structure: a) single and b) multiple*

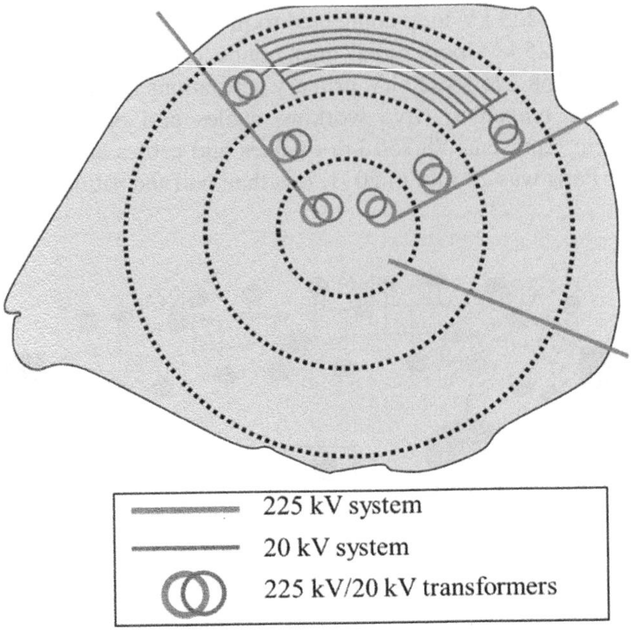

Figure 2.17. *Parisian power system*

2.2.1.1.8. Comparison of the different structures

These architectures can be compared because of the utilization coefficient (Coiffier 2013) described by equation [2.52]. This coefficient is generally lower than 1 in order to allow for the implementation of a resupply procedure in the event of a fault.

$$CU = \frac{\sum_{f=1}^{N_{feeders\,MV}} P_{max}(f)}{\sum_{f=1}^{N_{feeders\,MV}} P_{allowed}(f)} \qquad [2.52]$$

with:

– $P_{max}(f)$: maximum power delivered by the feeder k under normal operation to supply the load when it is at maximum consumption;

– $P_{allowed}(f)$: maximum admissible power of feeder k (thermal limitation).

To estimate the utilization coefficient for these structures, we assume that the maximum power of the feeders is identical and equal to P_{max}. Table 2.10 gives the value of the coefficients for the different structures.

Type of structure	Theoretical maximum utilization coefficient
Classic secured feeder, loops, simple double-branching and daisy chain	$CU = \frac{N_{feeders\ MV} \times P_{max}}{N_{feeders\ MV} \times 2 \times P_{max}} = 0.5$
Spindle, herringbone, tuples and multiple double-branching	$CU = \frac{N_{feeders\ MV} \times P_{max}}{N_{feeders\ MV} \times P_{max} + P_{max}}$ 1) In the case of a backup cable for six working cables, we have $CU = \frac{6}{7} = 0.86$. 2) In the case of Paris (five working cables for one backup cable), $CU = \frac{5}{6} = 0.83$.
Mesh	$CU = \frac{N_{lps} \times P_{max}}{N_{lps} \times \left(P_{max} + \frac{P_{max}}{3}\right)} = 0.75$ N_{lps}: number of loops
Grid	CU is not calculable

Table 2.10. *Utilization coefficients of the different structures with two connection points*

2.2.1.2. *Single point architectures (rural areas)*

In areas with low load density (rural areas), it is expensive to develop the structures presented in the previous section. Therefore, single point structures are used, as shown by the diagram depicted on the left side of Figure 2.18. Each load is connected to the primary substation by a single path and referred to as a *simple tree* structure. In this type of structure, there is no possibility of backing up the loads in the event of a fault, thereby degrading service quality when compared to that of urban systems. In this case, the utilization coefficient is equal to 1. In order to ensure good service quality, it is necessary to limit the length and power of these structures. In the case where the load density is somewhat higher, we can use a tree structure with a primary feeder (diagram on the right side of Figure 2.18). One or more main lines connect two primary substations and directly feed a portion of the loads, which are essentially those with high power, or those that are near a main line and for which a direct connection increases line length only slightly. Loads located far from main lines are fed by antennas that are connected to the main lines and can either be fed in radial or loop modes (we call these local loops). For rural areas with low load densities, these architectures are fairly conventional because they are simple to design and operate with a basic protection scheme and are relatively inexpensive (Schlabbach and Rofalski 2014). On the other hand, the service quality is poorer than those of urban systems because there is little or no backup between feeders and because this structure has little scalability.

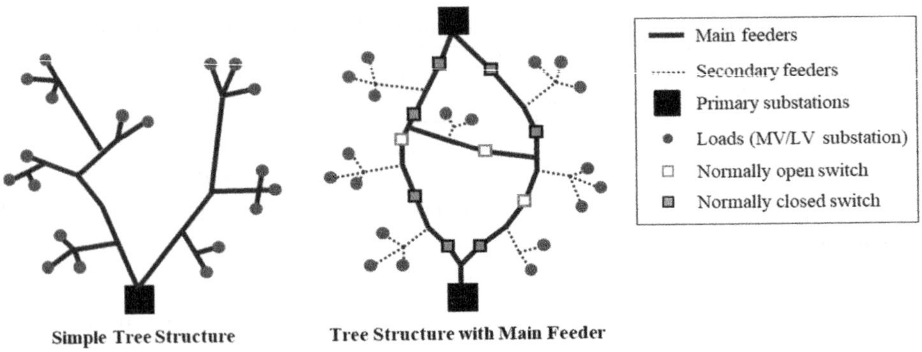

Figure 2.18. *Rural structures*

2.2.2. *LV system architectures*

The French low voltage systems are short, tree-like and unbalanced (single-phase supply because of the distribution of the neutral and three-phase). The rule of the shortest connection is employed by using single or decreasing sections along the feeder. Indeed, at this level of the system, looping or any structure that increases the length is not economically justified. With the increase of renewable energies such as photovoltaics and electric vehicles, the low voltage system could be tightly constrained. New strategies could emerge. The concept of LV and MV microgrids are currently being investigated by many research institutions, distributors (in the broadest sense) and manufacturers with different objectives. One of the major challenges for these systems is the development of DC systems or mixed AC/DC systems due to the high costs and losses of the power electronics devices required. Moreover, the protection equipment and logic applied to an AC setup cannot be implemented as a DC or mixed AC/DC system. It is therefore necessary to investigate new systems for the protection of goods and people, adapted to the design and the preferred operating mode of these microgrids.

In Germany, we can still find meshed LV systems for the supply of industrial or aging systems (Schlabbach and Rofalski 2014). These are robust to LV incidents due to the redundancies they offer. Depending on how these systems are connected to the MV system (Figure 2.19 shows three possible configurations), they can also be robust to MV incidents. For example, if they are fed by an MV feeder in the event of an incident on the MV/LV transformer, there is no interruption to the supply, which is done via the other transformers on the feeder. If they are fed by several feeders, they then become robust to incidents on the MV lines or feeders. These architectures therefore have excellent supply quality with reduced joule losses and a good voltage profile. However, the CAPEX is high due to the redundancy of the lines and required protection equipment, as well as due to the necessary use of directional protection devices in order to detect faulty LV currents, which could see a rise in MV in the event of an MV fault, potentially resulting in protection blinding (the circuit breaker of the feeder where the fault occurs does not trip because the fault current is lower than the protection tripping threshold) or an untimely trip (the circuit breaker of a feeder that is not at fault trips because a faulty current is flowing). In order to ensure the continuity of service in the event of a first phase to earth fault, these systems are also associated with an isolated or a "perfectly" compensated neutral earthing system. Earth voltage monitoring systems must be installed in order to guarantee the protection of people and property at the permissible contact voltages according to the European and French standard NF EN50522. It requires additional means of protection and qualified personnel. These systems are therefore reserved for cases where high reliability is the priority.

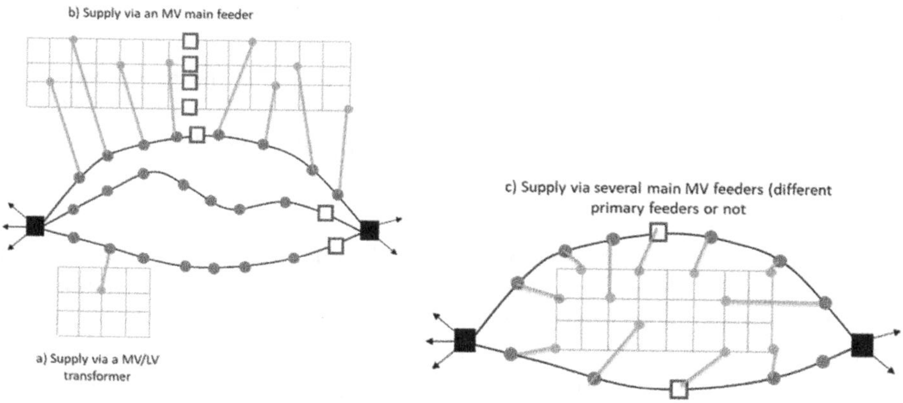

Figure 2.19. *German LV mesh systems*

2.3. Typical architectures of distributed neutral systems (North American system)

The North American system is quite different from the European system because it distributes the neutral at medium voltage and thus allows for the development of secondary feeders (laterals) in single, two or three phase. This structural choice is justified because these systems have both urban areas with very high load density and extremely rural areas for which the distribution of the three phases at medium voltage would be very expensive (see Chapter 1). The regular direct earthing of the neutral causes high fault currents which are therefore self-extinguishing when they are fugitive (e.g. a tree branch touching a phase burst into flame spontaneously and burns rapidly because the fault current is very high). The distribution system has a tree structure with overhead conductors that supply consumers between 4 and 35 kV in medium voltage and 208 or 480 V between phases in low voltage (Behnke et al. 2005). Unlike European-type systems, for which only the medium-voltage topology can be divided into several structures as per the service quality required, North American-type systems also have different low-voltage topologies. Ninety-seven percent of US consumers are supplied via fully radial systems because they are inexpensive and simple to operate and develop. The rest of the load, corresponding to the major US metropolises, is supplied by a radial MV system and meshed LV systems in two main types of systems: the area network or the spot system (IREC 2013). In this section, the different MV and LV system architectures that we may encounter in the US are described.

2.3.1. *MV system architectures*

There are four main types of MV system architectures in the North American system: *simple radial overhead, primary auto loop, underground residential distribution* and *primary selective system* (Settembrin and Fisher 1991).

2.3.1.1. *Simple radial overhead system*

This structure, historically developed to supply small residential areas, consists of several primary feeders (each loaded between 2 and 6 MVA; Settembrin and Fisher 1991) along which distribution transformers are connected. Secondary feeders may also be run from the primary feeders to supply other customers (laterals). As in the European rural system, reclosing circuit breakers are located at the head of each medium-voltage feeder and eliminate self-extinguishing faults by performing several tripping and reclosing cycles. The reliability of this type of system is the worst since a permanent incident on a line or in the primary substation puts all the consumers in the dark without any possibility of resupply. When the feeder is very long, we can sometimes find in the system sectionalizers, which allow for the isolation of the system located downstream of where the fault occurs. This improves service quality by not cutting off users upstream of the fault (Settembrin and Fisher 1991). Often, manual normally open switches are installed to connect feeders to other medium voltage systems so that consumers can be re-supplied in the event of a fault or maintenance. This structure, in terms of topology, is therefore similar to the structures shown in Figure 2.18. Figure 2.20 shows an example of a simple overhead radial system with three phases and a neutral to illustrate the different supply modes. Next, the single-wire diagram will be used.

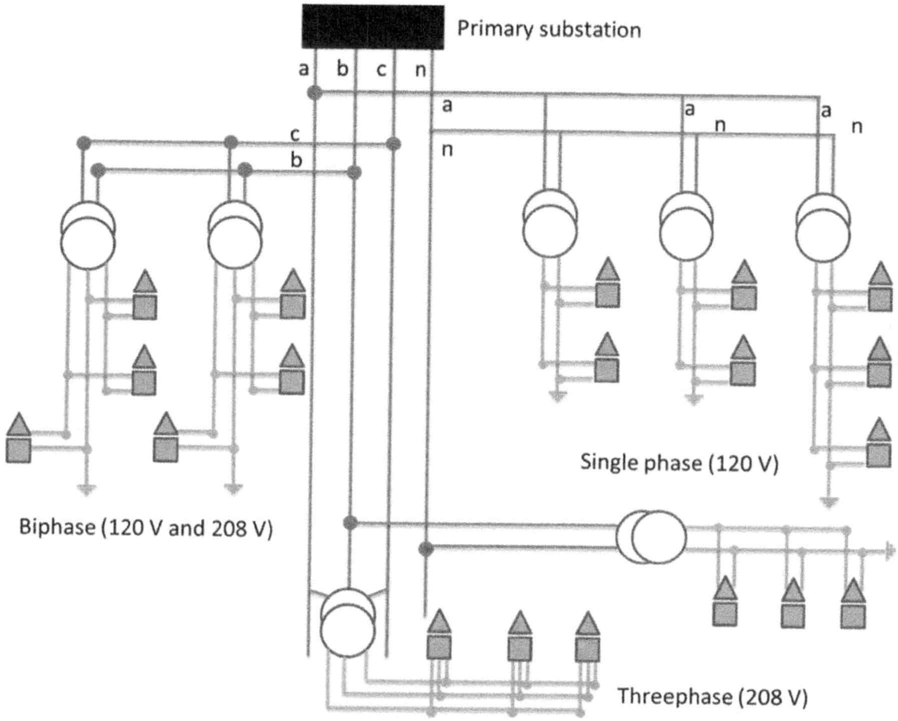

Figure 2.20. *Simple radial overhead system*

2.3.1.2. *Primary auto loop system*

This structure, used in rural and peri-urban areas, has loaded feeders between 6 MVA and 12 MVA. Two primary feeders for two different substations are looped by a normally open recloser switch, which in the event of a fault allows for the automatic restoration of part of the loads once the fault has been isolated. In addition to the two reclosers at the head of each feeder, a recloser is added in the middle of each feeder in order to feed more consumers in the event of a fault. Therefore, this structure reduces the number of long interruptions for customers, but increases the number of short interruptions (presence of reclosers). However, long outages may occur because the selectivity logic to be implemented is complex (Settembrin and Fisher 1991).

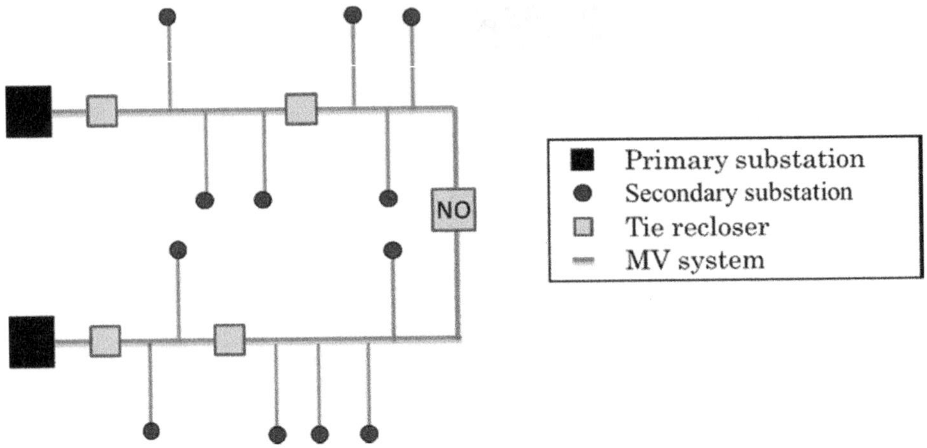

Figure 2.21. *Primary auto loop system*

2.3.1.3. *Underground residential distribution (URD) system*

This is the same conventional secured feeder system used in France, as shown in Figure 2.10, with the difference that all switches are manual and that there is not necessarily a fault passage indicator at all distribution stations. Therefore, permanent faults take longer than in France because of the time required to search for the fault. This system is used to supply residential areas mainly due to esthetics (buried lines) and reliability.

2.3.1.4. *Primary selective system*

This type of system is used to supply areas with a high load density, that is, where the sum of the power of the loads supplied by a distribution transformer is greater than 300 kVA (Settembrin and Fisher 1991). In this case, each transformer is connected to two main feeders from two different primary substations. The connection is made at a feeder break via two remote-controlled switches, one normally closed and one normally open. A current and voltage transformer and local intelligence make it possible:

– to disconnect the transformer from the primary substation that supplies it and to automatically connect it to the primary substation that backs it up in the event of a fault in the medium voltage part. There is no need to locate the fault beforehand because no other distribution transformer is connected to these feeders;

– to prevent automatic switchover in the event of a fault on the distribution transformer or downstream of it.

2.3.1.5. *Special-spare cable system*

Figure 2.23 shows the architecture of this structure more often employed in areas with low load densities, in North and South American systems, but also can be found in dense areas in Asia (Schlabbach and Rofalski 2014). Primary feeders (in blue) bring energy to consumption areas that are supplied by secondary feeders. All of these secondary feeders are connected to a dedicated primary feeder via t-connectors for overhead lines and t-joints for normally open underground cables. This is a back-up feeder that allows consumers to be re-supplied in the event of a fault on a secondary feeder. In normal mode, it is maintained under voltage but at zero load. This rather simple structure presents a good economic compromise, since only the primary feeders and the back-up feeder are oversized in order to allow for the resumption of service in the event of an incident on the radial feeders and possibly on a primary feeder. In general, the backup feeder is sized with the same cross-section as the primary feeders.

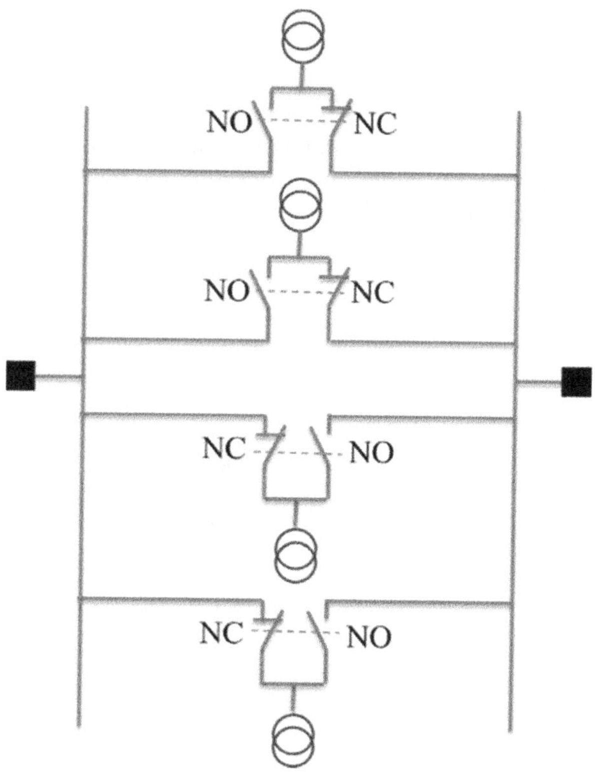

Figure 2.22. *Primary selective system*

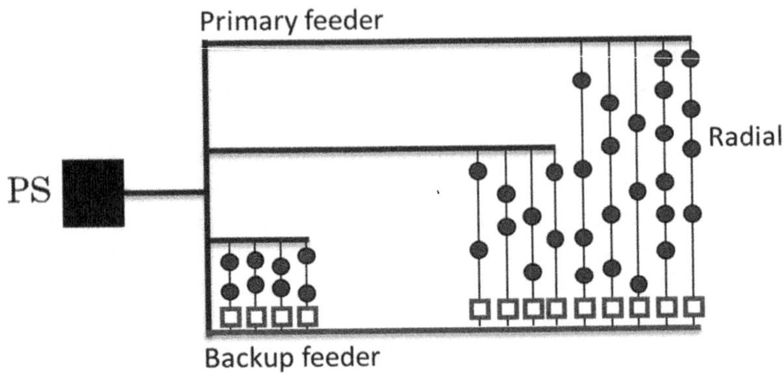

Figure 2.23. *Special-spare cable system*

2.3.2. *LV system architectures*

2.3.2.1. *Secondary selective system*

The logic behind this structure, commonly applied to industrial contexts, is to ensure the back-up by the LV system by connecting the transformers' secondaries through a normally open circuit breaker.

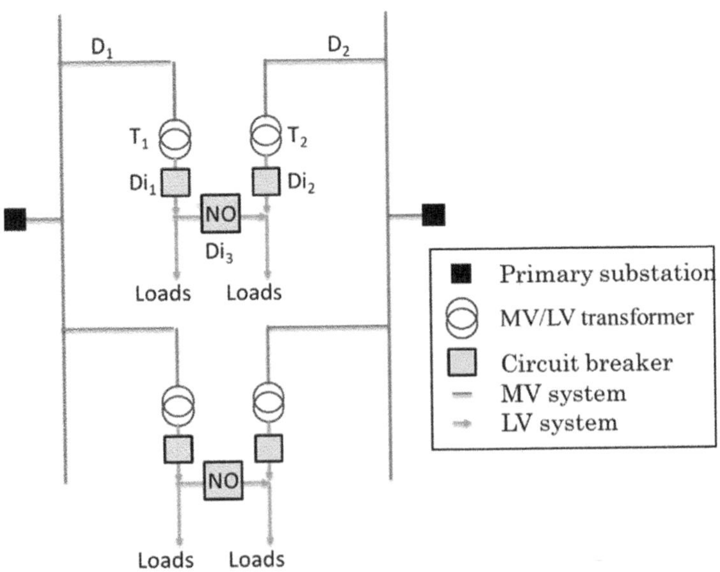

Figure 2.24. *Individual inverted secured feeder system*

As shown in Figure 2.24, in normal mode, each transformer is supplied by its own feeder. If a fault occurs in a primary substation or on a medium voltage feeder (e.g. D_1), then the circuit breaker placed downstream of the distribution transformer (Di_1) opens and the one linking the secondaries of the two transformers (Di_3) closes (mechanical interlocking), thus allowing the loads connected to the T_1 transformer to be supplied via the D_2 feeder. This structure is more reliable than the individual secured feeder because the loss of a distribution transformer does not cut off the consumers. Moreover, if an intervention is planned on the LV system by the planner (maintenance or a new connection, for example), it is not necessary to interrupt the current because it is possible to work in parallel in LV.

2.3.2.2. *Distributed grid system (also called area network)*

In very high load density city centers in which both high and low loads coexist (Settembrin and Fisher 1991), such as San Francisco and New York City, the low voltage distribution system is meshed to allow for good supply quality as shown in Figure 2.25 (Elliot et al. 2014). The number of medium-voltage feeders is generally greater than 6 and even reaches 24 for New York City (Settembrin and Fisher 1991). In the event of a fault at the MV system level, to prevent the other MV sources from feeding the fault through the LV system, directional protection devices called system protectors detect the currents flowing in the opposite direction and open the circuit. In Figure 2.26, if a fault occurs on the MV feeder D_{21} of primary substation PS_2, due to the LV system, both primary substations PS_1 and PS_2 will contribute to the fault current. One of two scenarios will then follow:

– the current flowing in the MV feeder D_{11} of substation PS_1 to supply the fault is strong enough to trip the circuit breaker of this feeder. We then speak of untimely tripping: the D_{11} feeder is cut off even though there is no fault on it;

– the current flowing in MV feeder D_{11} is low enough not to trip the circuit-breaker of this feeder. The fault on feeder D_{21} then continues to be energized, which creates a dangerous situation for the people going to repair the faulted zone.

These two cases are not desirable and the directional protection devices are designed to avoid them because they detect the so-called reverse flows, that is, going from the LV system to the MV system. On the other hand, these directional protection devices do not detect faults occurring in the LV system circuit, in which cable limiters allow the fault to be extinguished without exceeding the thermal capacities of the cables. In this structure, in theory, the consumer never experiences an interruption in service. The problem with this type of structure with regard to the development of renewable energy production is that the penetration rate of production is limited to the sum of the maximum consumption.

Indeed, in the opposite context, production is then exported on the distribution system. The energy flows being in the opposite direction, the directional protection devices are triggered.

2.3.2.3. *Spot network*

This type of structure, which can be found in New York for example, is used to supply areas with high loads such as airports, hospitals and shopping centers (Settembrin and Fisher 1991), as shown in Figure 2.27. All LV consumers are connected to a single busbar with a guaranteed continuity of supply. This busbar is fed by a circuit breaker-MV feeder-transformer-system protector assembly (between two and six assemblies for the largest systems, up to 25 MW). As for the meshed LV system, in the event of faults, the sources in parallel feed the fault and the *system protectors* detect the reverse fault currents. Similarly, in the event of a fault on the low voltage system, *cable limiters* allow for the elimination of the fault without interrupting the current.

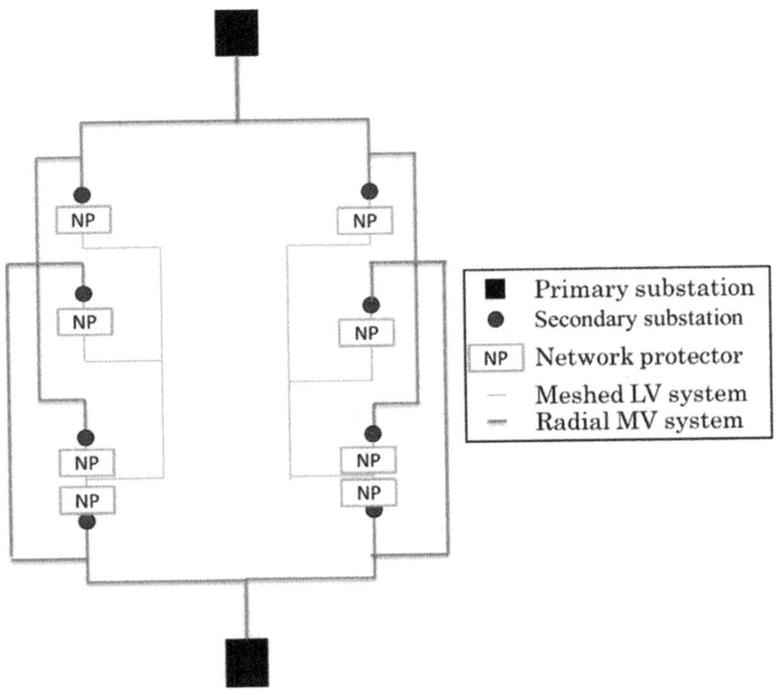

Figure 2.25. *MV and LV architectures for dense urban systems*

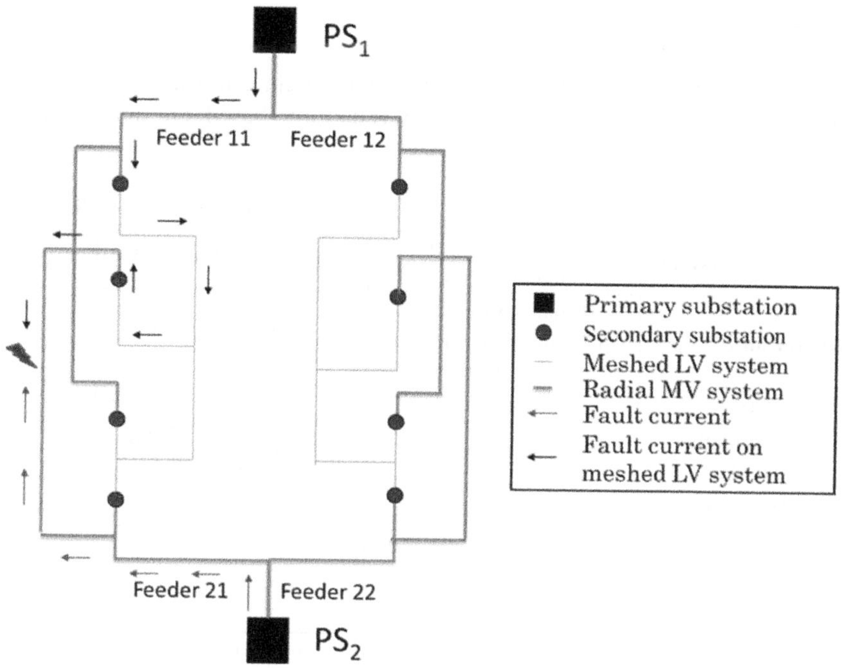

Figure 2.26. *Fault currents without directional protection devices*

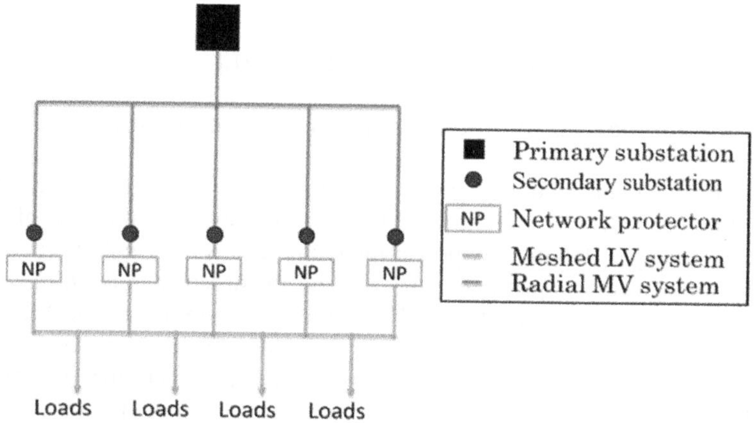

Figure 2.27. *Spot network*

2.3.3. *Comparison of architectures*

Table 2.11 compares the North American-type architectures, presented in this section, in terms of application areas and reliability based on the findings of Settembrin and Fisher (1991).

	Type	Power	Reliability
Simple radial system	Small residential and commercial	Feeders from 3 to 6 MVA	–
Primary auto loop	Small residential and commercial in rural and peri-urban	Feeders from 6 to 12 MVA	-
URD	Allotments		+
Primary selective system	Concentrated loads	Concentrated LV load clusters above 300 kVA	+
Secondary selective system	Industries		++
Distributed grid system	Large city centers (coexistence of high and low load density areas)		+++
Spot system	Very high densities concentrated loads (airports, hospitals)	Up to 6 feeders per primary substation from 17 to 25 MVA	+++

Table 2.11. *Comparison of North American architectures*

2.4. Other architectures encountered in the world

2.4.1. *Multi-divided and multi-connected structure (Japan and China)*

Japan is among the countries with very impressive average outage time per customer (annual national average of 20 min and 7 min for the city of Tokyo; TEPCO 2015) without having any power interconnections to neighboring countries. Its distribution system is operated at 6.6 kV and its structure is *multi-divided* and *multi-connected* (Miyata et al. 2017). This structure is similar to the French grid structure, as shown in Figure 2.28. Indeed, each MV feeder is partitioned into n sections consisting of n normally closed remote switches, except the last one which is open. Each section of each feeder is connected to a section of another feeder via a normally open switch. Additional manual normally closed switches may be installed on sections to improve service recovery. Each feeder must be oversized to supply $1/n$ additional section hence a utilization factor equal to $CU = \frac{N_{feeders} \times P_{max}}{N_{feeders} \times \left(P_{max} + \frac{P_{max}}{n}\right)} = \frac{n}{n+1}$. In the case of three sections per feeder, we get

CU = 0.75. This structure is also widely used in China for overhead systems (Cheng et al. 2014) when we seek good QoS.

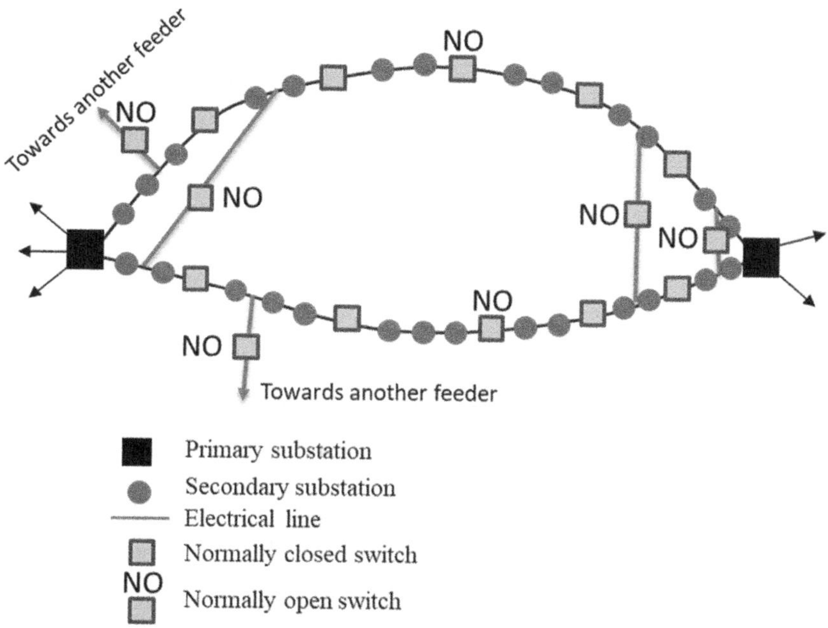

Figure 2.28. *Multi-divided and multi-connected system – case where the departure is divided into three sections*

2.4.2. *Loop and sub-loop system (Madrid, Berlin and China)*

Figure 2.29 shows a structure that we find mainly in China, but also in some old historical centers like Madrid and Berlin. Each feeder belonging to the same primary substation or to different primary substations feeds a reflection point. These reflection points are connected to each other by a normally open main loop (red loop). At each reflection point, sub-loops connect the MV/LV transformers. This structure offers a lot of redundancy and a lot of possibilities for resupply in the event of a fault. Indeed, it was designed in the 1970s when cables were of poor quality and therefore had a very high failure rate. This architecture tends to disappear in areas where the cables are renewed and therefore less likely to fail.

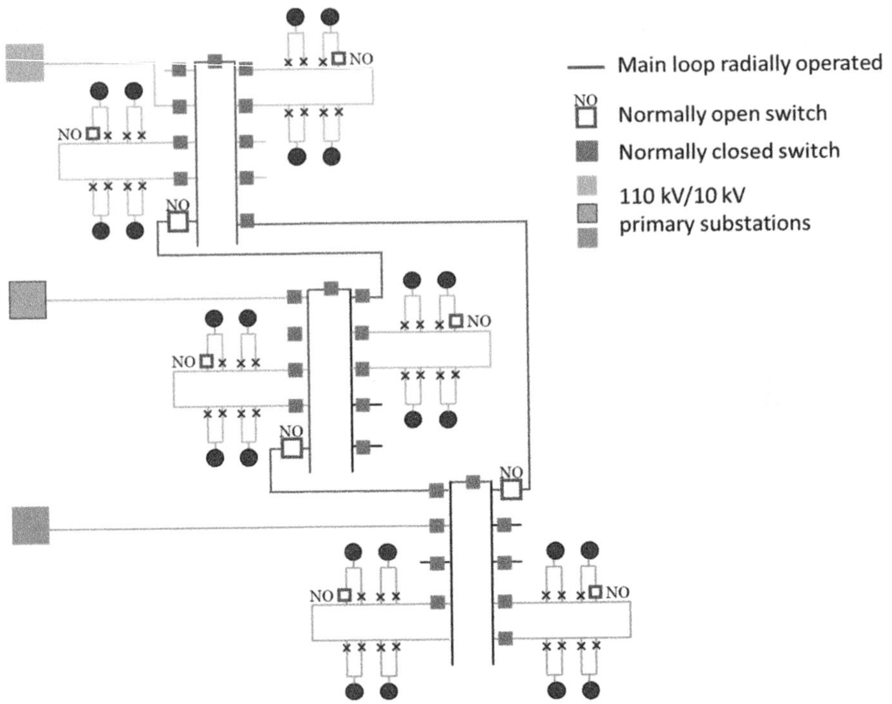

Figure 2.29. *Chinese loop and sub-loop structure*

2.4.3. *Two voltage levels, two types of distribution systems (Singapore)*

Singapore's distribution system consists of an English-influenced 22 kV grid and a 6.6 kV remote-controlled open loop system that was built after independence in 1965. The system is completely underground and its SAIDI of 15 s in 2016–2017 (SP Group 2018) is among the best in the world. Figure 2.30 provides a diagram of Singapore's power distribution system architecture. Approximately 10,000 distribution substations (6,600 in 22 kV and 3,420 in 6.6 kV) (SP Group 2021) are fully remotely conducted and those in the 22 kV system are equipped with pilot wire differential protection devices to properly detect faults in the meshed systems.

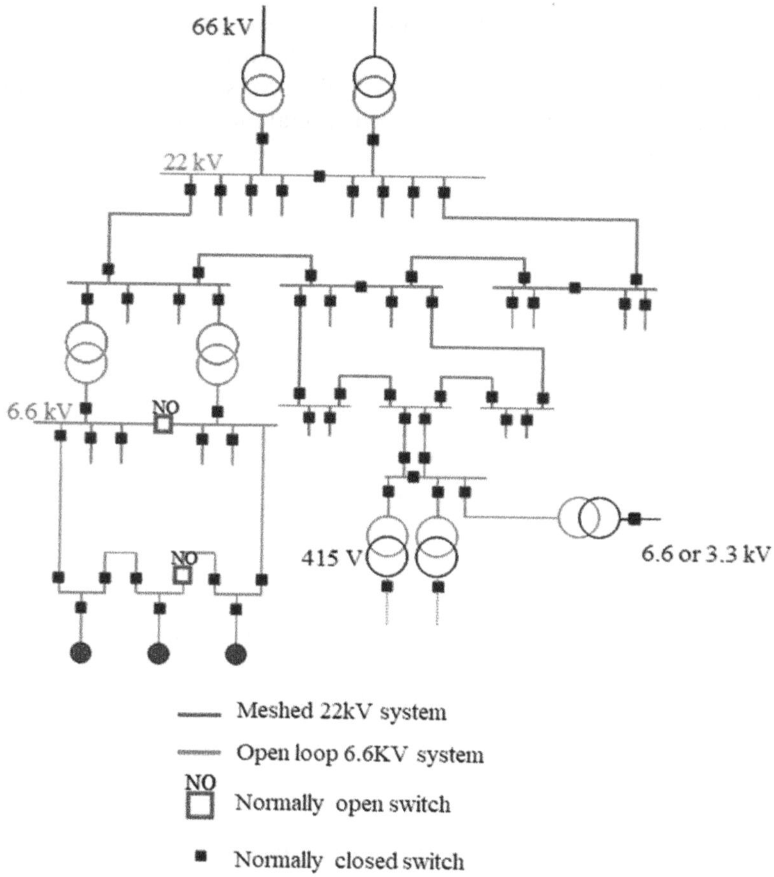

Figure 2.30. *Architecture of Singapore's power system*

2.4.4. *Secured feeder and spot network (Indonesia, Malaysia)*

The Indonesian system was built under French influence in the 1980s–1990s, thus according to the secured feeder, however the default rate of the system components is significant because they are designed according to Indonesian standards, which are far less demanding than French standards. Therefore in 2014, the Indonesian SAIDI was 493 min.

In Malaysia and Indonesia, we also find spot systems at the medium voltage level in some key economic areas, in order to ensure a reliable system and to attract foreign investors. Figure 2.31 shows the American spot system adapted to medium voltage. Each HV/MV primary substation has two transformers connected to two different HV systems upstream. The downstream part of each transformer is connected to a common bus via circuit breakers equipped with relays that are set to trip the circuit breaker when the power flows in the reverse direction, that is, from MV to HV. To this common bus are connected several radial feeders supplying the final consumers via MV/LV transformers.

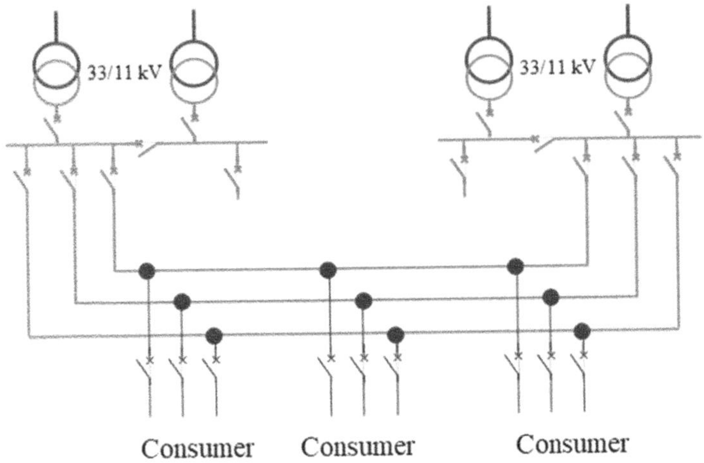

Figure 2.31. *Spot system in Malaysia and Indonesia*

2.4.5. *United Arab Emirates*

The typical distribution system in the United Arab Emirates (UAE) resembles the French secured feeder. However, as shown in Figure 2.32, the feeders are not connected in pairs to back up each other, as in France, but are instead all connected to a common switching station. Provided these are properly sized, each feeder can theoretically provide backup for another feeder. One of the feeders is called the "emergency cable" and no MV/LV transformer is connected to it. This structure can also be operated in a meshed manner, as in Dubai for example, that is, by closing the normally open breakers.

Foreign systems are therefore heterogeneous, which is sometimes the cause of incidents because the operating or maintenance teams are not trained in these different operating modes. It is very difficult to maintain systems with very different technologies. The structure of many foreign systems is mainly the result of colonial influence. In medium voltage, we find cases of meshed architectures in certain environments where we want to ensure excellent QoS with high economic stakes (to supply an important economic zone or to attract investors) or in the case of historical heritage.

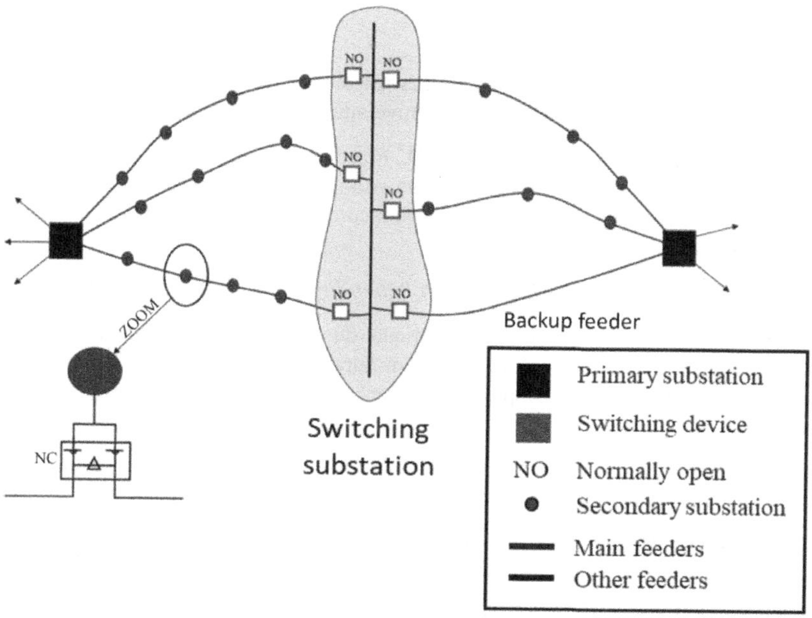

Figure 2.32. *Typical distribution system in the UAE*

2.5. Conclusion

Historically, distribution systems were built to provide access to electricity for consumers. In a second phase, electrical performance objectives were added because countries' economies now depend directly on the continuity and quality of the supply. In this context, technical and economic indicators have been defined, as well as system development strategies to optimize these, in order to better plan the evolution of the electrical distribution system. Depending on the technical choices made in the past, we can find different distribution system topologies, each with its own advantages and disadvantages. Each country has its own system planning doctrines and techno-economic strategies for investment choices. As the energy and societal context of electricity systems is changing, the current methods and choices of planners are evolving. Indeed, the massive arrival of renewable energies, electric vehicles and storage facilities, represents a major change in the assumptions underlying planning calculations. Therefore, strategies that were historically considered too costly have proven to be of interest in this new context. The objective of the following chapters is to explore the possible evolutions of planning methods and distribution system architectures in this new energy context.

2.6. References

Alvarez-Hérault, M.C. (2009). Architectures des réseaux de distribution du futur en présence de production décentralisée. PhD Thesis, Université de Grenoble, Grenoble.

Behnke, M., Erdman, W., Horgan, S., Dawson, D., Feero, W., Soudi, F., Smith, D., Whitaker, C., Kroposki, B., et al. (2005). Secondary network distribution systems background and issues related to the interconnection of distributed resources. Report, National Renewable Energy Laboratory, Springfield.

Brown, R. (2008). *Electric Power Distribution Reliability*, 2nd ed. CRC Press, Boca Raton.

Canepa, D. (2012). Approvisionnement électrique du Grand Paris [Online]. Available at: http://www.driee.ile-de-france.developpement-durable.gouv.fr/IMG/pdf/Seminaire_du_6_mars_2012_-_Presentations_Approvisionnement_electrique_du_Grand_Paris__cle2dcb2f-1.pdf [Accessed 2018].

Carrive, P. (1991). Réseaux de distribution : structure et planification. *Techniques de l'ingénieur Génie électrique*, 8, 1–32.

CEER (2008). 4th CEER benchmarking report on quality of electricity supply. *Proceedings of the Power Engineering International Conference*, Lisbon.

CEER (2016). 6th CEER benchmarking report on the quality of electricity and gas supply [Online]. Available at: https://www.ceer.eu/documents/104400/-/-/d064733a-9614-e320-a068-2086ed27be7f.

CEER (2017). CEER report on power losses. [Online]. Available at: https://www.ceer.eu/documents/104400/-/-/09ecee88-e877-3305-6767-e75404637087 [Accessed 2018].

CEER (2018). Energy quality of supply work stream (EQS WS) – CEER benchmarking report 6.1 on the on the continuity of electricity and gas supply – Data update 2015/2016 [Online]. Available at: https://www.ceer.eu/documents/104400/-/-/963153e6-2f42-78eb-22a4-06f1552dd34c.

Cheng, L., Chang, Y., Liu, M., Feng, H., Wu, Q. (2014). Typical medium voltage distribution system topologies in China: A review and a comparison of reliability. *Proceedings of the Institute of Electrical and Electronics Engineers International Conference on Probabilistic Methods Applied to Power Systems*, Durham.

Coiffier, A. (2013). Schémas directeurs de développement des réseaux électricité de distribution MT. *Techniques de l'ingénieur Génie électrique*, 1–28.

DGEC (2016). Programmation Pluriannuelle de l'Energie 2019-2023 2024-2028. Ministry of the Environment, Energy and the Sea, Paris.

Directorate-General of the Treasury (2016). Le secteur électrique au Nigéria [Online]. Available at: https://www.tresor.economie.gouv.fr/Ressources/File/430985 [Accessed 2018].

DOE (2015). Insights into advanced distribution management systems [Online]. Available at: https://www.smartgrid.gov/files/ADMS-Guide_2-11.2015.pdf [Accessed 2018].

Dfeeder, A. (2002). Le concept d'énergie non distribuée, outil d'aide à la décision dans la gestion des réseaux électriques. *Réalités industrielles*, 62–68.

Elliot, R., Lewis, K., Benjamin, R., Erickson, J., O'Donnell, A., Lee, D. (2014). The integration of distribution level generation & storage into the grid. Report, California Public Utilities Commission.

EMA (2015). Reliability [Online]. Available at: https://www.ema.gov.sg/cmsmedia/About-Us/Singapore-Energy-Story/PDF/03.%20Reliability.pdf [Accessed 10 August 2018].

Enedis (2008). Description physique du réseau public. Technical note, Enedis-NOI-RES_07E Version 3 01/03/2017.

Enedis (2018). Qualité de fourniture [Online]. Available at: http://www.enedis.fr/qualite-de-fourniture#onglet-duree-moyenne-de-coupure-bt [Accessed 2018].

Enedis (2020). Principes d'étude et règles techniques pour le raccordement au Réseau Public de Distribution géré par Enedis d'une Installation de Consommation en HTA. Enedis, Paris.

Enedis (2021a). Le réseau électrique intelligent ou smart grid [Online]. Available at: https://www.enedis.fr/greenlys [Accessed 2021].

Enedis (2021b). Chiffres clés 2020 [Online]. [Accessed 2021].

Enedis and ADEeF (2017). Comandabilité et actions : services de flexibilité. Study, Valorisation économique des Smart Grid : contribution des gestionnaires de réseau public de distribution. Enedis, Paris.

Energy Market Authority (2017). Smart energy sustainable future – Annual report 2016/2017. Rapport, Energy Market Authority, Singapore.

Energy Mix (2018). Community protesters shut down Port Harcourt transmission station. Report [Online]. Available at: http://energymixreport.com/community-protesters-shut-port-harcourt-transmission-station/ [Accessed 2018].

FTICL Energy (2016). Étude sur la régulation incitative de la qualité d'alimentation des gestionnaires de réseaux publics de transport et de distribution d'électricité. Report, CRE.

Garry, A. (2016). Intégration des incertitudes liées à la production et à son effacement sur les méthodes de planification des réseaux. PhD Thesis, Université Grenoble, Grenoble.

Gouin, V. (2015). Évaluation de l'impact du Smart Grid sur les pratiques de planification en cas d'insertion de production décentralisée et de charges flexibles. PhD Thesis, Université de Grenoble, Grenoble.

He, Y., Nilsson, A., Carlsson, F. (2011). Failure causes of distribution network components. *Proceedings of the 21st International Conference on Electricity Distribution*. CIRED, Frankfurt.

INSEE (2019). Les entreprises en France [Online]. Available at: https://www.insee.fr/fr/statistiques/4256020. [Accessed 2021].

IREC (2013). Connecting to the grid [Online]. Available at: https://www.energy.gov/sites/prod/files/2013/11/f4/connecting_to_the_grid_2009.pdf. [Accessed 2021].

Jorgenson, D. (1963). Capital theory and investment behavior. *American Economic Review*, 53(2), 247–259.

Khator, S.K. and Leung, L.C. (2007). Power distribution planning: A review of models and issues. *IEEE Transactions on Power Systems*, 12(3), 1151–1159.

Marglin, S. (1963). Approaches to dynamic investment planning. *The American Economic Review*, 54(4), 427–429.

Marignac, Y. and Legrand, V. (2003). Énergie non distribuée : apports et limites du concept d'électricité non distribuée pour une nouvelle approche de la sécurité énergétique française. Report.

Miyata, M., Koizumi, S., Watanabe, M. (2017). Case study of the distribution system planning for a multi-divided and multiconnected system. *Proceedings of the 24th International Conference and Exhibition on Electricity Distribution*. CIRED, Glasgow.

Mutialu, R. and Zafar, M. (2017). System efficiency of California's electricity grid (CPUC) [Online]. Available at: http://www.cpuc.ca.gov/uploadedFiles/CPUC_Public_Website/Content/About_Us/Organization/Divisions/Policy_and_Planning/PPD_Work/PPD_Work_Products_(2014_forward)/System_Efficiency_Report%20PPD_May_24_Final.pdf [Accessed 2018].

OCCTO (2016). Report on the quality of electricity supply. Report, Organization for Cross-regional Coordination of Transmission Operators, Tokyo.

Oddi, M. (2011). Plan de protection des réseaux de distribution à moyenne tension – mise en œuvre. *Techniques de l'ingénieur Génie électrique*, 1–26 [Online]. Available at: https://doi.org/10.51257/a-v1-d4812.

PG&E (2014). PG&E overview [Online]. Available at: http://www.pgecorp.com/corp_responsibility/reports/2016/bu01_pge_overview.jsp. [Accessed 2018].

PG&E (2015). Pacific gas ans electric company distribution resources plan: Opportunities and challenges for energy efficiency. *Proceedings of the ACEEE (National Conference on Energy Efficiency as a Resource)*. Little Rock, Arkansas.

PG&E (2017). 2016 Annual electric reliability report. Report, Pacific Gas and Electric, San Francisco.

PG&E (2019). Base interruptible program [Online]. Available at: https://www.pge.com/en_US/large-business/save-energy-and-money/energy-management-programs/demand-response-programs/base-inerruptible/base-inerruptible.page [Accessed 2021].

RTE (2018). Statistiques de l'énergie en France 2016 : statistiques de production, consommation et échanges [Online]. Available at: https://www.rte-france.com/fr/article/statistiques-de-l-energie-electrique-en-france [Accessed 2018].

Schlabbach, J. and Rofalski, K. (2014). *Power System Engineering: Planning, Design, and Operation of Power Systems and Equipment*. John Wiley & Sons, Weinheim.

Settembrin, R. and Fisher, J. (1991). Reliability and quality comparisons of electric power distribution systems. *Proceedings of the IEEE Power Engineering Society Transmission and Distribution Conference,* Dallas.

Sim, E. (2021). Quelques notions essentielles pour mieux comprendre et mieux exploiter les indicateurs économiques. Report.

SP Group (2018). Always here for you [Online]. Available at: https://www.spgroup.com.sg/poweringthenation/always-here-for-you.html [Accessed 2018].

SP Group (2021). Singapore power annual report 2015/2016 [Online]. Available at: https://www.spgroup.com.sg/wcm/connect/spgrp/ea560383-d6d8-4a92-9925-af27cce9462b/SP+Group+Annual+Report+FY1516.pdf?MOD=AJPERES&CONVERT_TO=url&CACHEID=ROOTWORKSPACE.Z18_M1IEHBK0MOUJ20ABQK7Q593U32-ea560383-d6d8-4a92-9925-af27cce9462b-lINZpPn [Accessed 2021].

Syndicat départemental d'énergie électrique de la Gironde (2015). Évolution des indicateurs de qualité et de continuité de fourniture. Report, AEC expertise & conseil, May.

Temraz, H.K. and Quintana, V.H. (1993). Distribution system expansion planning models: An overview. *Electric Power Systems Research*, 26(1), 61–70.

TEPCO (2015). Electricity supply facilities [Online]. Available at: http://www.tepco.co.jp/en/corpinfo/illustrated/electricity-supply/1253683_6499.html [Accessed 2018].

TEPCO (2016). Electricity supply facilities [Online]. Available at: http://www.tepco.co.jp/en/corpinfo/illustrated/electricity-supply/distribution-underground-e.html [Accessed 2018].

Wang, W. and Pinter, S. (2014). Dynamic line rating systems for transmission lines [Online]. Available at: https://www.smartgrid.gov/files/SGDP_Transmission_DLR_Topical_Report_04-25-14_FINAL.pdf [Accessed 2018].

Willis, H.L., Tram, H., Engel, M.V., Finley, L. (1995). Optimization applications to power distribution. *IEEE Computer Applications in Power*, 8(4), 12–17.

Willis, H.L., Tram, H., Engel, M.V., Finley, L. (1996). Selecting and applying distribution optimization methods. *IEEE Computer Applications in Power*, 9(1), 12–17.

3

Integration of Distributed Energy Resources in Distribution System Planning

3.1. Introduction

Chapters 1 and 2 presented in detail the historical evolution of power systems with a particular focus on distribution systems. This reminder of the technological, structural and regulatory choices is important because in the context of massive and rapid distributed energy resource (DER) development (defined in section 1.6.3 of Chapter 1), it is not economically and temporally possible to start from scratch. In this chapter, we first analyze the impact of DERs on distribution systems as well as governmental recommendations to foster their integration. We focus on the solutions that allow for the integration of DERs and in particular the flexibility solutions, also termed: non-wire alternatives. Next, we see that the evolution of DERs reveals a breaking point in the evolution of distribution systems. As long as the DER threshold remains low, the adoption of flexibility with simple remuneration mechanisms is sufficient. However, when the DER rate becomes significant, an evolution of the regulation is necessary, in particular, the development of a new structure for local markets (collaborative or not) that are coordinated with the wholesale markets.

For a color version of all the figures in this chapter, see www.iste.co.uk/alvarezherault/distribution.zip

3.2. Impact of distributed energy resources on the planning methods of distribution power systems

3.2.1. *Problems brought about by the appearance of DERs*

The appearance of DERs in the distribution system, which was not historically designed for this purpose, can lead to a number of constraints, each with varying degrees of intensity and occurrence. When these constraints appear, the current methods for integrating DERs into the power system are quite basic and have not involved any research into optimization. Moreover, regulations currently provide little incentive for performance and innovation, instead they are often solely based on the CAPEX costs of equipment. The cost of development, maintenance and modernization of the electrical distribution system is very important and in the United States represents about 20 billion dollars per year (AEE 2018) and, in France over the 2017–2020 period, 4.2 billion euros per year (CRE 2021). Therefore, if the share of DERs becomes non-negligible, then the existing tools will have to evolve in order to avoid a significant increase in power system investments. The main constraints caused by the appearance of DERs are voltage, temperature rise, stability, quality and protection against malfunctions.

3.2.1.1. *Voltage constraints*

The voltage profile decreases along the distribution system due to the load. At first glance, the introduction of generation, provided it remains low, could be beneficial to the system as it tends to raise the voltage level by injecting active power. However, the operator is still forced to size the system according to the consumption peak, without considering this generation because of its intermittent and uncontrollable nature. Beyond a certain threshold of generation penetration, overvoltage may appear in the system, as shown in Figure 3.1. Similarly, without a specific incentive mechanism, electric vehicle charging is superimposed on the consumption peak, which has the effect of aggravating the voltage drop (Figure 3.1). The HV/MV on-load tap changers or MV/LV off-load tap changers can then be used to reduce the voltage, but this action is limited because it only has a 'translation' effect on the voltage. Therefore, the resolution of overvoltage (respectively undervoltage) at one point in the system can lead to the appearance of one or many more undervoltage incidents (respectively overvoltage) at other points, as shown in Figure 3.3 (and Figure 3.2, respectively).

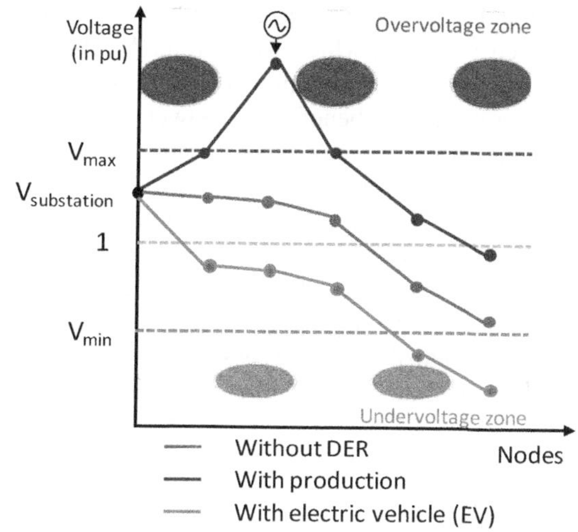

Figure 3.1. *Voltage profile on a feeder with and without DER (DER = production or electric vehicle)*

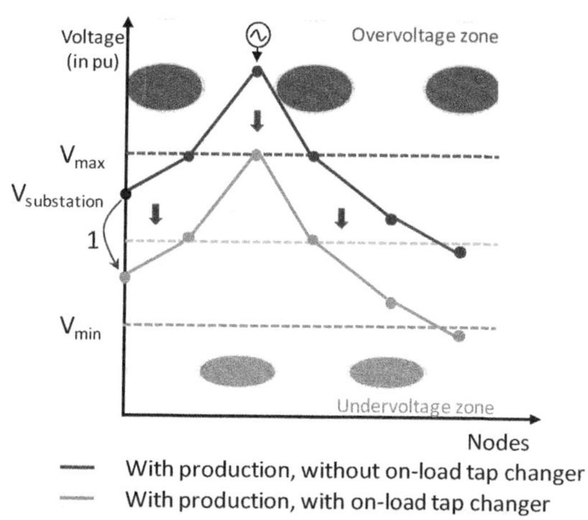

Figure 3.2. *Voltage profile with production – action of the on-load tap changer*

3.2.1.2. *Thermal constraints*

The presence of DERs in distribution systems changes the direction of power flows that were previously unidirectional, from the transformer to the load. These flows can be bidirectional due to a power export from the generation or storage systems, or even reversed from that which was originally planned. This can lead to more or less positive consequences on the power system, as illustrated in Figure 3.4. More local power generation can benefit the power system by limiting losses. However, local thermal stresses may also occur on some conductors and transformers if they are not sized for the new power extremes that may be different from the typical peak load. The distribution system can thus become a power exporter and potentially create constraints on the transmission system. Conversely, in urban areas, the development of electric vehicles can also create thermal constraints in the conductors if no load transfer mechanism has been provided. In general, the insertion and control of new types of uses at the distribution system level can create thermal constraints if no interaction between the system and the users takes place.

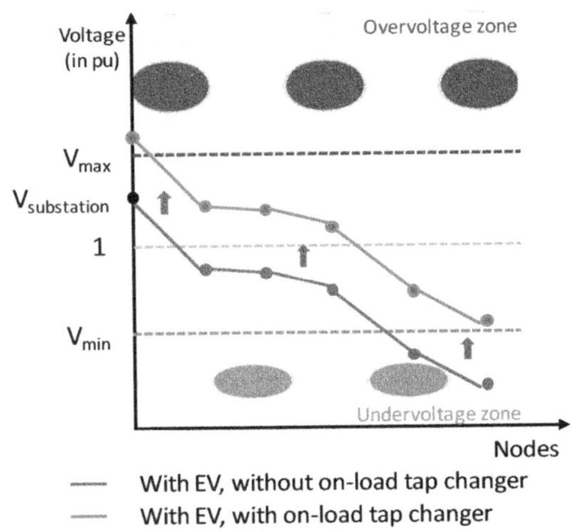

Figure 3.3. *Voltage profile with EV – action of the on-load tap changer*

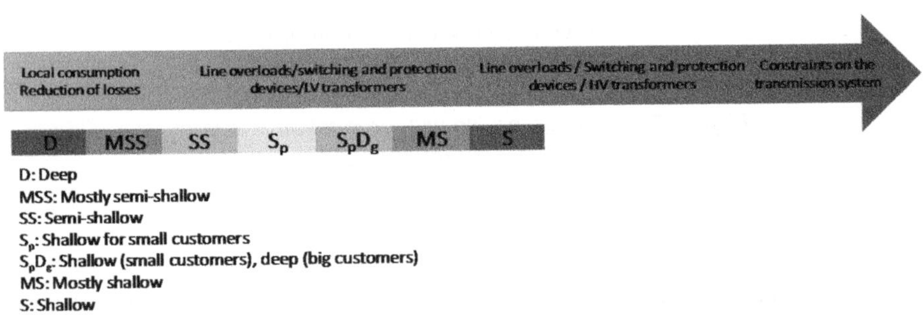

Figure 3.4. *Impact of DER increase on power flows*

3.2.1.3. *Stability constraints*

Distributed producers are equipped with decoupling protection that disconnects them from the power system when the frequency goes outside of the regulatory ranges defined in the grid codes. This precaution, taken in the context of marginal development for this type of generation, allows for system stability to be maintained. For example, in Germany the high decoupling frequency was set at 50.2 Hz, while in Italy and Denmark it was 50.3 Hz (Von Appen et al. 2013). When the frequency measured at the connection point of the distributed generation became higher than this maximum frequency, the generation was disconnected from the system, allowing the system frequency to be lowered before the primary control acted. However, a high penetration of distributed generation creates a risk of widespread outages as a consequence of these decoupling protection devices with this type of "on-off" control. A modification of these settings has already taken place in Germany, as summarized in Von Appen et al. (2013). During the blackout of November 4, 2006, the de-energizing of an HV line in Germany for the passage of a ship divided the European power system into three zones operating at different frequencies. At 50.2 Hz, Germany was part of an over-frequency zone. If the installed PV capacity had been the same as today (49.55 GW, or 23.4% of the German installed base), and if the incident had occurred at the time of the production peak, the European capacity would not have been sufficient to compensate for the loss of production caused by the decoupling of the PV production. The frequency would therefore have fallen sharply, which would have led to a generalized blackout. As a result of this analysis between 2012 and 2014, Germany changed the control of decoupling protection devices to introduce a gradual reduction of production rather than a total capping. This affected 315,000 PV producers at a total cost of €75 million. Similarly, a risk of outages can occur in underfrequency areas: if a significant number of generation producers are disconnected because the frequency falls below the decoupling protection threshold, it then becomes necessary to curtail more customers in order to return to a steady state, as the Australian Energy Market Operator (AEMO) explains in the AEMO 2016 Report. This was notably the case on August 9, 2019 in the UK, which led the Office of Gas and Electricity Markets (OFGEM) to outline a list of recommendations on the adjustment of protection devices for, among others, distributed producers, in its OFGEM 2020 Report. It is therefore necessary to coordinate the frequency settings of the global system and local generation at the level of the global power system.

A second problem related to stability is the expected decrease in the share of rotating machines in the electricity mix, in favor of the development of power electronics-based production. This loss of inertia, an intrinsic property of synchronous machines, must be compensated for either in a real way (the use of synchronous machines as static compensators, for example) or in a virtual way (thanks to storage, for example).

3.2.1.4. *Other constraints*

Other negative effects may also occur:

– *alterations to the quality of the voltage and current wave*: for example, rapid voltage variations, harmonics or flicker (see section 1.2.6.1);

– *malfunctioning of protection devices*: in radial systems, the protection devices are calibrated on tripping thresholds with few or no generation sources. In the event of a fault, the generation sources located in the vicinity of the fault contribute to the fault current and two phenomena can occur:

- *untimely tripping of the protection devices of a healthy feeder*: the generation sources supplying the fault to a neighboring feeder can cause the protection devices of the healthy feeder to be tripped,

- *blinding of the protection devices of the faulty feeder*: the impedance of the generation sources reduces the value of the fault current, which is subsequently not detected by the feeder's protection devices. Another example can occur on LV meshed systems in large American urban centers that have been equipped with directional protection: a fault situation can be detected in the event of an export onto the MV system of surplus power that has not been consumed locally (see section 2.4.2.2).

Therefore, the protection plan designed without DERs must be reviewed, or even completely changed, with the appearance of DERs.

3.2.2. *A need for an advanced planning tool that integrates DERs*

As explained in Chapter 2, power systems are historically sized according to load growth assumptions and system managers plan investments according to the constraints that appear or the quality of supply to be maintained. The main variable is therefore the evolution of the load, which is rather simple to forecast. DERs have already been present for a few years, but in a rather marginal way, which has not disrupted system planning too much. On the other hand, due to their present strong growth, connection studies have become necessary in order to size the connection of DERs to the system, and also to reinforce other parts of the system that could be constrained due to their presence. However, the trajectories and technological choices for the development of the electrical system in the long-term are not modified for the moment since the historical dimensioning case study remains based on the annual consumption peak.

In the future, DERs will experience significant development thanks to technological advances and, above all, to the expected decrease in costs. We can cite the example of batteries, whose installation costs according to the International Renewable Energy Agency (IRENA) are expected to drop by 50–66% by 2030

(IRENA 2017). This could lead to overinvestment situations and therefore a significant cost for the DSO, and consequently for consumers if these DERs are not integrated into planning studies in the same way as consumption is, and continue to only be considered a posteriori In a presentation by Pilo (2018), ISGAN[1] explains that this shift in planning significantly increases the complexity of investment decisions for several reasons, including:

– *the more complex identification of the cases that are important for the network*: bidirectional flow situations will multiply, which makes it more difficult to assess the maximum transit situations (location, time of year and value). For example, in the case of France which has a large share going to electric heating, 28% of residential consumption in 2016 in the RTE 2017 forecast, the network is sized according to the winter peak. As air conditioning systems are still underdeveloped in the residential sector, this sizing could shift in summer for areas with high photovoltaic production. For example, in Spain, the difference between the winter and summer peak consumption is decreasing year by year, and even saw a reversal in 2016 (RED 2020);

– *the multiplication of planning options*: the emergence of unconventional operational solutions, also known as flexibility services or non-wire alternatives (NWA), to defer or cancel a new investment onto the power system. These strategies can use existing equipment (such as distributed generation, storage and electric vehicles) combined with power system control algorithms, such as load curtailment or generation. As defined in SEPA (2010), an NWA should not degrade the quality of service and accessibility to the system. Therefore, it is important to be able to list and model these options and include them in the cost–benefit analyses used in planning;

– *a multi-temporal problem*: time scales must be shorter and take into account the different lifetimes of components. For example, the average lifetime of lines and transformers is 40–60 years (although they can reach 80 years), which is much longer than that of the information systems necessary for the implementation of NWAs, which is in the region of 10–15 years;

– *access to large amounts of data*: smart meters will allow us to obtain a great deal of information and to better predict power flow curves, but this requires the development of appropriate methods and algorithms (data analysis);

– *the increase in the degree of uncertainty*: the uncertainties surrounding the development of DERs, whether in terms of type, quantity or behavior (notably due to intermittency and consumer behavior), require the use of probabilistic methods associated with risk analysis;

1 International Smart Grids Action Network: international platform to support governmental concerns and actions around the world to accelerate the development of clean and smart grids.

– *a multi-target problem of a different nature*: the planning targets go beyond the traditional calculation of CAPEX and OPEX (see Chapter 2) but include other criteria that are not easily monetized, such as the price of a flexibility service like load curtailment.

Therefore, the electric distribution system, which until now has been looking for solutions to integrate DERs passively, will in the future have to use them to ensure proper operation at reasonable costs (EEA 2018). It is therefore necessary to evolve planning methods and tools because, as mentioned in the CIGRE C6 TB 591 working group report (CIGRE 2014b), a survey of 33 DSOs in 14 countries shows that most of them do not integrate flexibility into their planning methods due to a lack of planning tools, methodology and business cases.

3.2.3. *Government policy recommendations on the evolution of distribution system planning methods*

3.2.3.1. *In Europe: the Clean Energy Package*

The European Clean Energy Package (CEP) (Nouicer and Meeus 2019), presented in November 2016 by the European Commission, proposes measures to improve energy efficiency and the integration of renewable energy by 2030. Table 3.1 summarizes its recommendations for the evolution of planning methods.

Recommendations for distribution systems
Investment planning[2] *(Article 32, "Incentives for the use of flexibility in distribution systems" (Girouard 2015))*
– Carry out multi-year development plans: detail the necessary investments for a 5- to 10-year horizon, including loads, electric vehicle charging points, production and storage. – Include in these plans flexibilities as an alternative to traditional investments. Nouicer and Meeus (2019) state that DSOs can include up to 5% of generation curtailment in their development plans (with an exemption for countries with high renewable penetration). – Review and update these plans with the TSO. – Submit these development plans every 2 years to the regulator who may request their modification. – DSOs must consult with users affected by these investment plans and publish the results of the survey. – More involvement from regulators and TSOs in DSO investment plans.

2 The DSOs concerned by these recommendations must have more than 100,000 customers.

Recommendations for distribution systems
Development of flexibility (excluding frequency stability), including congestion management (Article 32, "Encouraging the use of flexibility in distribution networks" (Girouard 2015)) – Development of a regulatory framework to incentivize DSOs to use flexibility services that are cost-effective for them (which is very difficult to define). – DSOs or the regulator should define the specifications of flexibility services and the associated standardized products. – These flexibility services must be market-based unless the regulator exempts them. – Coordination with TSOs is necessary to avoid double use of flexibility at the transmission and distribution levels to solve the same problem.
Charging stations and storage The DSO cannot be the owner of recharging and storage infrastructures (market system) except under derogations. – Exemption for charging stations (Article 33 "*Integration of electromobility into the electricity grid*" (European Commission 2019)): - other parties cannot/will not own or the cost of the service provided would be too high (to be reevaluated at least every 5 years[3]); - agreement by the regulator; - same management mode of the charging points as the third-party access, as defined in Article 6. – two types of possible exemptions for storage (Article 36 *"Ownership of energy storage facilities by distribution system operators"* (European Commission 2019)): - Case 1 – storage is a fully integrated component of the grid[4]: the regulator's agreement is sufficient; - Case 2 – storage is not a fully integrated component of the grid: three conditions required: – other parties cannot/will not own or the cost of the service provided would be too high (to be reevaluated at least every 5 years); – agreement by the regulator; – need for storage to fulfill the DSO's mission.

Table 3.1. *Clean Energy Package recommendations (2019) regarding the evolution of planning methods*

3 In the event of new owners, the DSO would cease this activity and could recover the residual value of the investments from the charging infrastructure.

4 Definition from EU Directive 2019/944 (Girouard 2015): "Fully grid-integrated components: components that are integrated into the transmission or distribution system, including storage facilities, and are used for the sole purpose of ensuring the reliable and secure operation of the distribution or transmission system, but not for balancing or congestion management purposes."

The CEP analysis identifies three important points:

– *DSOs must publish their Distribution Network Development Plan (DNDP).* In Europe, only DSOs in Italy, Hungary, the Netherlands, Poland, the United Kingdom and Germany do so, but this only concerns HV equipment;

– *a coordinated approach is needed between transmission and distribution.* The investment decision process is less structured and transparent for distribution systems than for transmission systems. Therefore, an investment in the transmission system could also have a beneficial effect on distribution and thus avoid the need for an investment into distribution and vice versa;

– one of the obstacles to the development of DERs is that there is no regulation in most of the European countries that authorize DSOs to contract flexibility services. Only Germany, Belgium and part of the United Kingdom are authorized to do so, through contracts allowing the connection costs of the units to be reduced and authorizing the DSO to control them (survey conducted among European DSOs in the framework of the EVOLVDSO project). Some countries, such as the UK, the Netherlands, Norway, Germany, Australia and recently France, have also set up flexibility platforms, some of which allow for the contracting of services between DSOs and flexibility providers, while others allow for a connection between TSOs, DSOs and energy markets in order to relieve a local constraint (see section 3.4.4.2).

3.2.3.2. *In the United States: example of California*

In the United States, several states have begun to evolve their planning tools to incorporate the presence of DERs. In California, Assembly Bill No. 327 of October 7, 2013 requires three IOUs to submit Distribution Resources Plans (DRPs) by July 1, 2015 with the goal of evolving their planning methods by considering the services that DERs can provide, thereby reducing costs while providing technical benefits (FERC 2018). In the 2015 DRP, California's private system operators (PG&E, SCR and SDG&E) thus incorporated DERs into their planning method in three forms:

– implementation of methodologies to find the optimal DER location – application to two study cases in 2018:

- Case Study A – Integration Capacity Analysis (ICA): mapping the DER capacity of each line in the distribution system,

- Case Study B – Locational Net Benefits Analysis (LNBA): optimal DER location to defer or cancel an investment;

– implementation of demonstrations to validate the methodologies developed above:

- Demonstration Project C: using DER to defer system investments based on the LNBA tool,

- Demonstration Project D: operating a power system with a high penetration rate of DERs,

- Demonstration Project E: planning and operating a microgrid;

– evolution of current planning rules:

- DER growth scenarios and load,

- modernization of the system investment decision process,

- Distribution Investment Deferral Framework (DIDF): definition of a new methodology to identify DER connection areas, evaluate their interest and select the most suitable ones.

The methods presented in FERC (2018), to which we will return later, do not question the major historical choices such as architectures and associated operating modes, but instead aim at improving the various planning stages.

Another procedure that has been tested since 2015 and allows for the use of flexibility by the DSO is the demand response auction mechanism (DRAM) (PG&E 2021b). This allows flexibility aggregators to offer their service to the DSO directly through the wholesale market and CAISO.

3.2.4. *Transitioning to planning with DERs*

3.2.4.1. *New planning specifications*

Table 3.2 summarizes the recommendations for changes to traditional planning in order to integrate the development of DERs and to guarantee the sizing of the power system according to controlled costs and risks.

Traditional planning	**Planning with DERs**
Simple load models Based on telemetry at the level of the primary substation only	*Accurate load and production models* For example, deployment of communicating meters
Operation of the power system within the technical limits Operation of the system set at a given level of risk	*Operation of the power system beyond its technical limits* Integration of operational planning in long-term planning (use of flexibility Active Management System, AMS (TSO - DNO 2019))

Traditional planning	Planning with DERs
Deterministic calculations based on a worst-case scenario Overinvestment and oversized architectures associated with a very low level of risk	*Risk-oriented probabilistic calculations using time series*
A single target: cost optimization CAPEX, joule losses cost and reliability cost (see Chapter 2)	*Multiple targets* To be added: compensation for flexibility, DER insertion targets, environmental targets
Decoupled planning of energy systems	*System Coupling (Integrated Energy Network (Howard 2017))* Energy systems (electric-water-heat-mobility-data), measurement systems and information and communication systems

Table 3.2. *Evolution of planning methods with DERs*

COMMENT ON TABLE 3.2.– As explained in Chapter 2, the medium voltage architectures encountered around the world have loops that allow the system to be reconfigured in the event of a fault in order to ensure service recovery. In France, as the load is thermally sensitive, with a large proportion going to electric heating, the power considered in normal operation is calculated according to a predefined risk level for the occurrence of a certain temperature. The DSO considers a power called P_{tmb} (power at minimum basic temperature) corresponding to the power for the 30th lowest value of the daily average temperature observed over the last 30 years. In other words, the probability that the actual annual minimum power is higher than the P_{tmb} is about $\frac{30}{30 \times 365} = 99.73\%$ (i.e. a risk of 0.27%). On the other hand, in the event of an N-1 emergency scenario, the probability of losing an element that is already low, the DSO considers a less constraining case of power, that is, the one corresponding to the average temperature on 15 January over several decades.

3.2.4.2. *A two-stage transition*

The transition from current planning methods to new methods integrating DERs is complex and cannot be achieved quickly given the scale of the problem: very large distribution systems (1.4 million km for France alone), numerous players, complex regulatory structures, development history, country policies, different energy mixes and specificities of renewables (intermittent or predictable). As mentioned, some DERs have existed for a long time on the power system, such as marginal generation or the French on-peak/off-peak hours system, which can be assimilated to fixed *demand-side flexibility*. The evolution of the distribution system expected in the coming years will depend on the development dynamics of DERs (in terms of number, power and location) with, in particular, threshold effects requiring a technological and regulatory evolution of the power systems. De Martini and Kristov (2015) introduce three phases of power system development:

– Phase 1: low DER rate → need to upgrade the power system;

– Phase 2: medium DER rate → need to integrate DERs into planning;

– Phase 3: high DER rate → need to implement new market models incorporating DERs.

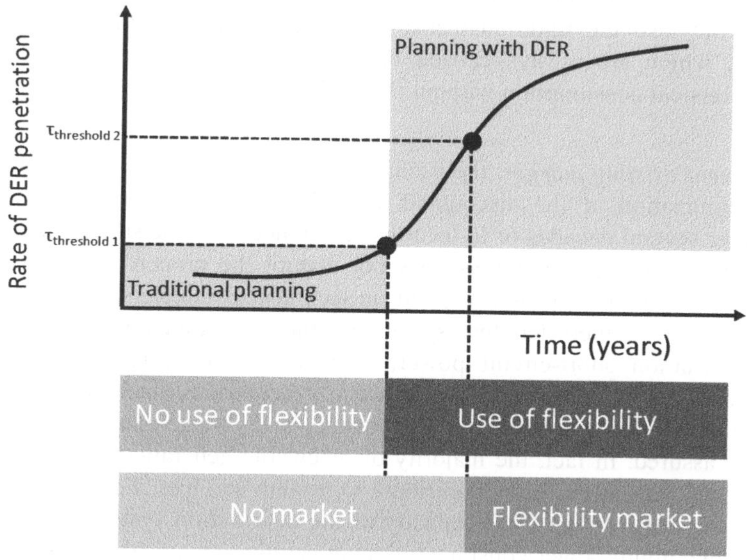

Figure 3.5. *Evolution of the distribution system according to development of DERs*

The DER transition rates between phases depend on a great many parameters, such as those mentioned above. In reality, the technological building blocks to control DERs so as to provide services to the power system are mature enough to be deployed. The real challenge is to determine the business models and operating procedures that are beneficial to all stakeholders (DERs by themselves, local energy communities and local markets). Therefore, we propose to replace the model of De Martini and Kristov (2015) with the one shown in Figure 3.5: as long as DER penetration remains below $\tau_{threshold\ 1}$, traditional planning methods remain possible; between $\tau_{threshold1}$ and $\tau_{threshold\ 2}$, it is possible to use flexibilities with relatively simple economic models; and when DER penetration becomes higher than $\tau_{threshold\ 2}$, the need for flexibilities as well as the number of flexibility providers increases, and it is necessary to define rules in the form of a framework of the local energy market and all the interactions between players as well as with the wholesale market. The rest of this chapter is thus structured according to this model: traditional planning versus planning with DERs using flexibilities with or without an associated market.

3.3. Phase 1: traditional "fit and forget" planning

In the first phase, the presence of DERs is sufficiently low so as to not call into question the main planning principles, that is, the choice of architectures, their operation, but also the traditional cost–benefit calculation methods presented in Chapter 2, which are always carried out on assumptions pertaining mainly to so-called classical consumption without DERs. In this phase, we can distinguish two sub-cases:

– *sufficient existing margin*: the sizing of the system's electrical equipment due to the minimization of the discounted TOTEX (CAPEX, losses and quality of supply) over several decades or to local policy choices may be sufficient enough to ensure that the presence of DERs does not disrupt the proper functioning of the power system. This margin will depend on technical parameters, such as line and cable technologies, their lengths, system architecture, transformer capacity and primary substation short-circuit power, as well as on climatic and regulatory parameters. In addition, as long as the share of synchronous machines remains high in relation to renewable generation, the frequency and voltage margins of the overall system are assured. In fact, the majority of renewable generation sources are static by design and do not currently contribute to system services. For example, in the United States, the average DER capacity of the distribution system is estimated at 5% of the maximum power consumed without creating technical constraints. In the case of California, 8% of feeders have more than 15% margin for their hosting capacity, 3% of feeders have more than a 30% margin, and 1% of feeders have a 100% margin (IRENA 2017);

– *insufficient margin*: some of the constraints listed in section 3.2.1 appear at one or more points in the system. The potential solutions are identical to traditional planning solutions, that is, with a high CAPEX: construction of lines, primary substations, secondary substations, switchgear, etc. The system operators then set up connection rules making it possible to select the various technical solutions and estimate their costs.

The integration of DERs is therefore treated on a case-by-case basis, and considered more as a passive element, that is, to be connected in the same way as a conventional load, rather than as an active element that can participate in the minimization of TOTEX if the rules are well defined. The objective is therefore to solve all the constraints of the worst-case scenario that still have a reasonable probability of occurring. In other words, DER power must be able to flow at any time of the year without causing technical constraints. In France, for example, this worst-case scenario corresponds to a situation where generation is at a maximum and consumption is at a minimum (estimated at 20% of maximum MV consumption) (CIGRE 2014a).

In both cases, the sizing is done by considering the worst-case scenario, which may be related to consumption or generation. In the US, this is called "fit and forget", and in the UK it is called "firm connection", since there is no questioning or adaptation of the final choice made. Therefore, CAPEX can be very high for potentially only a few constraints in a year.

3.3.1. *Allocation of DER connection costs*

The regulation reinforces this approach since the DSO is obliged to offer a connection solution to any customer wishing to install a DER. If the DER is a producer, the connection process is based on the queuing principle. The DSO conducts a technical study for the first producer in the queue and estimates the cost of the work through a technical and financial proposal. Traditionally, the costs of connecting a producer to the distribution system are defined by the regulations of each country. We distinguish four main levels of connection costs as presented in INOGATE (2009) and Furusawa et al. (2019): deep, shallow, semi-shallow (the case for France) and super shallow costs. Figure 3.6 summarizes the customer (consumer and producer) connection cost models for several European countries listed by the European Commission (2015).

As shown in Figure 3.7, the most constraining for the producer is the deep cost, since in this case the producer must not only pay for their connection to the system but also the possible reinforcement costs in the event their connection causes a constraint elsewhere along the distribution system. In the shallow cost, only the connection cost is borne by the producer, in the semi-shallow cost, the reinforcement cost can be shared between the DSO and the producer according to a distribution key that depends on the country, while in the super shallow cost, the DSO bears all costs. In the case of the shallow cost, producers have an incentive to connect and if their penetration rate becomes significant, the DSO may have to face a significant reinforcement of the system leading to unreasonable costs. On the other hand, in the case where the producer bears all the costs, this indirectly favors the development of producers in the "optimal" locations of the system but limits their penetration once the system has reached its limits (through construction) since the costs for the producers would become prohibitive. Therefore, these cost allocation models are limited in the case of significant generation development.

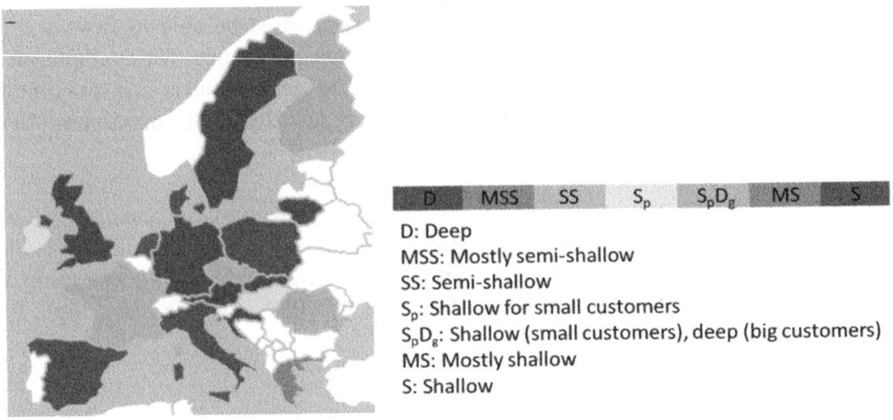

Figure 3.6. *Connection cost model for customers (producers and consumers)*

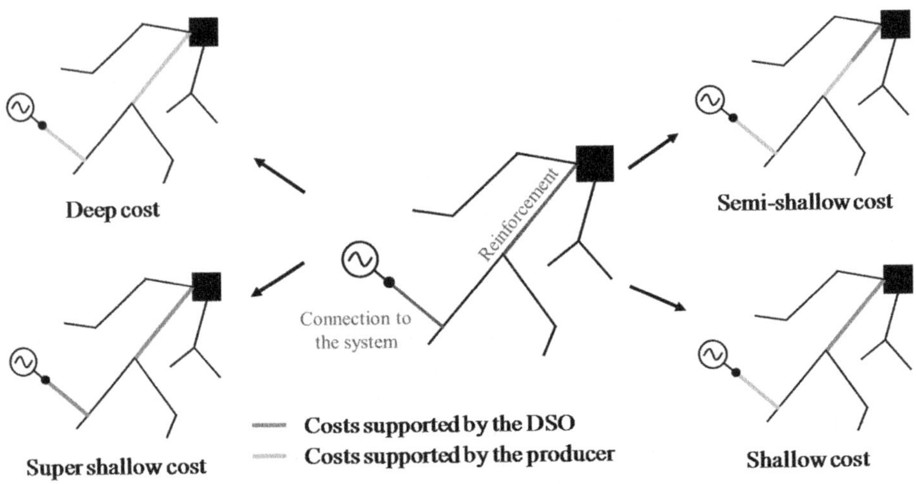

Figure 3.7. *Four levels of connection costs for renewable energies*

Therefore, DSOs are looking to develop tools to integrate DERs into their studies in order to better model electrical quantities and find the optimal locations for developing DERs.

3.3.2. *Estimated hosting capacity of the distribution system*

Some DSOs are developing tools to estimate the DER hosting capacity of their system, so that they can determine optimal DER locations, for example. De Martini and Kristov (2015) define the hosting capacity of a system as the additional DER power that can be added to each part of the system without having to reinforce it or change its equipment. It can also be defined globally for a given system.

3.3.2.1. *Example of tools developed in the US*

3.3.2.1.1. In California, the Integration Capacity Analysis developed by IOU

California's private system operators have developed a tool to assess the capacity of DERs in the distribution system called Integration Capacity Analysis (ICA) or Case Study A, as introduced in section 3.2.3.2, particularly to help determine optimal DER locations. This tool went live in late 2018 and is updated monthly. Capacity is calculated for each node and line independently of each other. It is requested that California's IOUs use a common methodology by avoiding the use of heuristics, considering different types of DERs and using time series. The scope studied is medium voltage, and the constraints considered are overtemperature, protection, quality and safety. The results are public and available online (California IDER and DRP Working Groups 2017) as well as on PG&E's website. In Newlander (2016), the generation hosting capacity for each medium voltage line is plotted on a map, allowing one to locate preferred areas for DER connection. One can see the generation hosting capacity for each medium voltage line thus making it possible to locate preferential areas for DER connection.

Two methods have been proposed and compared by the IOUs:

– *the iterative method*: it consists of performing successive load flow calculations by increasing/decreasing the capacity of the DERs;

– *the simplified method*: it first evaluates the initial conditions of the system using a load flow calculation to determine a set of dimensioning variables (such as the load level of the system components, the voltage profile and the faults). Next, the maximum DER capacity is evaluated independently for each dimensional quantity via classical electrotechnical equations, examples of which are given in Newlander (2016).

The time series employed are constructed from energy data measured by smart meters. In case study A, they include 288 points over 24 h, corresponding to a time step of 5 min for each of the 12 days, representative of each month of the year.

These maps allow for the visual location of sites of interest for DER deployment but also areas where system investments could happen in the future. These could also be integrated in the DER connection process.

3.3.2.1.2. EPRI's distribution resource integration and value estimation tool

EPRI (European Commission 2019) provides a tool to study the hosting capacity of a system through statistical simulations. In this kind of method, DERs are not integrated into the planning process but rather considered a posteriori in order to deduce optimal DER locations and to lessen the oversizing of necessary reinforcements.

3.3.2.1.3. WattPlan: DER Planning Software

WattPlan is a distribution system planning software with DERs developed by Clean Power Research and the Sacramento Municipal Utility District (SMUD) to integrate DERs into planning. It allows for the simulation of DER development scenarios and the investigation of the least expensive power system development solutions needed to incorporate them (Novotny 2018).

3.3.2.2. *In France: the S3RENR*

In France, following the Grenelle II law of 2010, the development of renewable energies was structured into two stages. First, the local authorities of each region defined renewable energy targets for 2030 in the Regional Climate, Air and Energy Plans (SRCAE). Then, the TSO RTE, in agreement with the DSOs, was in charge of distributing these objectives to its different substations by evaluating the cost of the new structures required (a new transformer for example) and reinforcements. These forecast costs make it possible to define a share per producer (in k€/MW), which is then shared rather than supported by the first one to arrive, which unlocks a significant system investment. These studies are carried out through informal meetings between TSOs, DSOs, producers and local authorities in order to define realistic and transparent scenarios for the development of renewable energies.

This information constitutes the S3RENR (*Schémas régionaux de raccordement au réseau des énergies renouvelables*: Regional power system connection schemes for renewable energies), which are made available on the Capareseau website in the form of a map and an Excel file to be downloaded (2,995 source stations were detailed in 2021). Figure 3.8 shows the variation in the amount of the co-payment for different electrical regions defined by Enedis and entered into Capareseau.

Let us take the example of the Égletons substation in Corrèze (Nouvelle-Aquitaine) with two transformers (15 MVA and 10 MVA), two MV voltage levels (15 kV and 20 kV) and one HV voltage level (90 kV). The capacity reserved under the S3RENR is 41.4 MW, of which 13.2 MW are ongoing projects. The works identified to increase the capacity of the substation are the reinforcement of HV works and the transformation of the 15 MVA transformer into 36 MVA.

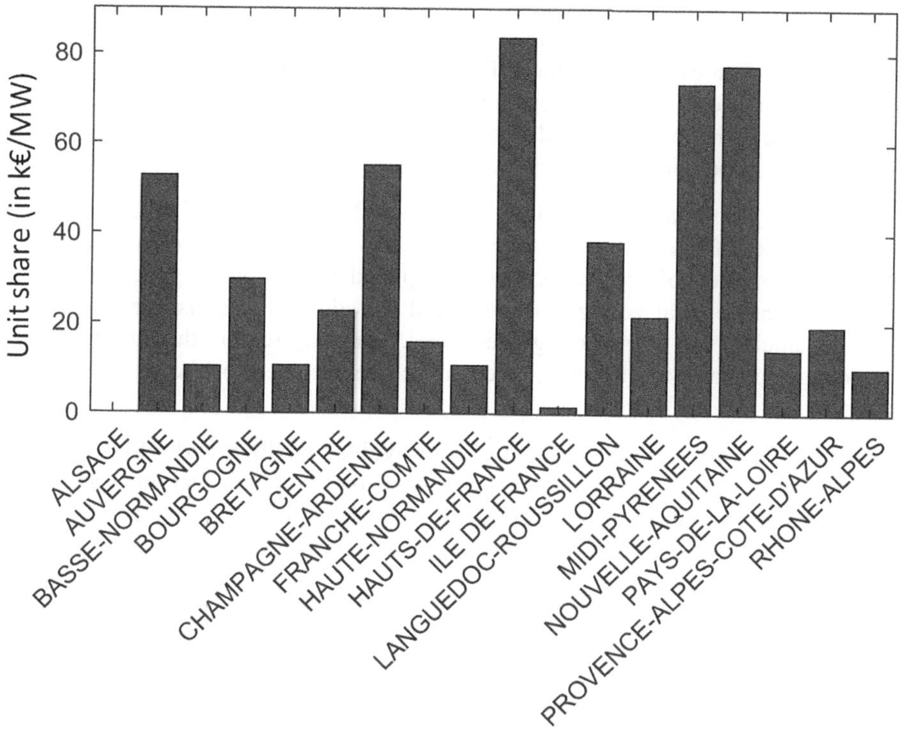

Figure 3.8. *Unit share by region*

3.3.3. *Locational Net Benefit Analysis*

In its 2015 DRP, PG&E proposes calculation methods to estimate the planning impact of DER insertion via the LNBA, as introduced in section 3.2.3.2. The optimal location of DERs can thus be determined by estimating their benefit as an alternative to the usual planning choices.

The LNBA is based on a method inspired by the Avoided Cost Model (ACM) developed by E3 (2020). It consists of calculating the additional or avoided cost of the future connection of DER to the power system. In particular, the developed method was applied to two areas, Escondido and Rancho San Diego (Case Study B), and showed that the use of DERs could effectively defer power system investments in some projects (SCE 2016). The main costs considered are as follows:

– *transmission and distribution system reinforcement*: additional or avoided CAPEX and OPEX for reinforcing substations and lines due to an increase in peak consumption. For example, if the DER is locally consumed generation, then the peak will decrease, which may negate reinforcement. On the other hand, if the DER is an electric vehicle charging during the peak, then the peak increases and the system needs to be reinforced;

– *voltage plan and quality*: additional or avoided CAPEX and OPEX to respect the admissible voltage variation ranges in the different operating regimes as well as the quality of the voltage wave. For example, the addition of generation to the system can cause local voltage increases and the addition of consumption (such as electric vehicles) can cause voltage decreases. This may require the modification of the voltage regulators' set points, or even their replacement or the addition of new ones;

– *reliability and resilience*: additional or avoided CAPEX and OPEX to ensure the reliability of the power system in the event of minor (reliability) and major (large-scale events) incidents. The DER mix can exacerbate the impact of a fault in the system, which may require new lines or reinforcements;

– *the frequency stability of the power system:*

- the additional or avoided costs of meeting the capacity required to maintain frequency stability. For example, in California, DSOs are required to have a generation margin equal to 15% of the peak demand (CPUC 2019). Therefore, the addition of electric vehicles, without taking into account the flexibility they can provide, causes an increase in peak demand and requires the purchase of additional capacity,

- the additional or avoided costs of being able to follow the larger 3-h consumption ramp. For example, if the ramp of the new net load curve (consumption minus generation) is greater, then an additional "fast" generation capacity will need to be purchased, that is, one capable of following the ramp,

- other costs (electricity, greenhouse gases, losses, ancillary services, renewable portfolio obligations, etc.): in California, selling prices are dynamic and can, for example, become negative if there is too much production compared with the capacity of the power system. DERs can therefore have a positive or negative impact on the price of electricity. A DER can also reduce losses by allowing local consumption, for example, or can on the contrary increase them, in the case of a large production at a time when consumption is low. These components are more or less similar to those of other private managers in California, and other states like Minnesota (ICF 2016).

3.3.4. *Distribution Investment Deferral Framework*

In California, the Distribution Investment Deferral Framework (DIDF) sets forth rules validated by the CPUC for the three IOUs to use services provided by DERs for the purpose of deferring or canceling power system investments as a result of thermal, voltage, or reliability constraints (SCE, PG&E and SDG&E 2016).

After targeting areas of interest for DER deployment, IOUs issue requests for offers (RFOs) for DERs to propose a project. For example, in 2021, PG&E selected seven locations where DERs could offset investments by being able to displace (consume or generate locally) up to 25.4 MW of power.

The various steps prior to the deployment of DER solutions are as follows for 2021 (PG&E 2021a):

– RFO publication (by November 01, 2021);

– DER proposals (by February 19, 2021);

– first selection of DERs (by May 03, 2021);

– response from selected DERs (by November 03, 2021);

– negotiations between DERs and IOU (by May 31, 2021);

– CPUC validation of selected DERs and proposed contracts (by June 14, 2021).

"Upstream meter" solutions					
Type of solutions	**Integration solution**	**Definition**	**Advantages**	**Disadvantages**	**Key actions**
Structural	Reinforcement of the lines or reconductoring	Replace existing lines using lines with larger cross-sections, change of voltage plan (long term)	→ Usual solutions → Make it possible to lift both current and voltage constraints → Reduce losses	→ Potentially very high investment cost → "Fit and forget"	Reduction of R and X lines
	Reinforcement of the substations	Replace transformers in HV/MV or MV/LV substations with higher capacity transformers or simply add an additional transformer			Reduction of R and X of the transformers and increase of the admissible power
	Change of topology	→ Adding new feeders → Creation of a primary or secondary substation → Addition of opening points (remote-controlled or not) → Addition of looping → Change of architecture	→ Allow the entire system to be redesigned and all constraints to be removed	→ Often high complexity → Cost is difficult to forecast → Depends on the parameters of the DER development scenarios	Change of L, R, X, V, P and Q
	Rebalancing of the phases	Rebalance the power on the systems by changing the connection phases of consumers/producers/DERs	→ Reduction of overloads → Reduction of losses	→ Potentially very high investment cost if the system is already unbalanced → Rebalancing at a given moment (usually a consumption or production peak) which is not optimal for all applications	Mastery of P

"Upstream meter" solutions					
Type of solution	**Integration solutions**	**Definition**	**Advantages**	**Disadvantages**	**Key actions**
Equipment	Reactive power management	→ Adding capacitors, FACTS or hybrid systems (e.g. capacitors combined with DER drive) in parallel to provide reactive power to solve low voltage constraints	→ Simple to set up	→ Only solves low voltage constraints → "Fit and forget" → Voltage jump (case of capacitor banks) → High cost of FACTS	Reduction of Q
Decentralized operations (without development of communication systems)	On-load tap changer (OLTC) with a fixed voltage setpoint	Modify the plugs of the transformer according to a local voltage measurement	→ Solution already industrialized in some countries and for certain voltage levels	→ High investment cost → Can only solve voltage problems	Control of V
	Dynamic phase rebalancing	Device making it possible to automatically choose the connection phase in real-time of producers or consumers according to constraints on the system	→ Industrializable solution	→ Economic gains for the moment not quantified	Control of P and V
Centralized operation with development of communication systems	Centralized voltage control	Controlling the active and reactive power and reactive powers of the producers in a centralized way	→ Optimized production	→ Higher cost and more complex architecture (sensors, communication, controller) → Requires good knowledge of the system or many sensors	Control of P and Q
	OLTC with one possible voltage band	Modify the transformer taps according to a template according to measurements of voltages on the system	→ Industrializable solution	→ High investment cost → Increased complexity (sensors, communication) → Can only solve voltage problems	Control of V

"Upstream meter" solutions					
Type of solution	**Integration solutions**	**Definition**	**Advantages**	**Disadvantages**	**Key actions**
	Reconfiguration	The state of the operating system elements is modified to find an optimal configuration of the system according to predefined targets	→ Makes it possible to solve both voltage and current constraints	→ Changing the configuration without interrupting the power supply requires temporary looping of the system, which is not always possible	Change of L, R, X, V, P and Q
Initial settings	Optimization of the off-load tap (in the case of an off-load tap changer)	Change the setting of the substation. For example, in the case of a light load, highly productive system reducing the tap of the off-load tap changer could solve all the voltage constraints	→ Very low cost	→ Interesting only if the adjustment is to be done once (or very rarely)	Mastery of V
Model improvement of system elements and input data	Thermal consideration of the transformers	Depending on the ambient temperature, it is possible to overload the transformer for a certain period of time without degrading its aging. This is called Dynamic Transformer Rating	→ Initial cost considered zero → Improvement of the models	→ Not suitable for all case studies → Depends on temperature conditions and DER development	Temperature control
Review of planning methods	Grouped application for generation connections	Optimization of production placement	→ Makes it possible to reduce voltage variations → At the same power, reduced losses	→ Feasibility to be studied → Additional complexity of connection studies → "Fit and forget"	Mastery of P and Q

"Upstream meter" solutions					
Type of solution	**Integration solutions**	**Definition**	**Advantages**	**Disadvantages**	**Key actions**
Decentralized operations (without development of communication systems)	Local control of the voltage	Controlling the active and reactive powers of the producers with local measurements	→ Solution already industrialized in some countries and on some voltage levels → Uses the available reactive power margin when the producer is not supplying its maximum active power	→ Local voltage measurements to be made → Can only solve voltage problems, not current problems → Risk of interaction between inverters (pumping phenomenon)	Control of P and Q
Centralized operation with development of communication systems	Dynamic curtailment of the production	Centralized control of active and reactive power of producers	→ Cost depending on the energy not injected	→ Extra cost (sensors, communication, controller) → Setting up a business model for the complex producer	Mastery of P
	Modulation of consumption via a third party	A third party, such as an aggregator or a supplier, can control certain uses at the consumer's premises before reducing the power consumed when the system operator needs it	→ Cost depending on the energy curtailed	→ Extra cost (sensors, communication, controller) → Control of the complex realization on the production of low powers not instrumented → The consumer can cancel the deletion request at any time	Mastery of P

"Upstream meter" solutions					
Type of solution	**Integration solutions**	**Definition**	**Advantages**	**Disadvantages**	**Key actions**
Initial setting	Limitation of production when installed	Limit the maximum production of producers	→ Feasible and simple solution → Cost dependent on the energy not injected	→ "Fit and forget" → Ensure that annual non-injected energy remains low	Mastery of P
	Fixed tan φ for producer control	Set tan $\varphi = \frac{Q}{P}$ of producers to a given value.	→ Feasible and simple solution	→ "Fit and forget" → Cost depends on the type of installation (e.g. for an inverter, the cost of the energy not produced)	Mastery of P and Q
	Load curtailment with a fixed tariff system such as on-peak/off-peak hours	Implement a system of constant pricing during certain time periods to encourage consumers to consume during preferred periods	→ Feasible and simple solution	→ Implementation of time slots and tariffs depend on the systems and the level of DER development	Mastery of P
Improvement of system element and input data models	Consumption/production modeling	Recalculate the value of the maximum power of consumers/producers over time according to their production observed in the past	→ Model improvement → Release of unused power margin	→ Feasibility and cost to be studied	Mastery of P

Table 3.3. *List of solutions used or with the potential to resolve current and voltage constraints due to the insertion of DERs on the power system (Bernier 2019); advantages of NWA-type integration solutions*

3.4. Phase 2: planning with DERs

Beyond a certain DER penetration threshold, their simple consideration in investment decisions is no longer sufficient. The role of DERs must evolve to provide services to the power system with economic compensation: we speak of non-wire alternatives, defined by EPRI as a DSO-oriented solution to solve a technical constraint in the distribution system making it possible to defer or cancel a traditional reinforcement (EPRI 2019). Not only do alternative operating modes need to be evolved or even created, but a regulatory framework needs to be defined to make the whole thing work optimally. In concrete terms, instead of replacing a line because of a voltage or current constraint or simply because a larger section would reduce its OPEX and therefore its overall cost over the long term (see Chapter 2), alternative operational solutions are used, such as those cited in Table 3.3, inspired by those proposed in Bernier (2019) and Pilo (2018).

3.4.1. *List of possible insertion solutions*

3.4.1.1. *System investment planning*

The cost of the new solutions mentioned above must be integrated in the overall economic balance sheet in order to determine the economic optimum for all the actors in the system, in particular for the DSO. Until now, its investment decisions were mainly based on the CAPEX, since OPEX currently represents only about 20% of the overall discounted cost over the lifetime of the system (Gouin 2015) (valuation of joule losses, costs related to the reliability of the system as explained in Chapter 2). However, if we integrate NWAs into the investment decision process, then the CAPEX/OPEX imbalance could be reversed. The main issue is therefore to correctly estimate the cost of the flexibility levers in order to be able to perform a cost–benefit analysis based on the TOTEX. This cost–benefit analysis then makes it possible to compare:

– business as usual solutions such as new lines and transformers;

– non-wire alternatives where the investment costs would be related to the equipment necessary for the use of operational solutions, such as communication equipment, software or storage (if the system operator is authorized).

One example is the Ringo demonstration currently under development, which tests the value of using storage instead of reinforcing the system. Since system operators are not allowed to sell energy, the principle of Ringo is to deploy several storage systems such that the amount of electricity stored at a given time and place is completely de-stored at the same time at another place in the system.

There are already studies based on surveys or calculations at a macroscopic scale that predict system investment reductions through NWAs.

– US and Europe: from October 2016 to January 2017, Accenture (2017) surveyed more than 100 companies across 23 countries making power system investments. Among the survey's findings was that by 2030, the system investments needed to address thermal and voltage profile constraints associated with the development of distributed generation represent:

- $20 billion for the US to accommodate 50 GW of PV and wind between 2017 and 2030;

- $58 billion for Europe to accommodate 139 GW of PV and wind between 2017 and 2030. The European Commission (2016) is even more pessimistic, announcing an overinvestment related to distributed generation that averages $11 billion per year until 2030.

The survey highlights the reduction of nearly 30% of these investments through the use of certain NWAs: the optimal placement of generation via incentives (around 2.5%), generation curtailment contracts (around 10.5–12.5%), storage or demand-side flexibility (14–15%).

– France: in its report on the economic valuation of Smart Grids, Enedis (2017) conducted a cost–benefit analysis of six Smart Grid solutions on the MV perimeter: predictive management, dynamic self-healing after an incident, centralized voltage regulation, local dynamic control of MV producers' reactive power, MV generation capping, and flexibilities to resolve withdrawal constraints. The main conclusion is that the use of flexibility is beneficial for all the actors involved in the power system and could reach 24 k€/MW/year depending on the case studies.

– Australia: Energy Network Australia (ENA) reports that optimal DER location and control could save consumers $16 billion by 2050 (ENA 2017). This new consumer service purchase market could be worth $1.1 billion per year within 10 years.

– UK: in the report by Sanders et al. (2016) from Imperial College of London, a cumulative reduction of £4–13 billion by 2050 through distribution system-side investments could be achieved through the deployment of demand-side response and storage.

3.4.1.2. *Improved system operational planning*

In its report, Enedis (2019) highlights the benefit of NWAs in improving the reliability of power systems, and in particular, the time it takes to resupply customers and the energy not distributed in the event of:

– scheduled work on the system;

– exceptional climatic events that can be anticipated, such as cold spells (D-4 to D-1);

– resupply of customers during the period when a faulty zone is isolated but not yet repaired.

In the first case only, a capacity reservation with penalties if the flexibility is not available would be considered. According to Enedis (2019), the estimated gains could vary between 0 and 20 k€/MWh.

3.4.2. *Planning without flexibility markets*

3.4.2.1. *Plans to use NWA to defer system investments in California*

In California, in the 2017 DRP, the CPUC requires private system operators to test the services that DERs can provide:

– El Nido substation (Demonstration Project C): there was no need to reinforce the substation thanks to the local generation provided by the DERs;

– Huron substation (Demonstration Project D): curtailment offers for the problem of high generation penetration.

Another example is Assembly Bill 2514 of September 29, 2010, which requires the Californian regulator, the CPUC, to define a project for integrating storage into the power system.

Therefore, in rulemaking 10-12-007 (Commission of the State of California 2013), the CPUC sets a goal of 1.325 GW of storage installed on the systems of the three major Californian distributors (PG&E, SCE and SDG&E) by 2020 at all levels, transmission, distribution, and consumer, as shown in Table 3.4.

	2014	**2016**	**2018**	**2020**
SDG&E (165 MW)	*20*	*30*	*45*	*70*
Transmission (MW)	10	15	22	33
Distribution (MW)	7	10	15	23
Consumption (MW)	3	5	8	14
PG&E and SCE (580 MW)	*90*	*120*	*160*	*210*
Transmission (MW)	50	65	85	110
Distribution (MW)	30	40	50	65
Consumption (MW)	10	15	25	35

Table 3.4. *CPUC's storage development of storage requested by the CPUC*

Flexibility is provided to move 80% of the storage capacity to be installed at the transmission level to the distribution system and vice versa. Table 3.5 provides an example of PG&E's end-of-2018 balance sheet with 80% of the planned storage to be installed at the distribution to transmission level.

	Transmission	Distribution	Consumer
Target end-2018 (370 MW)	200	120	50
Revised target	296	24	50
Realized	692.5	36.5	50.2
Remaining until 2020	Exceeded	0.5	34.8

Table 3.5. *PG&E update as of November 16, 2018 (PG&E, Advice 5427-E, 2018)*

3.4.2.2. *Production curtailment to defer system investments*

3.4.2.2.1. Germany: the 3% rule

In 2017, Germany implemented the 3% rule, which, in order to reduce system investments, allows the system operator to cut up to 3% of the annual energy produced by a producer by compensating for the loss of revenue. The BMWi, Ministry of Economic Affairs and Energy determined the "3%" value on the basis of scenarios for the development of renewable energy production and concluded that the reduction of annual energy production by 3% would reduce power system reinforcement by 40% and investment by at least 15% (Furusawa et al. 2019). Another study conducted by E-Bridge Consulting, OFFIS and RWTH, on behalf of the Ministry of Economy and Innovation, found that only 3% of annual DER generation is controlled for the purpose of rendering system services. This study predicts a reduction in investment of nearly 40% if all production were controlled (enerquire 2017).

3.4.2.2.2. UK: flexible non-firm connection

In the UK, flexible non-firm connections are several alternative connections that offer reinforcement and are detailed below.

Timed capacity connections

This is a limitation of production that is contracted according to a pre-defined schedule (SP Energy Networks 2017; Furusawa et al. 2019). At certain times of the day and based on demand, a limitation set point is allocated to the producer. SP Energy Networks recommend using this alternative for producers below 0.5 MVA (below 1 MVA per Western Power Distribution (Western Power

Distribution 2015)) and connected to voltage levels less than or equal to 11 kV (SP Energy Networks 2017). The user is expected to comply with the contract; the DSO is only required to place conventional power sensors. This system is inexpensive because it does not require a communication system but only manages local constraints.

Export limiting devices

An automatic piece of equipment measuring the apparent power at the generation connection node. If a current and/or voltage constraint occurs, a reduction in generated power and/or a modulation of consumption takes place. SP Energy Networks (2017) explains that this solution can be used for any type of generation at any voltage level and is attractive to customers who want to increase their generation capacity so as to reduce their electricity consumption bill. They recommend adding directional relays to shut down generation to in case there is a communication fault, making sure they are not under stress.

Local management scheme

This solution makes it possible to manage local constraints at a producer level. A voltage and/or power measurement is performed at the connection node. In the event of a constraint, the producer's power is reduced. SP Energy Networks (2017) recommends this solution for simple networks with little generation and ideally operating at 33 kV.

Intertrip

A type of switch making it possible to cut off generation in case of constraints on the network. This solution is recommended for non-interconnected HV systems with few producers and only one stressed element such as a line or a transformer, since the element to be monitored must be instrumented, and which can be quickly complicated in an interconnected system.

Active Network Management

A centralized or decentralized real-time control of a given area consisting of several producers and several elements that may be in contraction. Based on measurement points, a maximum production capacity is allocated to each customer. Producers above these values are cut off. Two rules can be used:

– *the last in first out (LIFO) method*: the last producers to arrive are cut off first; according to ENA (2015) on good practices in the ANM drafted by Energy Network Association (ENA), this method is to be preferred in the case of first projects with curtailment contracts due to its simplicity of implementation and because the quantity of energy to be curtailed by the producer is precise;

– *the pro rata method*: all producers are cut in proportion to their installed capacity. ENA (2015) recommends this method in the case where the DSO wants to maximize the generation connection to its power system by implementing quotas.

An example of an ANM has been installed since 2016 on the Isle of Wight. According to Scottish and Southern Electricity Networks (2016), it has connected an additional 45 MVA of generation without reinforcement, representing a saving of about €2.7 million. However, other sources explain that in practice, the cost of developing ANMs is high and they have communication issues (Pratt 2017). Wight Community Energy is one of the Isle of Wight's producer associations representing a total capacity of 3.95 MW. The load curtailment required to avoid power system constraints led to cut all of its generation during the month of June 2017, which represented an 8.1% decrease in their annual energy output for 2017, corresponding to a financial loss of €55.7k. The expected cuts over the following months do not allow the producer to make a profit.

ENA (2015) specifically emphasizes the following recommendations:

– contracts and connection procedures:

- methods, tools and the current and future power system status must be communicated to producers (even if some anonymization may be necessary);

– DSOs must notify producers of their eligibility for deletion contracts;

– DSOs must notify producers in advance of upcoming load curtailment as well as of any changes to the load curtailment schedules provided for in the contract;

– even if the ANMs concern the DSOs, the TSOs must not reject the ANMs, particularly if long and costly work is planned on the transmission system;

– postponement of investment: DSOs must be able to determine the need for system curtailment in order to postpone reinforcements. Good communication between the planning and connection teams is important to determine which producers are eligible for load curtailment;

– in all cases, whether it is a temporary solution or a deferral of long-term investment, ENA (2015) recommends a systematic cost–benefit analysis to conclude on the interest of load curtailment according to each situation.

Consortium formation (Scottish and Southern Electricity Networks 2021)

The customers form a consortium and agree on the rules for sharing capacity and therefore costs.

3.4.2.2.3. Experimentation phase of the economic model

France: ORI (*Offre de Raccordement Intelligente*: "Smart Connection Offer")

The French DSO Enedis has been testing two smart connection offers since 2017 as part of the *Smart Grid Vendée* demonstration. Instead of connecting a producer via a dedicated feeder, which represents a very high CAPEX in turn compromising the connection project, the DSO proposes to connect it in full system at the nearest point with two possible counterparts:

– *guaranteed maximum power*: the producer's maximum power is limited to a certain value which may be exceeded if network conditions allow;

– *guaranteed maximum cut-off*: the producer agrees to reduce the power according to the needs of the network so that the annual energy cut-off does not exceed a certain threshold contracted with the DSO.

Table 3.6 shows the characteristics of the two producers testing Enedis' connection offers.

	Guaranteed maximum power	**Guaranteed maximum energy cut**
Name	Wind farm of five turbines, Chauché	Solar power plant, Talmont Saint-Hilaire
Maximum power	Five wind turbines P_{max} = 11.75 MW	5 MWp
Planned connection	Dedicated 10 km start	Dedicated start of 3.3 km
Cost of connection	€1.1 M	€494,000
Terms and conditions of the IRO	Limitation of the power to 98% of P_{max}, that is, 11.5 MW	Maximum total energy erased of 250 MWh/year for 3 years
New connection	Connection to the existing network by a 300 m cable	Connection to the existing network by a 40 m cable
Cost of the new connection	€100,000	€179,000
Cost saved by the producer	€1 million (91% savings)	€315,000 (64% savings)
Financial losses related to IROs	Production loss of less than €500,000 over 15 years	Loss of revenue of €21,500

Table 3.6. *ORI tested through the framework of the* Smart Grid Vendée *demonstration (Thual 2018)*

The DSO estimates that these smart offers could reduce the cost of connecting producers by 90 k€/MW and reduce delays by 7 to 10 months. In the offers tested, the DSO does not directly control the power injected at the output of the producer's installation but communicates with the producer to ask them to reduce their production. In the event that the system is under stress and the producer does not respond, the DSO can still disconnect the producer via the de-coupling protection that connects it to the distribution system (Furusawa et al. 2019).

Belgium

Belgium is currently experimenting with a connection offer similar to the maximum guaranteed power in France. The difference is that the DSO directly controls the output of the installation and can thus modulate the maximum power produced by the installation. For the time being, load curtailment is carried out in proportion to the power of the producers.

Table 3.7 shows the advantages and disadvantages of the connection options described above. Germany is the only country to have a model based on a producer remuneration system. While it is complicated to quantify a producer's unproduced energy in order to compensate it fairly, other models do not necessarily encourage producers to opt for these alternative offers. Given the strong uncertainties of the future, both in terms of the evolution of DERs and the evolution of regulations, the benefits estimated by producers from a connection offer may not be realized in the future.

Advantages	Disadvantages
Faster producer connection procedure May be simple to implement Better control of production Reduced CAPEX Few regulatory changes DSO and producer win	Definition of fair rules for the producer (pro rata versus LIFO) Quantification of the optimal curtailment rate (3% rule) Flexibility used locally (distribution) without interaction with the transmission system Long time before industrialization: first ideas in 2007 and implementation at the moment in the UK and in Europe only Germany

Table 3.7. *Advantages and disadvantages of the existing and testing phase connection offers*

3.4.3. *Planning with flexibility markets*

To make the most of DER flexibilities, it may be interesting to set up a market system between system operators and DERs similar to the capacity market via flexibility platforms called Distributed Energy Resources Management Systems. Smarter Grid Solutions (2021) is an example of a flexibility platform provider that manages 308 DERs representing 1,040 MW of generation across 404 business use cases in 23 countries, mainly in Europe, North America and Australia.

3.4.3.1. *Principle*

BDEW, the German Association of Water and Energy Industries, and USEF (Universal Smart Energy Framework), founded in 2014, bringing together various players in the Dutch electricity chain, are proposing a market model at the distribution system level that allows for the use of flexibility services to defer or avoid investments into the system.

Power systems traditionally have two main regimes (see Chapter 2):

– *the normal regime (or N regime) in green*: the power system is functioning correctly, that is, the production/consumption balance is assured, the thermal limits of the conductors are respected as well as the regulatory voltage ranges. The system is therefore sized so that all consumption and generation can flow through the system even under the maximum conditions that can potentially be reached;

– *the fault regime (N-1 or more) in red*: when the normal operating conditions are not respected due to an unexpected overload or a fault, for example.

The USEF (respectively BDEW) proposes to incorporate two (respectively one) intermediate regimes between the normal and fault regimes (USEF 2015; BDEW 2015). In the case of overload constraints, the DSO can activate flexibilities on demand or generation through a local market. In particular, this intermediate regime can reduce investments in the power system and improve reliability, thereby avoiding the fault conditions due to the disconnection of the overloaded lines. This intermediate level, represented by the yellow in Figure 3.9, is a distribution capacity market similar to the one done at the national level. A second level, represented by the orange in Figure 3.9, is introduced in the event that the capacity market is not sufficient to resolve congestion. A consumption/production load curtailment system based on priority customers is carried out directly by the DSO. This is the same principle as the load shedding carried out by RTE, except that the advantage here is that the DSO can control the placement of load curtailment. The flexibility described here is "one-way", that is, a reduction in generation or consumption to cancel a local constraint. However, it must also be considered in the opposite direction, that is, increasing consumption or generation via storage, for example, to resolve low voltage constraints (e.g. Ringo on the transmission system).

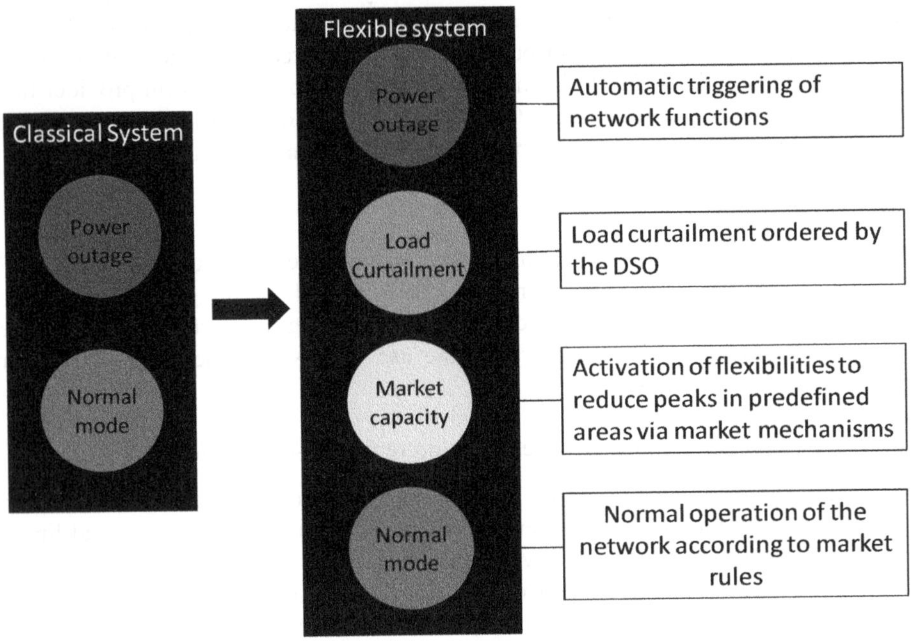

Figure 3.9. *New regimes introduced by the integration of flexibility into planning (diagram inspired by (USEF 2015))*

Several experimental projects have been developed in recent years to test flexibility market platform concepts:

– SINTEG project (Germany): 2016–2020, Smart Energy Showcases – Digital Agenda for The Energy Transition, is funded by the BMWi in order to carry out flexibility market experiments in five different areas of Germany at different scales to address different challenges;

– SMILE project (France): *Smart Ideas to Link Energies*, development of a flexibility platform accessible to all communities;

– nodes (Norway): this is a proposal for a European decentralized flexibilities market that would also be connected to other markets and could thus value flexibility even at times when it is not useful for the DSO (EnergiNorge 2021).

3.4.3.2. *Existing flexibility markets*

Two types of flexibility platforms already exist or are being tested. As Delta-EE (2019) suggests, these can be classified into two categories:

– long-term platform between the DSO and the flexibility suppliers;

– short-term platform between the DSO and the suppliers, but also between the DSO and the TSO and the energy markets (wholesale and capacity market).

3.4.3.2.1. Long-term platforms (UK)

Two examples of long-term flexibility platforms currently exist – Flexible Power offered directly by the DSO Western Power Distribution (WPD) since 2018 and Piclo Flex offered by an independent player. These interfaces connect buyers, DSOs, and sellers who can be any type of flexibilities provider. The principle is that DSOs post their needs on the platform and then DERs have a certain period to propose their offers, that is, the power and its selling price. These are then selected, or not, by algorithms incorporated into the platform.

Flexible Power (WPD)

Current and future constrained management zones are mapped on a web interface allowing all flexibilities providers to offer their services, regardless of their size and type. Flexibility requirements are categorized into three types of services (Godfrey and Sawdon 2020):

– *secure:* to solve constraints related to peak consumption or production;

– *dynamic*: to supply customers in the event of specific work on the system, such as maintenance work;

– *restore*: to supply customers in the event of system faults (e.g. short-circuit).

Depending on the type of service, the remuneration varies from £300 to £600/MWh for contracts ranging from 1 to 5 years. It also depends on the number of suppliers offering their service: if supply barely matches demand, flexibilities suppliers are remunerated with a fixed cost. If the number of suppliers is large and the overall supply exceeds the demand, then a market system is set up.

In 2020, WPD identified 334 MW of flexibility needs in 42 locations across their system (Western Power Distribution 2020).

Piclo Flex

Figure 3.10. *The Piclo Flex platform: an example of the need for flexibilities (source: UKPN)*

An example is shown in Figure 3.10 (Flex 2021). The UKPN system operator has a need for increased generation and/or reduced consumption on its 11 kV system in the Brandon area over four periods: from December 1, 2020 to February 28, 2021, 2021/2022, 2022/2023, and 2023/2024. It may or may not indicate the purpose of this requirement which is to defer investments. The tender was open from October 31, 2019 to November 3, 2019.

For each period, the DSO enters its requirements: power volumes to be produced/reduced, time slots and size constraints and the minimum operating time of DERs, as shown in Figure 3.11.

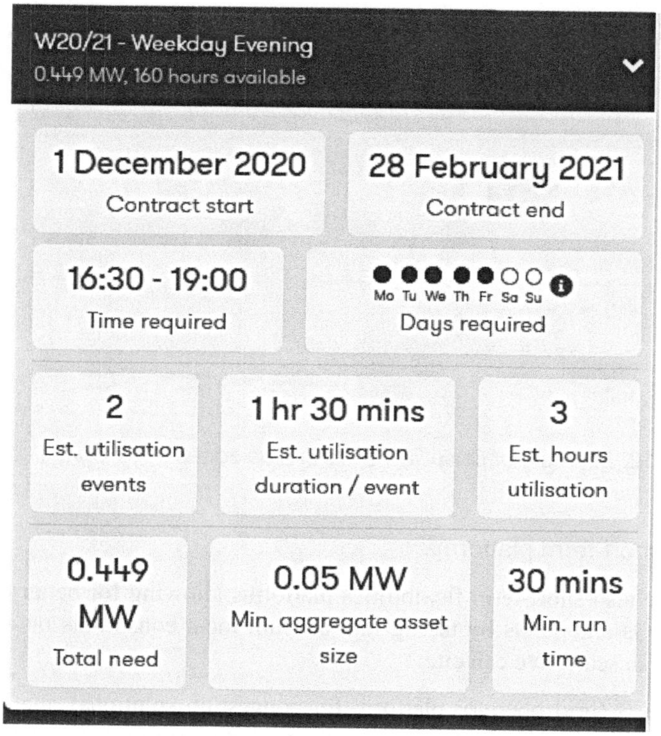

Figure 3.11. *The Piclo Flex platform: detail on the flexibility needs over the period 2020/2021*

Experimentation in progress in France

In November 2019, Enedis launched flexibilities tenders to defer its system investments, in six case studies outlined in Enedis (2021). Figure 3.12 shows the example of Flex Cantal, where flexibilities are needed in the five blue zones, corresponding to five MV feeders. The LV connected photovoltaic generation causes overvoltage constraints between 11:00 am and 4:00 pm. The DSO requires at least 0.5 MVAr for each MV feeder to cancel the necessary reinforcement works.

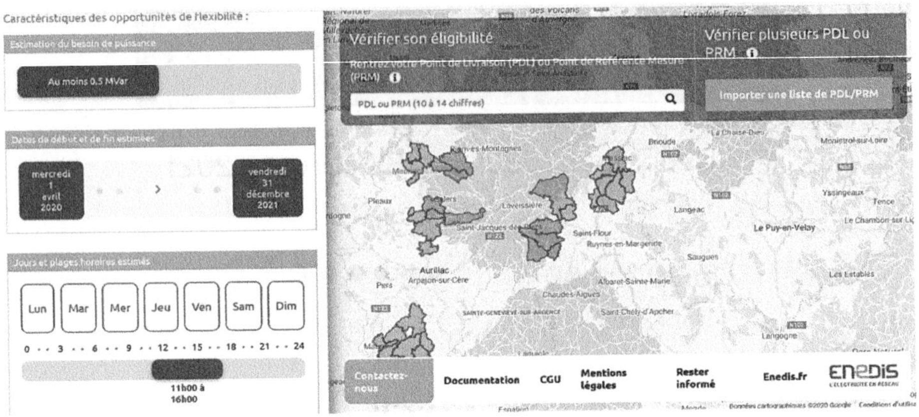

Figure 3.12. *Example of flexibility needs in blue zones (Flex Cantal case study)*

3.4.3.2.2. Short-term platforms

There are also short-term flexibilities platforms allowing for better coordination between DSOs and TSOs by taking into account local constraints on a national or even European scale. We can cite:

– GOPACS (grid operator platform for congestion solutions), the Netherlands, January 2019; it is connected to a wholesale market platform, EPTA and operates in four steps:

- Step 1: the distribution system operators estimate the locations and dates of future congestion on their system and enter them into GOPACS,

- Step 2: GOPACS generates a signal on the markets to which the flexibilities suppliers can respond (reduction/increase in production or consumption). The imbalance generated by this modification is compensated for by other offers outside the congested area,

- Step 3: these offers are transmitted to GOPACS, which checks:

– respect for local constraints;

– the non-degradation of the global system;

- Step 4: the system operators then pay the difference between the two previous costs to resolve the congestion;

– Enera demonstration, Germany, February 2019: local flexibility markets interact with the *epexspot* wholesale market by issuing tariff signals incorporating local constraints (epexspot 2019);

– Centrica LEM, UK, 2018: demonstration project in Cornwall in collaboration with WPD's Flexible Power platform to identify areas in constraints.

3.4.3.2.3. Short-term and long-term platforms

Two platforms allow both DSOs and flexibility operators to interface for better planning and operation of the distribution system, and also to take into account local constraints at the national level: NODES, Norway, 2018 (NODES 2021) and *Decentralised Energy Exchange*, GreenSync, Australia, 2018 (deX 2018).

In France, the ATLAS flexibility platform is currently being tested as part of the SMILE project (SMILE 2008).

3.4.3.3. *Emerging market model*

If DER penetration becomes too high, it becomes necessary to define a market structure that not only allows DER-to-power system exchanges but also DER-to-DER exchanges that do not necessarily belong to the same primary substation. The transmission system is then impacted more, which requires stronger coordination between DSOs and TSOs. Howard (2017) suggests the need for two types of structures: one structure allowing for bilateral energy transactions and another spot market-type structure for residual energy. In any case, the evolving role of the DSO will require further coordination to be defined between the DSO and the TSO. Accenture (2016) analyzes different market models incorporating DERs in different ways to propose possible future developments in market models. At the end of the report, two models in particular seem promising in the future: the Distribution Platform Optimizer (DPO) and Energy Supranet.

3.5. Conclusion

The integration of DERs into planning is complex due to the lack of regulation at this time. Even if the technologies are present and mature, power system operators are reluctant to rely on flexibility to change traditional planning decisions. Moreover, the current regulation on the remuneration of system operators encourages them to invest rather than choose solutions that reduce their costs (Girouard 2015). They therefore prefer to wait for the regulation to change before opting for solutions that increase the level of risk. However, regulation needs concrete case studies via full-scale deployments to obtain feedback and evolve. It is therefore a complex situation in which few actors dare to take the first step.

3.6. References

Accenture (2016). Electricity network transformation roadmap [Online]. Available at: https://www.energynetworks.com.au/resources/reports/insights-from-global-jurisdictions-new-market-actors-evolving-business-models/.

Accenture (2017). Power surge ahead – How distribution utilities can get smart with distributed generation [Online]. Available at: https://www.accenture.com/_acnmedia/pdf-64/accenture-power-surge-ahead-how-distribution-utilities-can-get-smart-with-distributed-generation.pdf.

AEE (2018). Distribution system planning – Proactively planning for more distributed assets at the grid edge. Coley Girouard and Ryan Katofsky.

AEMO (2016). Response of existing PV inverters to frequency disturbances, Australian Energy Market Operator (AEMO), 2016, [Online]. Available at: https://www.yumpu.com/en/document/read/56347416/response-of-existing-pv-inverters-to-frequency-disturbances/9.

BDEW (2015). Smart grid traffic light concept: Design of the amber phase. Discussion paper, Berlin.

Bernier, M. (2019). Solutions smart grid innovantes pour l'intégration massive de la production photovoltaïque au réseau public de distribution en zone rurale. PhD Thesis, Université Grenoble Alpes, Grenoble.

California IDER and DRP Working Groups (2017). California distribution resources Plan (R.14-08-013) integration capacity analysis. Report, ICA Working Group.

CAPARESEAU (2021). Capacités d'accueil pour le raccordement aux réseaux de transport et de distribution des installations de production d'électricité [Online]. Available at: https://www.capareseau.fr/ [Accessed 2021].

CIGRE (2014a). Capacity of distribution feeders for hosting DER. Working document, CIGRE Working Group C6.24.

CIGRE (2014b). Planning and optimization methods for active distribution systems. Working document, CIGRE Working group C6.19.

Commission of the State of California (2013). Decision adopting energy storage procurement framework and design program [Accessed 2019].

CPUC (2019). 2018 Resource Adequacy Report. Report, California Public Utilities Comission Energy Division, San Francisco.

CRE (2021). Le réseau public de transport d'électricité [Online]. Disponible available at : https://www.cre.fr/Electricite/Reseaux-d-electricite/Investissements [Accessed 2021].

De Martini, P., Kristov, L., Schwartz, L. (2015). Distribution systems in a high distributed energy resources – Future planning, market design, operation and oversight. Report, Berkeley Lab, Berkeley.

Delta-EE (2019). *Delta-EE webinar highlights: Demand-side flexibility for DNOs – Everything you need to know*. Film.

deX (2018). *Creating the grid of tomorrow, today*. Film.

E3 (2020). ACM: Avoided Cost Model [Online]. Available at: https://www.ethree.com/tools/acm-avoided-cost-model/.

ENA (2015). Active network management good practice guide. *Baringa Partners*.

ENA (2017). Are we there yet? [Online]. Available at: https://www.energynetworks.com.au/news/energy-insider/are-we-there-yet/ [Accessed 2021].

Enedis (2017). Economic assessment of smart grids solutions. ADEEF (Association des distributeurs d'électricité en France).

Enedis (2019). Les flexibilités au service de la transition énergétique et de la performance du réseau de distribution. Working document, Enedis, Paris.

Enedis (2021). Les flexibilités locales au service de la transition énergétique [Online]. Available at: https://flexibilites-enedis.fr/ [Accessed 2021].

EnergiNorge (2021). Nodes European market place for decentral flexibility [Online]. Available at: http://nodesmarket.com [Accessed 2021].

enerquire (2017). Where do the DNOs in Europe see their future? Survey results discussed [Online]. Available at: https://www.enerquire.com/blog/future-of-the-DNO-data-hubs-demand-response-balancing-responsibility [Accessed 2021].

epexspot (2019). Local flexibility markets – A market-based solution to solve local grid. *Proceedings of the 2nd Annual Nordic Electricity Market Forum*. Oslo.

EPRI (2019). Panel session: Incorporating non-wires alternatives into the planning process. *Proceedings of the IEEE General Meeting 2019*. Atlanta.

European Commission (2015). Study on tariff design for distribution systems. Report.

European Commission (2016). Impact assessment support study on: "Policies for DNOs, Distribution Tariffs and Data Handling". Report.

European Commission (2019). Directive (UE) 2019/944 du parlement européen et du conseil du 5 juin 2019 concernant des règles communes pour le marché intérieur de l'électricité et modifiant la directive 2012/27/UE (refonte). *Journal officiel de l'Union européenne*.

FERC (2018). Distributed energy resources technical considerations for the bulk power system. Report, Docket No. AD18-10-000.

Flex, P. (2021). The independent marketplace for trading energy flexibility online [Online]. Available at: https://picloflex.com/ [Accessed 2021].

Furusawa, K., Brunekreeft, G., Hattori, T. (2019). Constrained connection for distributed generation by DNOs in European Countries. Working document, Jacobs Univeristy Bremen, Bremen.

Girouard, C. (2015). How do electric utilities make money? [Online]. Available at: https://blog.aee.net/how-do-electric-utilities-make-money [Accessed 2021].

Godfrey, B. and Sawdon, H. (2020). Flexible power webinar – Routes to participation WPD 2020 Procurement cycle 1 [Online]. Available at: https://www.westernpower.co.uk/news-and-events/latest-events/flexible-power-routes-to-participation.

Gouin, V. (2015). Évaluation de l'impact du Smart Grid sur les pratiques de planification en cas d'insertion de production décentralisée et de charges flexibles. PhD Thesis, Université Grenoble Alpes, Grenoble.

Howard, M. (2017). The integrated energy network: The times demand it. *EPRI Journal* [Online]. Available at: https://eprijournal.com/the-integrated-energy-network-the-times-demand-it/ [Accessed 2021].

ICF (2016). Integrated Distribution Planning, 2016, prepared by the ICF International [Online]. Available at: https://www.energy.gov/sites/prod/files/2016/09/f33/DOE%20MPUC%20Integrated%20Distribution%20Planning%208312016.pdf

INOGATE (2009). European best practice regarding to the connection to the grid, connection tariffs. Report, INOGATE.

IRENA (2017). Electricity storage and renewables: Costs and markets to 2030. Working document IRENA, Abu Dhabi.

Newlander, J. (2016). Demonstration Projects A & B Final Reports of San Diego Gas & Electric Company (U 902-E). Report, San Diego Gas & Electric, San Diego.

NODES (2021). A fully integrated marketplace for flexibility. Working document, NODES Market.

Nouicer, A. and Meeus, L. (2019). The EU Clean Energy Package. Report, European University Institute, Fiesole.

Novotny, G. (2018). Why WattPlan Grid: Planning for DER adoption [Online]. Available at: https://www.cleanpower.com/2018/wattplan-grid-planning-der-adoption/ [Accessed 2021].

OFGEM (2020). 9 August 2019 power outage report. Report, OFGEM, London.

PG&E (2018). Erik Jacobson, Advice Letter 5427-E, Notice of Cancellation of PG&E's Energy Storage Request for Offers, Targets and Revised Procurement Strategy Pursuant to Ordering, Paragraph 6 of Decision 18-10-036, 2019, https://www.pge.com/tariffs/assets/pdf/adviceletter/ELEC_5427-E.pdf. [Accessed 2021].

PG&E (2021a). 2021 Distribution Investment Deferral Framework Solicitation Protocol [Online]. Available at: https://www.pge.com/en_US/for-our-business-partners/energy-supply/electric-rfo/wholesale-electric-power-procurement/2019-didf-rfo.page?WT.mc_id=Vanity_rfo-didf&ctx=large-business [Accessed 2021].

PG&E (2021b). Demand response auction mechanism [Online]. Available at: https://www.pge.com/en_US/large-business/save-energy-and-money/energy-management-programs/demand-response-programs/2020-demand-response/2020-demand-response-auction-mechanism.page?WT.mc_id=Vanity_dram [Accessed 2021].

Pilo, F. (2018). ISGAN – WEBINAR planning of distribution systems in the era of smart grids [Online]. Available at: https://www.slideshare.net/sustenergy/planning-of-distribution-systems-in-the-era-of-smart-grids?from_action=save [Accessed 2021].

Pratt, D. (2017). Isle of Wight curtailment reduced following intense negotiations with SSE. *Solar Power Portal.*

RED (2020). El sistema eléctrico español 2019 [Online]. Available at: https://www.ree.es/es/datos/publicaciones/informe-anual-sistema/informe-del-sistema-electrico-espanol-2019 [Accessed 2021].

Sanders, D., Hart, A., Ravishankar, M., Brunert, J. (2016). An analysis of electricity system flexibility for Great Britain. Report, Imperial College, London.

SCE (2016). Southern California Edison Company's (U 338-E) Updated demonstration projects A and B final reports. Report.

Scottish and Southern Electricity Networks (2016). Environment report. Report, Perth.

Scottish and Southern Electricity Networks (2021). Flexible generation connection application guide. Working document, Perth.

SEPA (2010). Non-wires alternatives: Incorporating NWAs into your grid modernization progam [Online]. Available at: https://sepapower.org/resource/non-wires-alternatives-nwa-incorporating-nwas-into-your-grid-modernization-program/thank-you/ [Accessed 2021].

SCE, PG&E and SDG&E (2016). Distribution infrastructure deferral framework (DIDF) & distribution deferral advisory group (DDAG) [Online]. Available at: https://slidetodoc.com/panel-2-distribution-infrastructure-deferral-framework-didf-distribution/ [Accessed 2021].

Smarter Grid Solutions (2021) [Online]. Available at: https://www.smartergriDNOlutions.com/ [Accessed 2021].

SMILE (2008). Atlas [Online]. Available at: https://smile-smartgrids.fr/fr/projets/atlas.html [Accessed 2021].

SP Energy Networks (2017). Flexible Connections and Principles Of Access Policy. ESDD-01-009 Issue No. 2.

Thual, F. (2018). Enedis lance les premières offres de raccordements intelligents en Vendée [Online]. Available at: https://www.latribune.fr/regions/pays-de-loire/enedis-lance-les-premieres-offres-de-raccordements-intelligentes-en-vendee-769826.html [Accessed 2021].

TSO-DNO (2019). An integrated approach to active system management with the focus on tso-DNO coordination in congestion management and balancing. Rapport, CEDEC/ENTSO-E/Geode/E.DNO/Eurelectric, Brussels.

USEF (2015). USEF: The framework explained [Online]. Available at: https://www.google.com/url?sa=t&rct=j&q=&esrc=s&source=web&cd=&ved=2ahUKEwirkJDoncv0AhUB3xoKHWAPDIYQFnoECAQQAQ&url=https%3A%2F%2Fwww.usef.energy%2Fapp%2Fuploads%2F2016%2F12%2FUSEF_TheFrameworkExplained-18nov15.pdf&usg=AOvVaw3M0Yp-A3nAinKC4Fy3TYdj [Accessed 2021].

Von Appen, J., Braun, M., Stetz, T., Diwold, K., Geibel, D. (2013). Time in the sun: The challenge of high PV penetration. *IEEE Power and Energy Mag.*, 11(2), 55–64.

Western Power Distribution (2020). *Flexible Power and Flexibility webinar*. Film.

Western Power Distribution (2015). Flexible Connections Offered.

4

Planning Case Studies

4.1. Introduction

As explained in the previous chapters, historically there have been several planning cultures in the world. Indeed, power systems have been designed and developed as technology has advanced, according to different development objectives. Access to affordable and safe electricity was the first technological challenge. Some countries in the world continue to operate at the limits of the system, in terms of allowable currents and voltages, and only invest in cases of extreme need; take India for example, where power companies only reinforce the system in the event of unforeseen incidents (World Bank Group 2014).

Figure 4.1 shows the exterior and interior of an MV/LV substation in India, whose lack of maintenance suggests a high failure rate.

The development of electrification thus allows for a significant economic boom, to which further targets pertaining to supply quality (in terms of its voltage and current, but also its continuity) have been added. These motivations have led some system operators around the world to establish key principles for the development of their power systems, based on historical data and assumptions. In addition, they have defined technical and economic indicators to evaluate the targets to be attained with respect to constraints. In Figure 2.1 of Chapter 2, we detail the four different planning time horizons ranging from real time (<1 h) to long term (20–30 years). In this chapter, we focus on the first three building blocks, that

This chapter was coauthored with Oana IONESCU RIFFAUD, Aurel GARRY for section 4.3.2.2, Mahana BERNIER for section 4.4.1, Ahmed HADJSAID for section 4.4.2, and Vannak VAI and Kanika YON for section 4.5.1.

is, the very short term, the medium term, and planning for the long term. Figure 4.2 summarizes the traditional distribution system planning methodology used by system operators. In recent years, additional objectives concerning the decarbonization of electricity and reductions in consumption have emerged, leading to the surge in the development of decentralized energy resources[1].

As described by the Smart Electric Power Alliance (SEPA 2020), the conventional planning method is very linear, without any interaction between the different players of the electric power system, chiefly because the latter have traditionally few in number and with very clear functions: production or consumption. The functions of each player are becoming more and more blurred because the same actor can now consume, produce, store and curtail, all at the same time. This change in the electrical landscape is leading to technological and societal ruptures. The power system, at the very least, demands an evolution of the planning methods, specifically those at the distribution level, as shown in Figure 4.3.

a) b)

Figure 4.1. *MV/LV substations in India: a) exterior of the substation; b) inside the substation. For a color version of this figure and all subsequent figures in this chapter, see www.iste.co.uk/alvarezherault/distribution.zip*

1 See definition in Chapter 1, section 1.6.3.

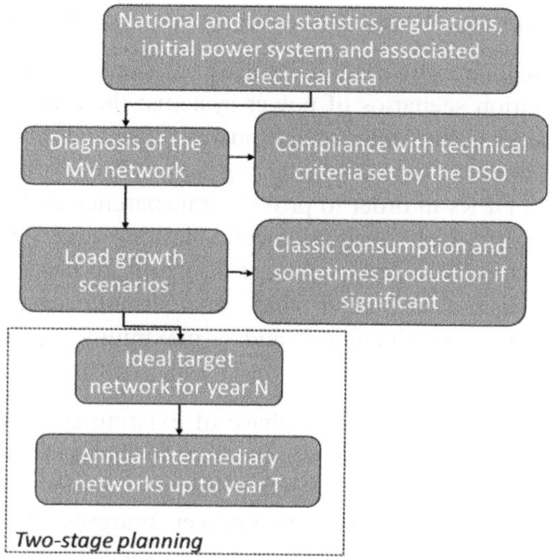

Figure 4.2. *Conventional planning*

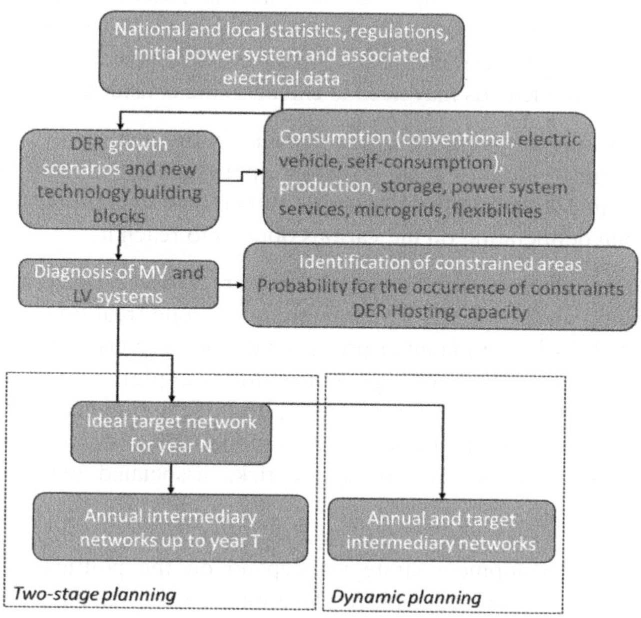

Figure 4.3. *Planning with DER*

Therefore, the major developments that need to be integrated are as follows.

– Taking into account new technological solutions, DERs and flexibilities (NWAs) in the evolution scenarios of power systems: the different forms of DERs and their control have a direct influence on power flows and therefore on the sizing of the systems that connect them. Data exchanges must therefore take place between system operators and DERs in order to provide transparency and to optimize system operation (SEPA 2020). New technological solutions and flexibilities must be modeled as an alternative to reinforcement. It is therefore necessary to concentrate on time-dated power curves rather than on extreme cases (e.g. maximum consumption–minimum production and maximum production–minimum consumption).

– Taking into account the diagnosis phase of existing systems (see Figure 4.3) prior to the modeling of scenarios: it is necessary to simulate the evolution of the existing power systems under these different scenarios in order to be able to estimate the evolution of the voltage and current/power margins. The probabilities of occurrence for different scenarios must also be modeled (using probabilistic rather than deterministic approaches) in order to be able to arbitrate between encouraging an investment or waiting for more information.

– The integration of low voltage into the planning studies: up until now LV has been, correctly, considered without apparent complexity to be the end of the system, that is, the direct connection of end-users. With the development of DERs, especially low voltage, planning methods may need to extend to the consideration of LV.

– Dynamic planning option: given the increase in the number of uncertainties, it may be interesting to study a more dynamic approach that models the arrival of events and possible actions on the system, rather than imagining a target to be attained, and then deliberating on the various options to reach it.

– Acceleration of time constants: in Chapter 2, we defined different time constants affecting planning, ranging from the millisecond (real time) to annual. The integration of NWAs in the planning process amounts to considering real-time and short-term phenomena (such as voltage control, for example), to long-term decisions (the decision to reinforce a line, for example). Therefore, it is necessary to redefine coherent temporal horizons that allow one to model the global system in the best way possible, all the while controlling the risks associated with each of these uncertainties.

Power system development strategies depend on the political and economic context as well as on the geographical location. Therefore, an area with a low population density will have long and lightly loaded power systems, while areas with a high load density will have more heavily loaded power systems. In the first case, the predominant constraints will be related to voltage, while in the second case, these will be related to current and power. In addition, the choice of operating modes and the mix of energy generation will vary according to the interconnection possibilities with other territories (continental or island system). We therefore differentiate the following three case studies:

– dense urban interconnected systems: high load density;

– rural interconnected systems: low load density;

– off-grid microgrid systems (see definition in Chapter 1): in geographical areas where the electrical system does not yet exist (developing countries), in areas that are difficult to access (such as islands) or in areas where electrical access would be very expensive (mountains, national parks). A microgrid can therefore have several scales: a building, a district, an LV feeder, an LV system, or even an MV system.

In the case of existing systems, a preliminary diagnosis of the system is necessary to guide planning decisions.

In this chapter, we present concrete methodologies applied to real-world examples that allow conventional planning to evolve toward plans that integrate DERs. In order to allow for two different levels of reading, these methodologies are described in a succinct manner so that the reader can quickly understand the principle. Real-world examples help to illustrate the benefit of the proposed methods. Then, in Chapter 6, more detailed explanations such as the mathematical formulation for the various demonstrations and the detailed description of the algorithms introduced here in Chapter 4 are given for readers wishing to replicate the algorithms and/or adapt them to their own case study.

4.2. State of the art of distribution systems with DERs

4.2.1. *New diagnostic criteria for distribution systems*

As we have already explained, electrical distribution systems are very heterogeneous around the world. When the DSO plans the system, a first step is to diagnose the system and identify weaknesses. For this purpose, realistic assumptions of consumption evolution are made, and the usual criteria, such as voltage drops, maximum currents and poorly supplied customers are estimated. In this methodology, two parameters need to be changed: it is necessary to model a general evolution of DERs, not only consumption, and it is necessary to define one or more criteria related to the goals for decarbonization. Therefore, the planner needs a mapping tool to visualize the weaknesses of the system and also a more accurate estimation of its margins, more commonly referred to as its hosting capacity (see Chapter 3). We recall that De Martini and Kristov (2015) define the hosting capacity of a system as the additional DER power that can be added to each part of the system without having to reinforce it or change its equipment, that is to say, without imposing technical constraints on the system.

In this work, we express the Hosting Capacity (HC) as a percentage of the maximum power consumed by conventional loads[2] (denoted as $P_{max}^{conso\ conv}$) using equation [4.1], where P_{max}^{DER} is the maximum power of DERs[3] that can be connected. The capacity can also be expressed directly in units of power or energy.

$$\mathrm{HC} = \frac{P_{max}^{DER}}{P_{max}^{conso\ conv}} \times 100. \qquad [4.1]$$

We also define the DER penetration rate as the share of installed DERs (denoted $P_{installed}^{DER}$), independent of system constraints, at a given time, which can also be defined in units of power, of energy, or as a percentage of $P_{max}^{conso\ conv}$.

It is the latter definition that we consider in this work. The penetration rate, denoted τ_{DER}, is given by equation [4.2].

$$\tau_{DER} = \frac{P_{installed}^{DER}}{P_{max}^{conso\ conv}} \times 100. \qquad [4.2]$$

We can emphasize that there is no correlation between the penetration rate and the hosting capacity. Depending on the case study, the hosting capacity can be estimated:

– for different spatial scales: the capacity can be calculated for each HV/MV or MV/LV substation, for each MV or LV feeder or line, or even for a "single point". For the latter, it is possible to define, for each node taken independently or globally, the quantity of DER that can be connected without imposing constraints;

– for different time scales: for example, the total photovoltaic power that can be injected without imposing constraints on the system, is greater at 13h00 than that at 23h00.

4.2.2. General principle for estimating the maximum DER power without imposing constraints on the system

As mentioned in Bollen and Rönnberg (2017), there are three steps for determining hosting capacity.

2 Conventional loads are the classic historical loads, such as electric heating, household appliances and lighting.

3 DERs can be generation, electric vehicles, demand side management, etc. The complete definition of DERs is given in Chapter 1.

– Step 1: definition of power system performance indicators according to the system state being considered:

- steady state: if the share of production responsible for maintaining the stability of the system is sufficient, then it is possible to be satisfied with an analysis of a steady state (e.g. in the case of a low voltage system). The technical quantities to be checked are the maximum values for the currents and voltages in the steady state;

- steady and dynamic state, provided stability constraints are likely to appear (in the case of an island, for example). It will then be necessary to monitor, in addition, the dynamic behavior of the frequency as well as that of the existing protection devices.

– Step 2: definition of the limits to be respected by the performance indicators. The admissible current, defined for each component of the system, must in general remain lower than the maximum admissible currents. However, it is possible to play on the thermal inertia of the components by allowing a certain amount of overload for a short duration. In this case, it will be necessary to evaluate the impact on the aging of the components. The voltage at all points of the system must be between x% of the nominal system voltage (x varies according to the DSO and the systems, see Chapter 1).

– Step 3: setting up a methodology to calculate performance indicators according to the amount of DER on the system:

- Step 3a: the definition of prospective scenarios for the evolution of the studied area is as follows: growth of conventional loads and DERs, creation of new areas (electrical substations, districts, etc.) and regulatory constraints, for example;

- Step 3b: the calculation of performance indicators obtained through classical load flow calculations in steady state completed by the frequency and voltage stability studies from dynamic state. The maximum DER power that can be connected to the system without imposing technical constraints is such that:

– all system performance indicators are within the allowed ranges,

– if we increment this power, then at least one of the performance indicators is outside the acceptable ranges.

This methodology is subject to significant uncertainties, with particular respect to the evolution scenarios (Step 3a). There are several approaches to modeling them, as summarized in Figure 4.4:

– Method 1 – Deterministic Approach: if the planner has good knowledge of the future and is able to build a few realistic scenarios, they can adopt a so-called deterministic approach, that is, they make assumptions on the evolution of variables through one or more scenarios (Step 3a) and then perform the various calculations (Step 3b) using power system simulation software. This is the method historically used since the penetration rate of DER is low, or even zero, and the only uncertain variable is the future conventional load, whose evolution was relatively predictable over several years;

– Methods 2 and 3 – Stochastic Approach: in the current context of significant future development of DERs, the planner only knows the trends or the orders of magnitude of their evolution at the national, or at best, regional level. In the latter case, so-called stochastic approaches are necessary, that is, an uncertain parameter is no longer assumed to be precisely known, but may have specific variation characteristics, such as a mean value and a standard deviation (Method 2 in Figure 4.4) or may vary between a high and a low scenario (Method 3 in Figure 4.4), according to a uniform distribution law, for example. One of the most widely used stochastic methods for taking uncertainty into account, due to the simplicity of implementation, is the Monte Carlo method.

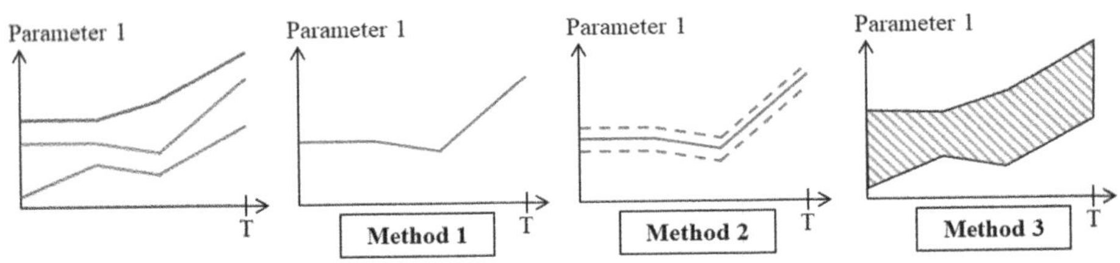

Figure 4.4. *Modeling of future changes in consumption and production*

Figure 4.5 shows an example of a single point photovoltaic generation hosting capacity calculation obtained through a real-world French rural LV system, anonymized with Method 1. In this example, the hosting capacity is calculated for each node independently. We can see here that it is possible to insert on this system between 0 and 36 kVA of photovoltaic generation depending on its location on the system: less than 6 kVA at the end of the line, because of significant voltage constraints, and up to 36 kVA around the distribution substation because the voltage and current constraints are respected.

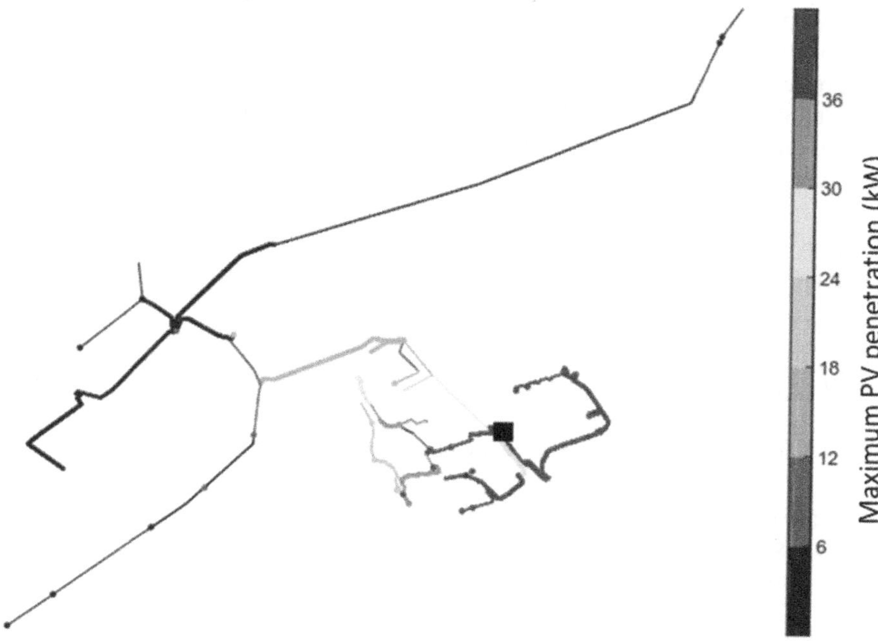

Figure 4.5. *Photovoltaic hosting capacity on an LV system (single point study) – the black square represents the MV/LV secondary substation*

4.2.3. *Decision support tools under uncertainty based on the Monte Carlo method*

4.2.3.1. *Principle illustrated via a simple example*

The Monte Carlo method is a classical numerical method commonly used in electrical engineering, but also in many other fields. It consists of performing a number of random draws in order to estimate a deterministic value when it is not possible to find an exact solution. To illustrate the principle of this method, let us consider the following simple example: "we seek to estimate the value of π". Let the circle center be (0; 0) and the radius r reside inside a square with center (0; 0), with a side denoted as l, such that $2r \leq l$. Then, we have the equality given by equation [4.3]. To estimate the value of π, it is sufficient to estimate the ratio $\frac{Area_{circle}}{Area_{square}}$.

$$\frac{Area_{circle}}{Area_{square}} = \frac{\pi r^2}{l^2} \rightarrow \pi = \frac{Area_{circle}}{Area_{square}} \times \frac{l^2}{r^2}. \qquad [4.3]$$

The Monte Carlo method can be used here to estimate the ratio of the areas of the circle and the square. To do this, we have to randomly draw of a point P inside the square several times, and see if this point is inside or outside the circle. Let N_{draws} be the total number of random draws, and N_{inside} the number of these draws located inside the circle, then the probability that a random draw for point P is inside the circle is given by equation [4.7].

$$p = \frac{Area_{circle}}{Area_{square}}. \tag{4.4}$$

We then have:

$$N_{inside} = \sum_{k=1}^{N_{draws}} X_k, \tag{4.5}$$

$X_k = 1$ if the point drawn is inside the circle and 0 otherwise.

$X_1, X_2, \dots, X_{N_{draws}}$ are thus independent random variables with identical Bernoulli distributions for parameter p, and N_{draws} is a random variable following a binomial distribution. We therefore have:

$$E\left[\frac{N_{inside}}{N_{draws}}\right] = p, \tag{4.6}$$

where E is the mathematical expectation.

Therefore, if the number of draws is sufficient, $\frac{Area_{circle}}{Area_{square}}$ can be approximated by $\frac{N_{inside}}{N_{draws}}$.

To converge faster, let us choose a square of minimal size, that is, such that $l = 2r$. Finally, the value of π is given by equation [4.7].

$$\pi = \frac{N_{inside}}{N_{draws}} \times \frac{l^2}{r^2} = \frac{N_{inside}}{N_{draws}} \times \frac{4r^2}{r^2} = 4 \times \frac{N_{inside}}{N_{draws}}. \tag{4.7}$$

Therefore, there is a trade-off between the number of draws, the computation time, and the accuracy of the desired result. Figure 4.8 shows the evolution of the value of π as a function of the number of draws performed.

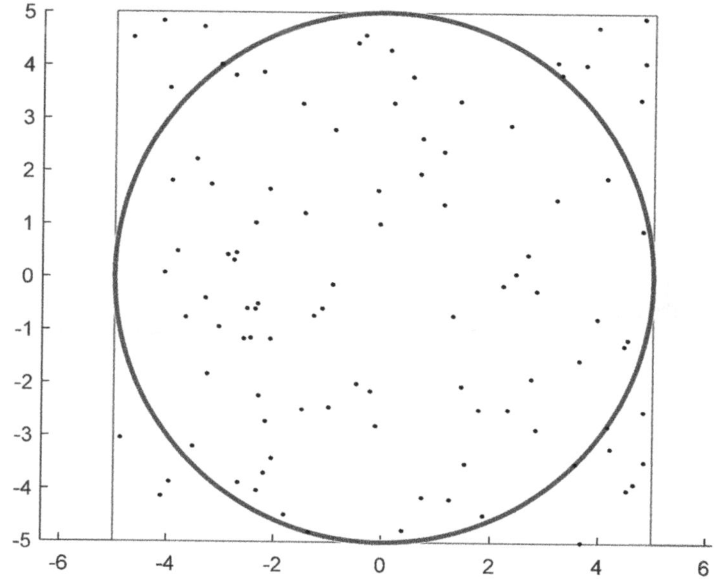

Figure 4.6. *Example 100 draws, 0.087 s, π = 3.08*

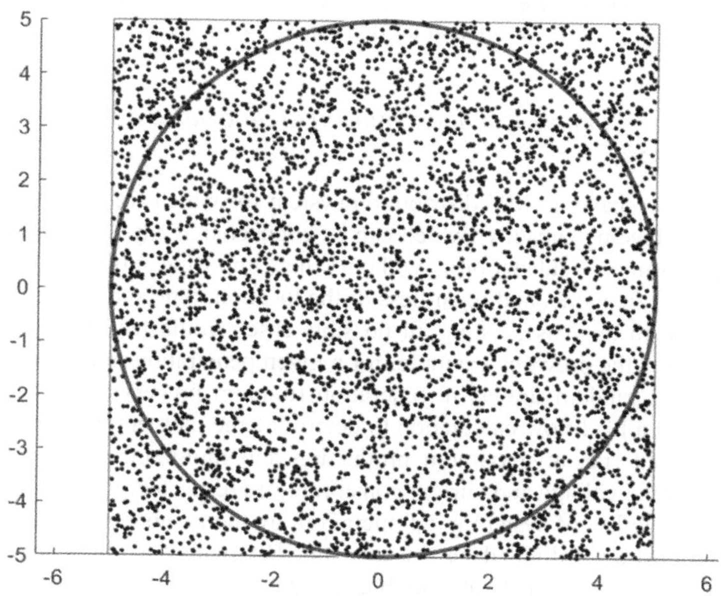

= 3.1304

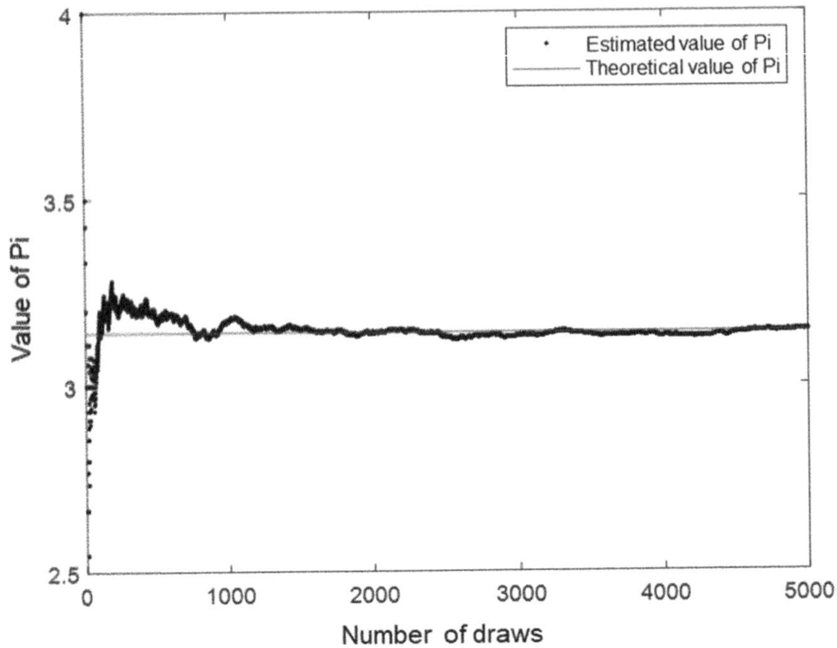

Figure 4.8. *Evolution of the value of π calculated as a function of the number of draws*

4.2.3.2. Applied example for the estimation of renewable generation capacity for an MV system

Let us say we have an MV distribution system for which we want to determine the generation hosting capacity (which can be connected at any point on the system outside the primary substation), without this imposing voltage and current constraints in static mode, and regardless of the location, type and installed power of the generation. In this case, listing all possible cases would lead to a combinatory explosion. Using the previous example, for a given penetration rate, it is possible to perform several random draws of generation (random powers and locations) and then to perform a load flow calculation to determine the moment when generation is most constrained (i.e. at maximum production and minimum consumption), and check whether the voltage and current constraints are respected. It is then possible to define the probability of system good operation, by the number of draws respecting the system constraints divided by the total number of draws. The capacity of the system can then be defined as the highest penetration rate of generation that guarantees a 100% probability of system operation.

Let us use as an example a branch of the Grenoble distribution system, consisting of three primary substations. These primary substations supply six lines in a secured feeder structure, which supply 136 MV/LV substations. This is a system with a very high load density (it includes a city center) and is therefore rather short: 38.70 km of lines for 61.19 MW, with conductors that have a large cross-section (240 mm^2). We place ourselves in a scenario of minimum consumption, equal to 20% of the maximum consumption, in the year 2030, considering the hypotheses of average evolution of consumption of 0.96% per year (the *Grenelle de l'environnement*[4] framework). The production penetration rate varies from 0% to 200% in 1% intervals. For each penetration rate, 1,000 random draws are made. The maximum power of a generation facility connected to the system is limited to 5 MW because beyond this value, it is generally directly connected to the primary substation.

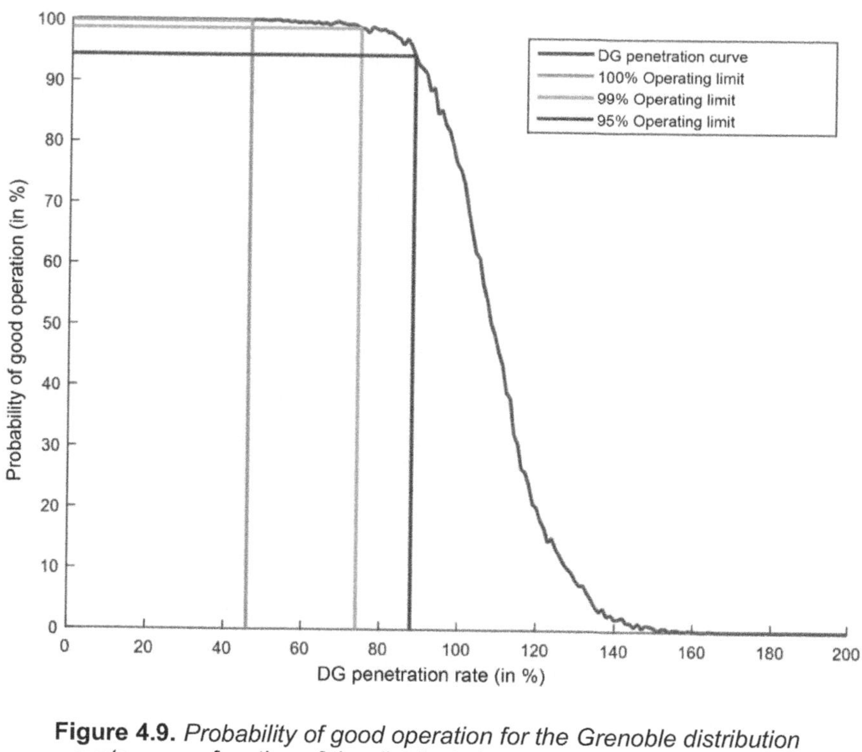

Figure 4.9. *Probability of good operation for the Grenoble distribution system as a function of the distributed generation penetration rate*

4 The *Grenelle de l'environnement* is a set of political meetings that took place at the end of 2007 and led to the drafting of the Grenelle I and II laws, in August 2009 and July 2010.

The blue curve in Figure 4.9 shows the probability of system good operation as a function of generation penetration. The hosting capacity of the Grenoble system reaches 44.6%. This method also relates the capacity of the system to a notion of risk. For example, if the operator knows that they have other mechanisms, such as generation curtailment, they can decide not to expand their system with no risk. In this example, if the operator accepts a malfunctioning risk of 1%, then they can increase the hosting capacity of the system by 26.6%, and almost double that if they take a risk of 5%.

4.2.3.3. *Visualization of system constraints*

As part of the GreenLys demonstration project (2012–2015), scenarios for the evolution of conventional loads, electric vehicles (EVs) and production were defined on the basis of the measures undertaken by the *Grenelle de l'environnement* project, and the 100% renewable energy (RE) scenario of NégaWatt (2013), which are summarized in Tables 4.1 and 4.2. These two scenarios are applied on the real-world distribution system of Grenoble over a study period from 2015 to 2030. The parameters for the two scenarios, Grenelle and 100% RE, are used as bounds between which the variables are randomly generated according to a uniform distribution. The evolution of the parameters for the two scenarios can be seen in Figure 4.10. Photovoltaic generation is the only type of generation considered for this study. For each system, 100 scenarios are randomly generated for the ranges of variation from the parameters for the Grenelle and 100% RE scenarios. The systems are studied under normal operating conditions.

	Grenelle and 100% RE scenarios			
	2015	**2020**	**2030**	**2050**
GDP growth rate	2 in %/year	2.1 in %/year	1.8 in %/year	-
National population **Lyon population** **Grenoble population**	64.514 million 505,094 161,071	65.962 million 523,695 163,629	68.532 million 538,249 168,706	72.275 million 590,933 179,336
Demand (energy)	100%	104.9%	115.4%	135.4%
Growth rate (energy)	0.96%/year	0.96%/year	0.8%/year	-
Demand (power)	100%	104.9%	114.4%	133.5%
Growth rate (power)	0.96%/year	0.87%/year	0.77%/year	-

Table 4.1. *Demand data for the Grenelle and 100% RE scenarios*

	Grenelle scenarios (noted S_G) and 100% RE (noted S_{100})							
	2015		2020		2030		2050	
	S_G	S_{100}	S_G	S_{100}	S_G	S_{100}	S_G	S_{100}
Share of RE	17%	17%	24%	33%	27%	55%	40%	100%
Share of HEVs	0%	0%	3%	0.4%	16%	6.1%	20%	-
Number of HEVs in Grenoble	0	0	2,700	360	14,846	5,660	19,727	-
Number of HEVs in the area of study	0	0	475	63	2 610	995	3,468	-

Table 4.2. *Data from the Grenelle and 100% RE scenarios on the development of RE and hybrid and electric vehicles (HEV)*

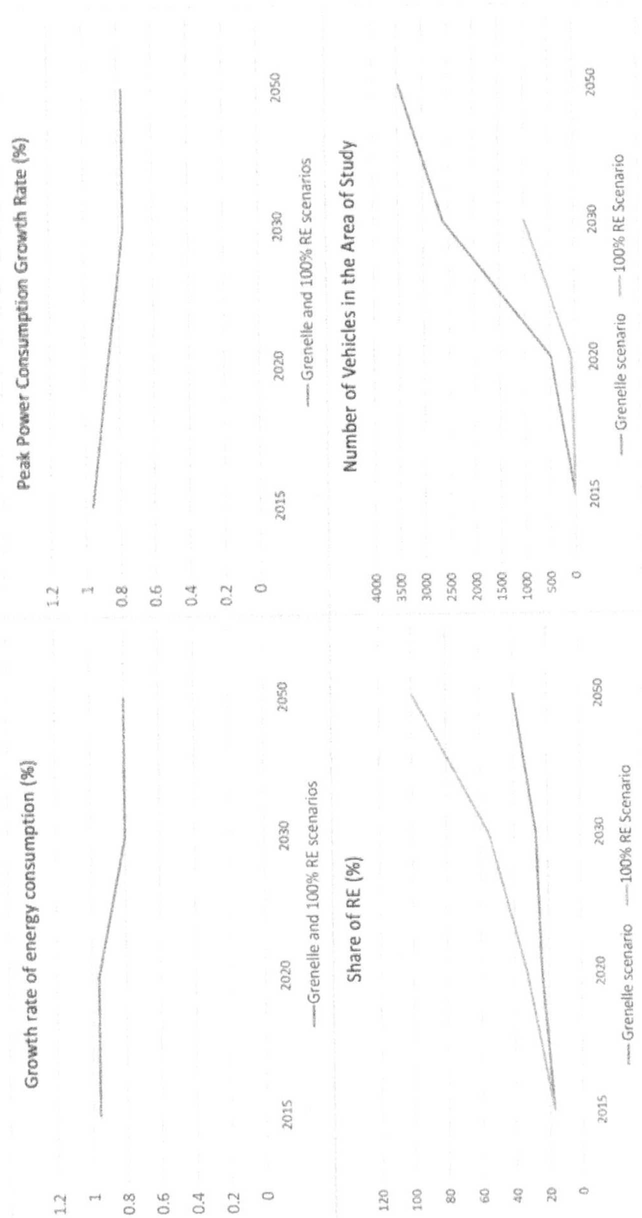

Figure 4.10. *Evolution of scenario parameters for the case studies*

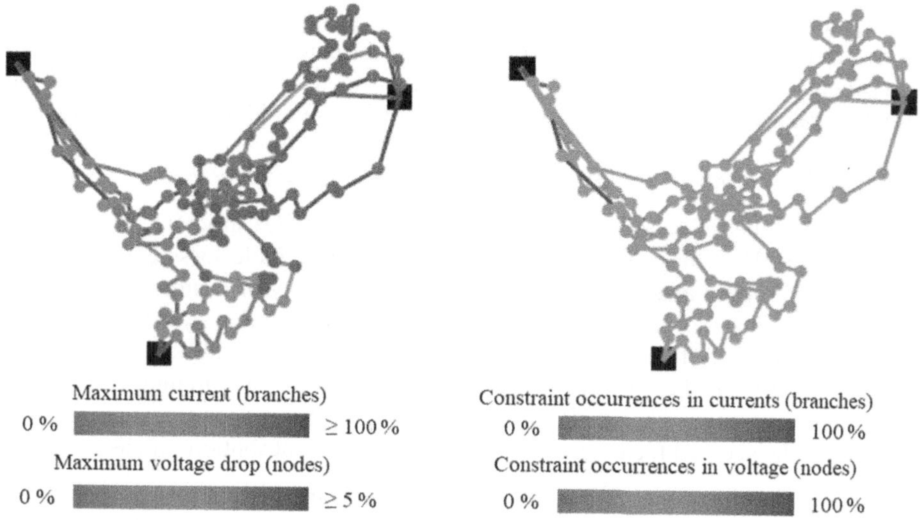

Figure 4.11. *Constraints on the real system of Grenoble at year 15*

The results are examined in the 15th and final year of the study period as this is when constraints are at their maximum. Figure 4.11 shows, graphically, the location, depth and occurrence of voltage (represented by nodes) and current (represented by branches) constraints. The depth corresponds to the maximum value of the constraint (in voltage or in current) among all the randomly generated scenarios. The occurrence corresponds to the proportion of scenarios for which a constraint appears, compared to the total number of scenarios; this indicator is therefore similar to the Monte Carlo-type approach presented above.

4.3. Dense urban interconnected systems

4.3.1. *Structural solution: topological optimization of electrical distribution systems*

The arrival of distributed energy resources in the distribution system introduces bidirectional power flows. The voltage plane is therefore no longer decreasing but may present local overvoltages that the HV/MV on-load tap changers cannot resolve. Therefore, a system sized with unidirectional flows and decreasing voltage assumptions can be at best sub-optimal in the case of high DER penetration, and at worst can be highly constrained (voltage, current) and require significant investments to resolve them. A question that then arises is: Are traditional architectures well optimized and still suitable for the case of high DER penetration?

4.3.1.1. *Case study 1: optimizing existing traditional architecture*

In France, the reference architecture is the secured feeder structure, as presented in Chapter 1. This structure is easy to operate, has low joule losses, allows for a good quality supply and is reliable, by securing the supply of customers served by the HV/MV substation in the event of faults in the distribution systems. It is also easier to upgrade. A first approach that favors the insertion of DERs advocates for the optimization of the existing construction methods for this architecture, before proposing new ones. Therefore, in this case study, we propose an optimization method for the secured feeder architecture whose main contributions are as follows:

– jointly optimize the various costs of the DSO: investment costs, operational costs, costs related to the continuity of the supply and the integration of DERs;

– take into account all the topological constraints inherent to the construction of an urban interconnected power system: configuration (imposed by the cities' architecture), costs and street congestion limits (only a limited number of electrical conductors can pass through a street);

– efficiently explore the solution space for the construction of the architecture: this is subject to a combinatorial explosion (large numbers of lines and nodes), nonlinearities (electrotechnical equations) and discrete variables (the presence of a remote-controlled switching device, or not) preventing the use of traditional optimization methods – these aspects are discussed in greater detail in Chapter 5;

– propose several solutions, from which the DSO can select the one that best meets the different objectives;

– applied to different use cases, that is, both to "green field planning", for which the initial system is non-existent, and to "expansion planning", for which there is an initial system whose removal, reuse or reinforcement can also be optimized.

4.3.1.1.1. Proposed algorithm

The principle of the proposed method is to divide the planning problem into three sub-objectives, and to process these one after the other, integrating the constraints as they arise while minimizing the degradation of the solution of the previous step. The different steps are shown in Figure 4.12.

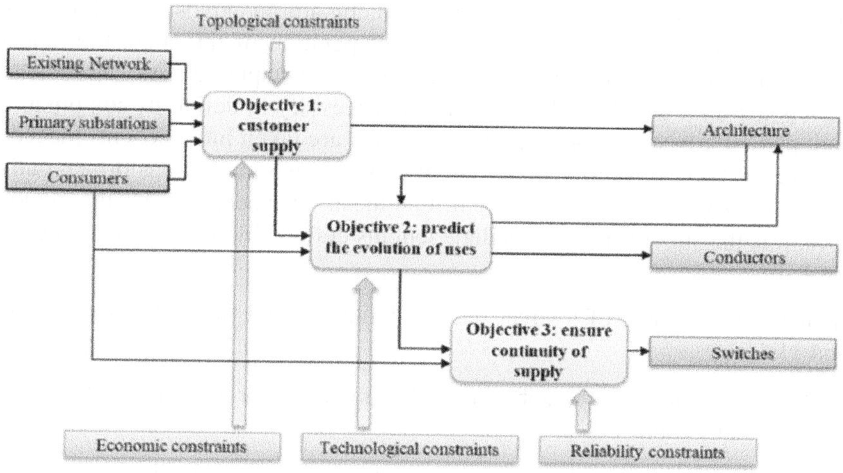

Figure 4.12. *Optimization steps of the secured feeder architecture*

Sub-objective #1: customer supply

Sub-objective No. 1 is to supply the customers of the power system (consumers and producers) and thus define the structure of the system that will carry or evacuate the energy between the customers and the primary substations connected to the transmission system. The objective function chosen to minimize is the total length of the system (named L_{tot}). This minimizes:

– the cost of conductors and civil works, proportional to the length of the branches;

– the probability of faults along the lines and therefore the number of RCSs required to ensure a given continuity of supply, in accordance with the DSO's objectives;

– the cost of joule losses;

– voltage drops and rises at any point in the system.

The solution is purely combinatorial and consists of partitioning the secured feeders between the primary substations and the MV/LV substations between the secured feeders. This double partitioning leads to a combinatorial explosion in terms of the number of solutions. Therefore, the selected algorithm is a Simulated Annealing algorithm which, compared to other meta-heuristic methods (Genetic Algorithm, Tabu Search and Ant-Colony), offers the best performance, low variance of the results and a low number of requests to the objective function (which reduces the computing time). The justification of the combinatorial explosion and the choice of the method are detailed in Chapter 5. The operation of this algorithm and the mathematical modeling are likewise described in Chapter 5. As the objective function is the total length of the system, it is necessary to define its computational method, particularly when topological constraints are taken into account. Indeed, in the absence of these constraints, the length between two given transformer stations (HV/MV or MV/LV) is the classical Euclidean distance, but in the opposite instance, it is the sum of the intermediate lengths defined by the streets, as shown in Figure 4.13:

a) shows an example of a street with its buildings and two MV/LV substations (10 and 11);

b) shows the equivalent graph obtained to model the electrical and topological elements listed in Table 4.3;

c) shows the shortest path to connect substations 10 and 11, obtained using the Dijkstra Algorithm (Dijkstra 1971);

d) shows the power line obtained between 10 and 11.

In this length calculation, road constraints such as property lines and municipal boundaries are not taken into account since they require finer modeling as well as access to these data.

Graph vertices	Graph edges
Street intersections HV/MV and MV/LV substations Connections of the substations to the streets	Edges connecting two intersections Edges connecting a substation to the nearest street

Table 4.3. *Composition of the graph in Figure 4.13*

In addition to taking into account the nature of the streets, other systems (telecoms, gas, water, transportation) as well as potential road system expansions, corrective coefficients are applied to the cost of the edges. The route of a secured feeder to connect several MV/LV substations is solved via the traveling salesman problem (Applegate 2006).

The result of the simulated annealing algorithm is a set of solutions that are optimized from the perspective of total length, but that do not respect all the technical constraints imposed, namely current and voltage constraints. The obtained solutions are then optimized according to a second criterion, namely the maximum power per secured feeder, which is used a lot in planning, together with the PL product (see Chapter 2), since it makes it possible to ensure that the thermal limits of the conductors and the authorized voltage drop and rise limits are not exceeded. This second optimization phase is a greedy algorithm and consists of exchanging MV/LV substations between secured feeders in order to balance the total power allocation and minimize the degradation of the previous objective function L_{tot}. The degradation of L_{tot} is limited by the use of topology, as illustrated in the example of Figure 4.14.

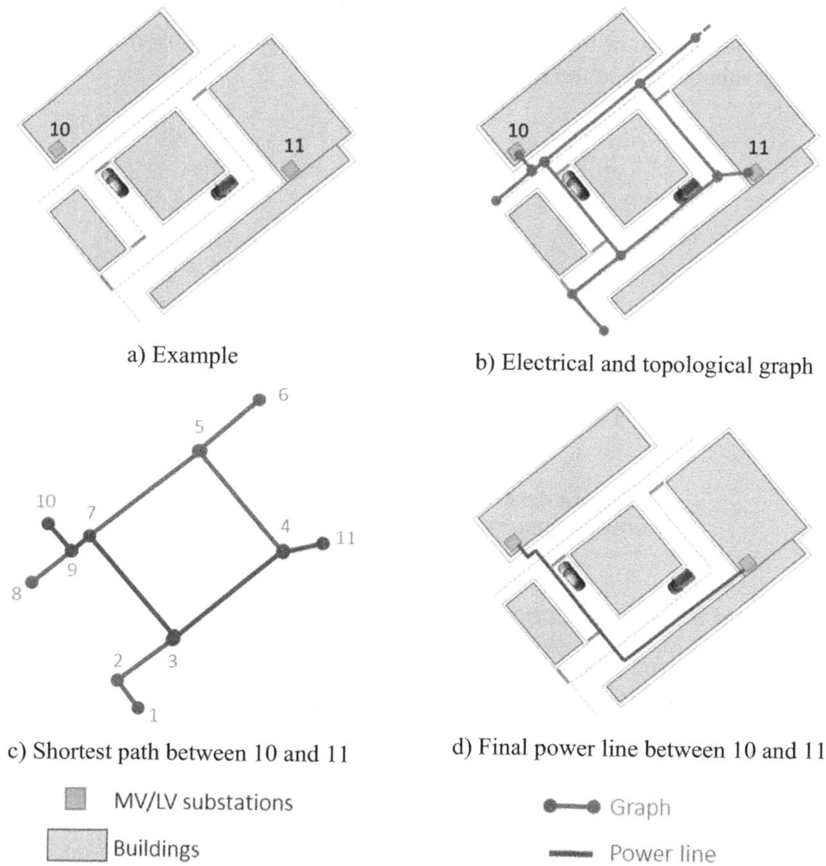

Figure 4.13. *Modeling topological constraints as a graph*

Sub-objective #2: predict the evolution of uses

Sub-objective No. 2 is to forecast the evolution of the uses on the system (consumption and production). During this second step, the conductors of the lines selected during the first step are sized via a techno-economic analysis making it possible to minimize the cost of the joule losses and to respect the technical constraints, which are the admissible current and voltage limits. The minimization of the total length of the system during the first step makes it possible to limit the cross-section of the conductors, ensuring voltage drops and rises remain within the authorized limits, and that the rebalancing of the loads between secured feeders limits the cross-section of the conductors, thereby ensuring the power flows between the primary substation and the MV/LV substations

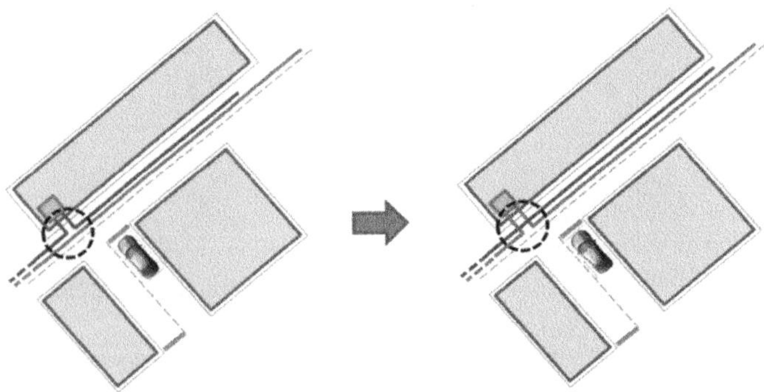

Figure 4.14. *Example of load exchange between secured feeders to rebalance power*

Sub-objective #3: ensure continuity of supply

Sub-objective No. 3 is to guarantee the continuity of the supply by placing a minimum number of RCSs along the secured feeders, in order to meet the objectives set by the DSO. The number of RCSs per secured feeder is first minimized by a techno-economic study based on a statistical analysis of the average outage time according to the number of RCSs. The RCSs are then placed on the secured feeders in such a way as to maximize the equality of the PL products between feeders and between the branches delimited by two RCSs, in order to guarantee an equitable continuity of supply for all the customers of the system.

4.3.1.1.2. Case study 1a: "green field" planning

The applied example is a portion of the Grenoble distribution power system (presented in section 4.2.3.2 and shown in Figure 4.15(a)), operated by *GreenAlp*, composed of six secured feeders fed by three primary substations, and serving 136 secondary substations for a total power of 61.19 MW. The total length of the actual system is estimated at 37.16 km. The topology of the area served is modeled in a graph with 480 vertices and 772 edges. After connecting the secondary substations, the graph has 830 vertices and 1122 edges. In this example, the distribution system for this area is constructed as for a "green field" planning case in order to compare it to the current solution (Figure 4.15(b)). Table 4.4 summarizes the performance metrics of the actual and optimized system:

– after the first stage, the optimized power system had a total length of 31.66 km, that is, a gain of 18.19% compared to the initial system;

– two secured feeders are identified as being overloaded by about 20% when compared to the maximum authorized power. The greedy algorithm of load exchange between secured feeders allows for better power distribution by increasing the total length to 741 m, or only 2.34% when compared to the initial solution, after 11 successive load exchanges;

– the conductors are sized in the second stage to meet the techno-economic balance, and the RCSs are positioned in the third stage to meet the reliability objectives. The observed gain on the total system cost is 16.70%.

a) b)

Figure 4.15. *Case study of Grenoble: a) the actual system; b) the optimized "green field" system*

The previously proposed solution had the same secured feeder configuration as the initial solution shown in Figure 4.16. Sixteen other secured feeder structures were studied (10 structures with six secured feeders and 6 structures with five secured feeders) for which the simulated annealing algorithm was simulated 10 times in order to examine the variability in the resulting solutions. Figure 4.16

shows another example of a six-secured-feeder structure and an example of a five-secured-feeder structure. These configurations always have at least one secured feeder for each pair of primary substations. It is also verified that for five secured feeders, the maximum allowable power per secured feeder is still respected, since the total power to be supplied is distributed over five and not six secured feeders.

Parameters \ Case study	Actual system	System of minimum total length	Final selected system	Final/actual gain[5]
Conductor length	38.70 km	31.66 km	32.41 km	16.3%
Trench length	22.69 km	21.91 km	22.48 km	0.9%
CAPEX	**3,566**	**3,022**	**3,092**	**13.3%**
Conductor cost	774 k€	633 k€	641 k€	17.2%
Trench cost	2,672 k€	2,341 k€	2,403 k€	10.1%
Switch cost	120 k€	48 k€	48 k€	60%
Updated OPEX	**589 k€**	**392 k€**	**369 k€**	**37.4%**
Discounted losses costs	552 k€	365 k€	342 k€	38%
Discounted ENS costs	18 k€	13 k€	13 k€	27.8%
Discounted outage costs	19 k€	14 k€	14 k€	26.3.%
TOTEX	**4,155 k€**	**3,414 k€**	**3,461 k€**	**16.7%**

Table 4.4. *Summary of the performance indicators of actual and optimized system*

The 160 solutions obtained are compared according to the four criteria shown in Figure 4.17: total system cost (CAPEX and discounted OPEX), generation hosting capacity as presented in section 4.2.3.2, power balancing between secured feeders and average SAIDI of the system. This multi-objective approach provides the DSO

5 $Profit = 1 - \frac{final\ value}{initial\ value}$.

with a set of solutions from which they can choose the one that best suits their selection criteria.

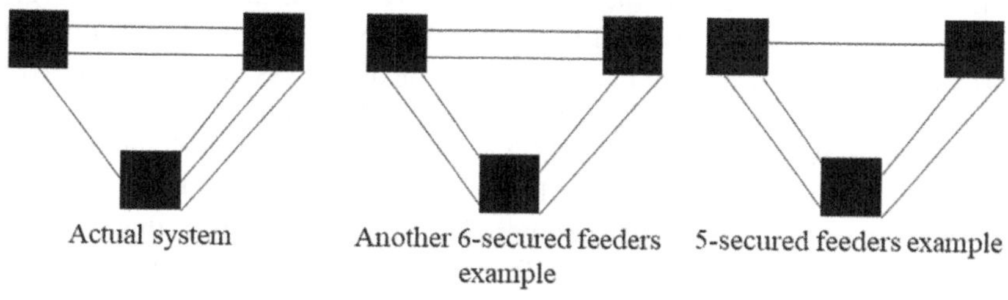

Figure 4.16. *Examples of secured feeder configurations between primary substations*

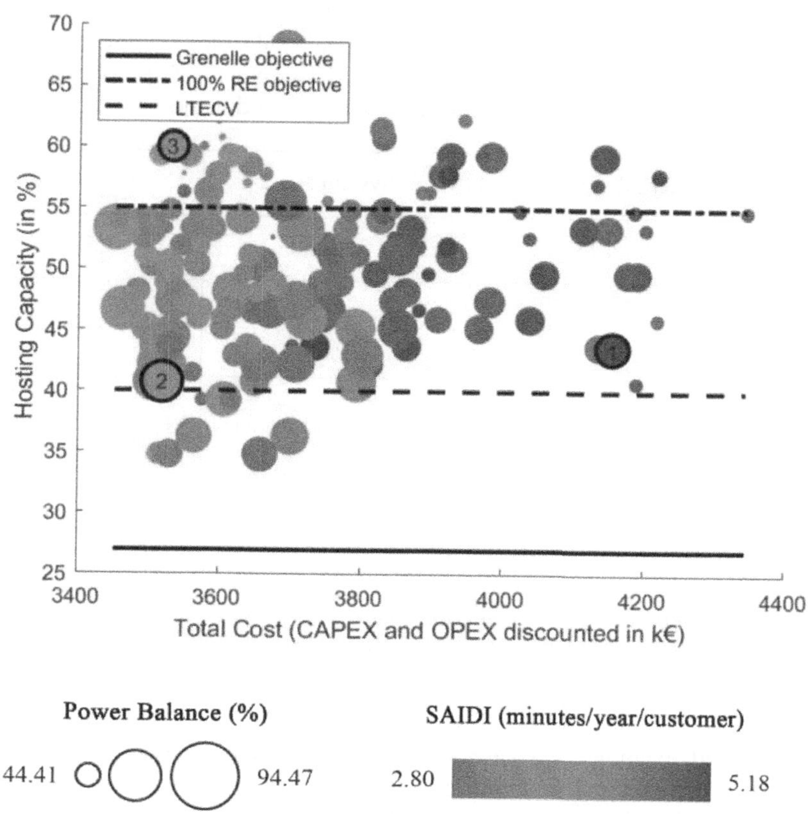

Figure 4.17. *Multi-objective approach to a set of solutions for 2030*

The three dotted lines represent the 2030 RE penetration rate objectives for the Grenelle (27% RE) and NégaWatt (55% RE) scenarios used by the *GreenLys* project, and the objectives for the LETCV (*Loi relative à la transition Energétique pour la Croissance Verte*: Law on Energy Transition for Green Growth) of August 17, 2015 (40% RE by 2030). Ninety five percent of the systems meet the LTECV target for RE, 100% of the systems meet the Grenelle targets, but only 21.5% meet those of the NégaWatt scenario. The DSO can then decide to prioritize one criterion over another. Let us take as an example the case where the DSO does not wish to give priority to one indicator over another. The "best" solution in this case will be the one with the highest average of the four normalized indicators: this is the No. 2 circle in Figure 4.17. This solution, in addition to having the best average, meets the minimum RE penetration targets for the Grenelle and LTECV scenarios. If the objective is to meet the objectives of the "100% RE" scenario, the best solution that meets this objective while minimizing the other three indicators is the one represented by circle No. 3. For comparison, we have also represented the performance of the current Grenoble system (circle No. 1). The detailed results of these three solutions are given in Table 4.5. We can already see that the development of an optimal planning methodology makes it possible to improve the different indicators.

	Actual system (No. 1)	**System No. 2**	**System No. 3**
Total cost	4,155 k€	3,519 k€	3,533 k€
HC hosting capacity	43.60%	40.73%	59.99%
Maximum SAIDI	5.18 min/year/customer	3.05 min/year/customer	3.16 min/year/customers
Power balance rate	60.44%	48.96%	62.09%

Table 4.5. *Values of the objectives for the selected solutions*

4.3.1.1.3. Case study 1b: system extension planning

The second example is a case of system expansion planning: the same MV system as in the previous example is considered (20 kV, six secured feeders and 136 MV/LV substations), to which is added the 5.5 kV MV system serving 36 MV/LV substations in the same area and with a maximum power of 11.84 MW. We use the methodology described above to minimize the transition work on the 5.5 kV system, taking into account the existing infrastructure. We keep the same number of six MV secured feeders, as they have sufficient power margin to host the additional 11.84 MW of consumption, and each 20 kV load remains connected to the same secured feeder as in the original solution. Figure 4.18 shows the location of the 5.5 kV substations relative to the 20 kV substations.

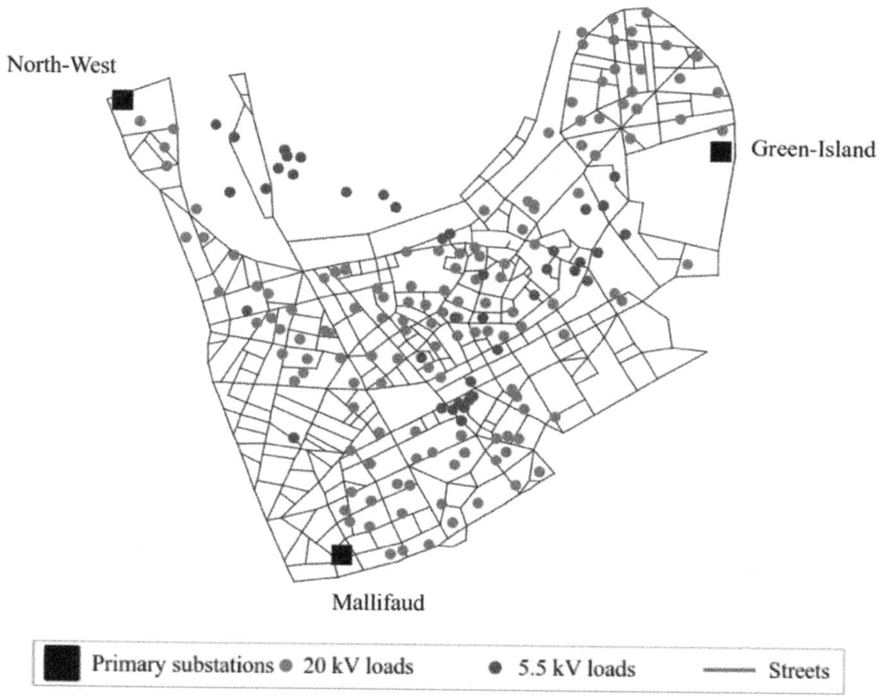

Figure 4.18. *20 and 5.5 kV systems*

The works to be considered are:

– keep a cable on the same secured feeder as the initial solution (20 kV system);

– transfer a cable from one secured feeder to another;

– lay a new cable;

– remove a cable.

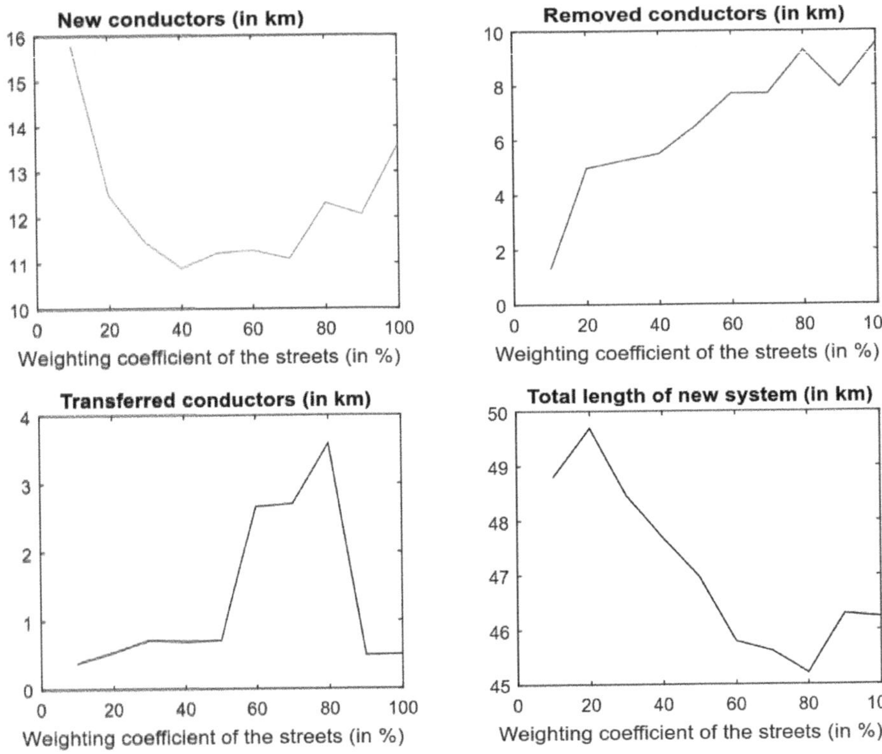

Figure 4.19. *Evolution of lengths for the transition from 5.5 to 20 kV*

To optimize the use of the existing 20 kV infrastructure, we assign a weighting factor α_{work} between 10 and 100% to the corresponding edges in the power system graph in order to "encourage" the use of the already existing cables. A zero weighting leads to unrealistic results as a secured feeder could then carry on indefinitely on a street considered to have zero work costs.

Figures 4.19 and 4.20 show the evolution of lengths and work costs for different values of the weighting factor α_{work}, with the results averaged over 100 simulations for each value of α_{work}. An optimal value of α_{work} of 80% minimizes the total length of the new system, with minimal new conductor installations (a plateau between the 40 and 80% values of an α_{work} is visible) and maximization of transfers from existing conductors between secured feeders, despite a total length of conductors removed that is greater than the results for the other values of α_{work}. The total cost of transferring the 5.5 kV system to 20 kV is also minimized for an α_{work} of 80%.

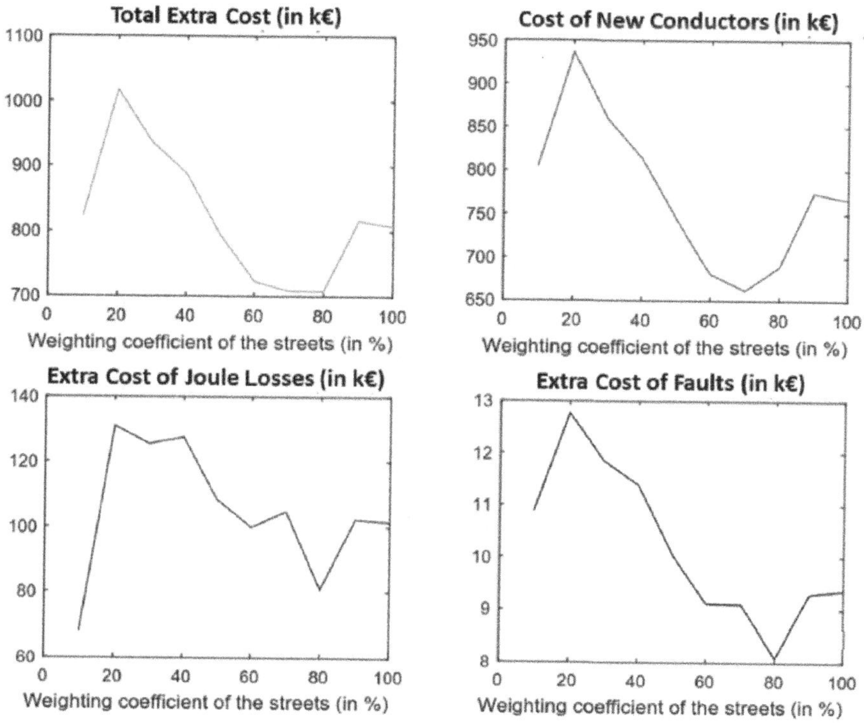

Figure 4.20. *Evolution of system costs for the transition from 5.5 to 20 kV*

These results are evaluated with the consideration that the cost of removing existing conductors is zero, that conductor transfer also has zero cost and that the cost of laying new conductors does not depend on the already existing trenches in the streets where the works are to be performed. These different parameters are adjustable by the DSO, who can then use this method for different values of α_{work} to identify a solution that satisfies both the cost requirements and the feasibility of the work in terms of potential road system expansions, impact on system users, etc. Indeed, Figures 4.21 and 4.22 show that while the total cost is more important for an $\alpha_{work} = 10\%$ than for $\alpha_{work} = 80\%$, there is less work to be done since there is no conductor transfer to be performed, and few cables are removed.

Figure 4.21. *Works to be carried out for an* α_{work} *= 80%*

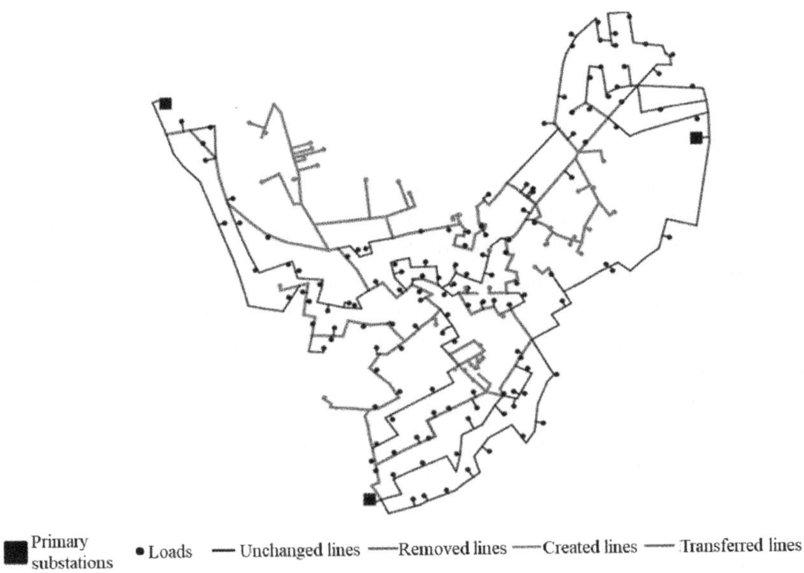

Figure 4.22. *Works to be carried out for an* α_{work} *= 10%*

4.3.1.2. *Case study 2: new distribution system topologies*

The transmission system that transports energy from the generation plants to the consumption areas is meshed, that is to say, there are several paths to direct the electricity flows. This historical choice is linked to reliability. Indeed, in the transmission system, each node represents thousands of MW and it is therefore necessary to ensure that the decommissioning of a line does not cause a power outage. However, system meshing at the distribution level is not considered economical in terms of the benefits it would bring. On the other hand, if DERs are connected to the distribution system, adding well-positioned alternative paths to evacuate energy not consumed locally could be more economical than increasing the cross-section of lines and cables already present in the system. Moreover, the mesh allows for a better distribution of the current within the lines, which are then less loaded. Since losses are proportional to the square of the current, meshing a system makes it possible to reduce losses. Finally, depending on the protection system envisaged (which depends on the way the system is meshed), the cost of the energy not supplied and of the power cut can also be reduced. Therefore, the meshing of systems allows for a reduction in OPEX but, in the absence of DER, the CAPEX of meshed systems is greater than the OPEX benefits it brings.

The radial system solution is therefore more economical. On the other hand, if the DER penetration rate becomes significant, since reinforcements will become necessary in the distribution system, it is possible that the CAPEX of the reinforced radial system becomes more significant than the CAPEX of a smartly meshed system. Therefore, in this section, we propose new meshed architectures to increase the penetration rate of DERs. Given the high cost of system infrastructures, we have imagined how to mesh smartly the existing structures in order to avoid modifying the architecture too much to make these systems mutable in the long term. Indeed, in the case of industrialized countries where the system already exists, completely rethinking the architecture and the operating modes of the systems would mean starting from scratch, and the transition cost of these systems would be considerable.

In the case of low DER penetration, the traditional secured feeder architecture is a good technical and economical compromise as it enables good reliability indices to be obtained with a limited investment (a few loops and a few remote-controlled switching devices). How do we efficiently mesh this structure in order to increase the penetration rate of renewable generations without totally changing the protection plan? Two solutions have been studied: a mixed mesh/radial operation in the meshed secured feeder structure (Alvarez-Herault et al. 2015) and a permanent meshed operation in hybrid loops (Alvarez-Herault et al. 2010).

4.3.1.2.1. Case study 2a: hybrid loops (meshing on the same transformer)

Description

The hybrid loop structure is a secured feeder structure (see Chapter 2), but the secured feeders connect consumer loops and not the consumers one by one. The dual advantage is that the lengths are significantly reduced and the local mesh allows for better power flow distribution and reduced joule losses. Figure 4.23 shows the evolution of the secured feeder into the hybrid structure. One secondary substation per loop, called a "loop head substation", is connected to a main secured feeder via two normally closed remote-controlled switching devices. In order to ensure the radial operation of each main secured feeder, one loop head substation per secured feeder is equipped with a normally open switching device. Therefore, in the event of a fault on the main secured feeder, the fault detection and isolation procedures remain the same, and the protection equipment remains unchanged. Figure 4.24 shows the connection method for a loop to a main secured feeder as well as the protection equipment. Each "loop head" substation is connected to the primary secured feeder via two switches and a circuit breaker. The other secondary substations are connected to the loop via normally closed automatic switches.

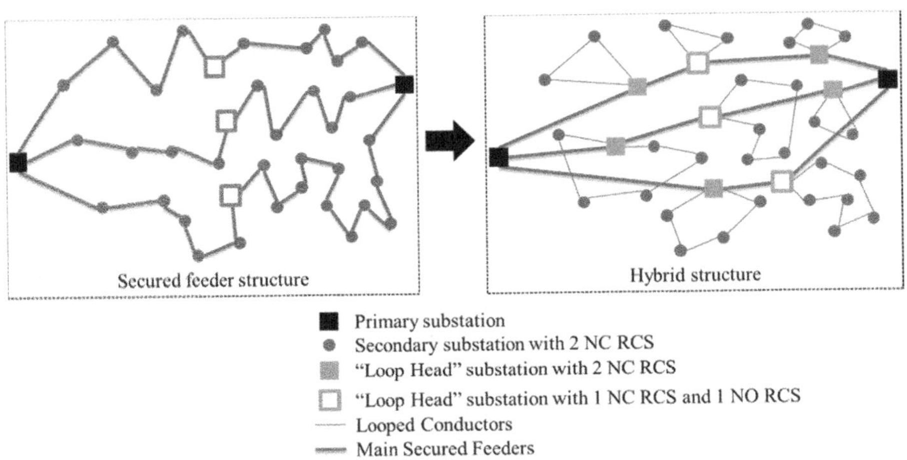

Figure 4.23. *Hybrid structure*

In the event of a fault in a loop, the circuit breaker in the "loop head" substation trips first, thus isolating the loop from the main secured feeder that remains connected to it. Then, all the switches of the secondary substations in the loop are opened, as shown in Step No. 1 of Figure 4.25, and the circuit breaker of the "loop head substation" closes. The switches then close successively and automatically in one direction with a 10-s delay until the fault is restored. The circuit breaker reopens, as well as the last closed switch that caused the fault to be fed. The state of the system is then that of Step No. 2 in Figure 4.25. The same procedure is repeated in the other direction. The fault is then automatically detected and isolated (Step No. 3 in Figure 4.25). The cost of the protection and switching equipment is then more significant than the cost of the feeder breaker but should be compensated by the savings on reinforcements and OPEX.

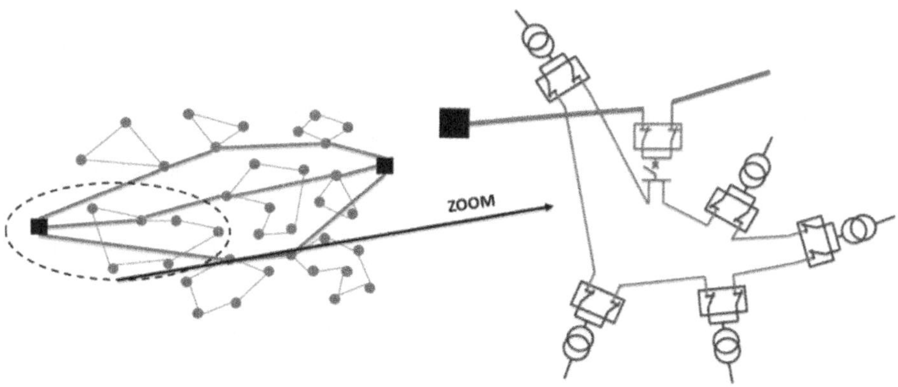

Figure 4.24. *Close-up of the loop protection plan*

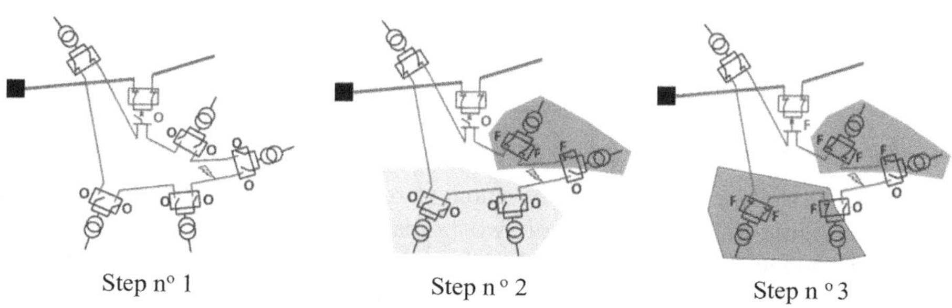

Figure 4.25. *Procedure for detection, localization and resupply in the event of a fault in a loop*

For an application with French criteria, it is then necessary to define the zones of equivalent PL product. As in the case of the secured feeder, the normally open switch of the main secured feeders is placed so that the PL product is equal on both sides. It follows that each loop must have the same PL product.

Construction algorithm

In the context of a long-term master development plan where all lines will have to be replaced, the objective is to find the hybrid loop structure that minimizes the total length of the system while guaranteeing good quality supply, with respect to the PL criterion. Therefore, minimizing the total length of the system will not necessarily lead to a solution that minimizes the TOTEX. Indeed, a system with very few loops and feeder breakers will have a very low cumulative length of conductors, but potentially more losses because the lines will be very heavily loaded. In addition, the PL products will be unbalanced. Solving this problem is complex and requires the use of heuristics. Indeed, the objective is to minimize the total length of the system and to maximize its quality of supply, that is, to minimize the SAIDI, SAIFI, ENS and to balance the PL criterion. For a given number of loops, the retained algorithm consists of:

– building an initial solution by distributing the number of loads fairly between the loops while minimizing the loop lengths;

– exchanging loads between loops to balance the PL criterion;

– connecting the loops via main secured feeders;

– sizing the conductors of the final architecture;

– performing a load flow calculation to evaluate the voltage and current plan;

– calculating the performance indices of the architecture: SAIDI, SAIDI, ENS, PL product, costs and capacity.

Applied example

The case study considered is the anonymized French MV real urban system shown in Figure 4.26, consisting of two primary substations and 984 secondary substations supplying 85,239 MV and LV customers for a total power consumption of 151.4 MW. The term "loads" will be used to refer to the secondary substations.

In a long-term planning logic, it is assumed that all conductors are to be replaced (Figure 4.27), but it would likewise be possible to fix a set of conductors that would remain unchanged. Two architectures are then determined: the traditional secured feeder structure and the hybrid loops in order to compare fairly the performance of the new proposed structure. An example of a long-term secured feeder structure consisting of 15 secured feeders is shown in Figure 4.28, and an example of a 10-loop hybrid loops structure is shown in Figure 4.29.

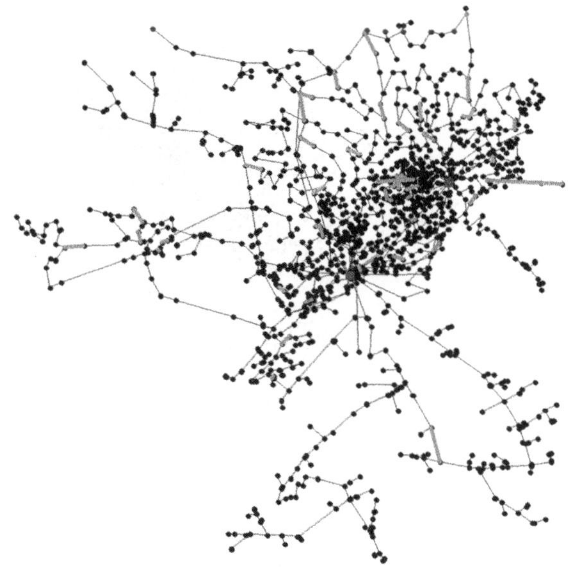

Figure 4.26. *Actual urban MV secured feeder structure*

Figure 4.27. *Input data*

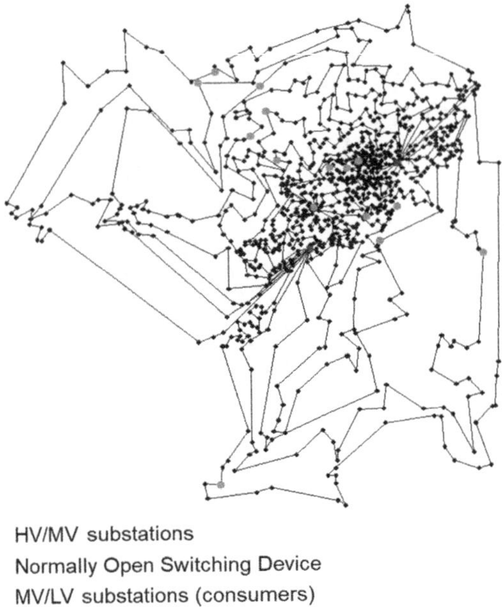

Figure 4.28. *Long-term secured feeder structure (15 main secured feeders)*

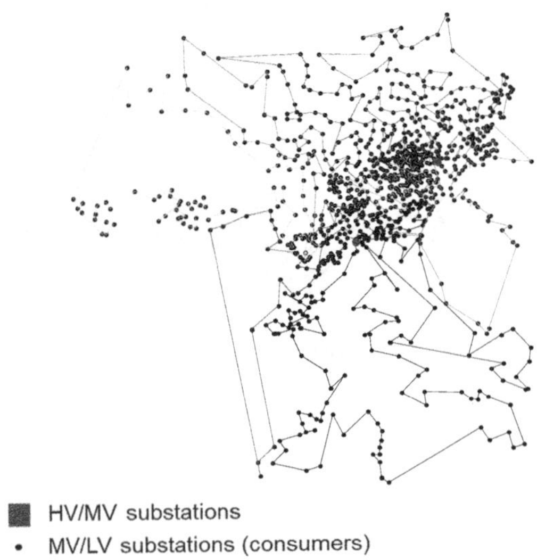

Figure 4.29. *Long-term hybrid structure (10 loops, primary main secured feeders not shown)*

The performance indicators (Figures 4.30, 4.31 and 4.32) were calculated by varying the number of main secured feeders in the secured feeder structure case and the number of loops in the hybrid loop case (with the number of main secured feeders fixed at 10).

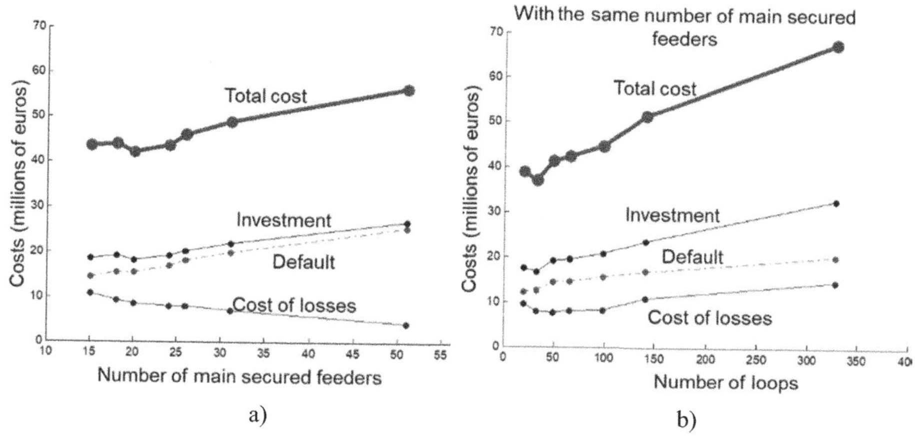

Figure 4.30. *CAPEX and OPEX updated according to the number of primary secured feeders (for the secured feeder structure: a) and according to the number of loops (for hybrid loops: b)*

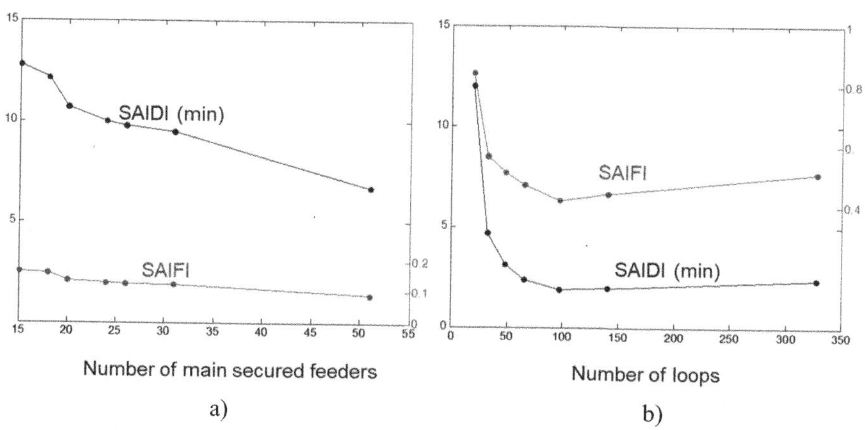

Figure 4.31. *SAIDI and SAIFI per feeder updated according to the number of main secured feeders (for the secured feeder structure: a) and according to the number of loops (for hybrid loops: b)*

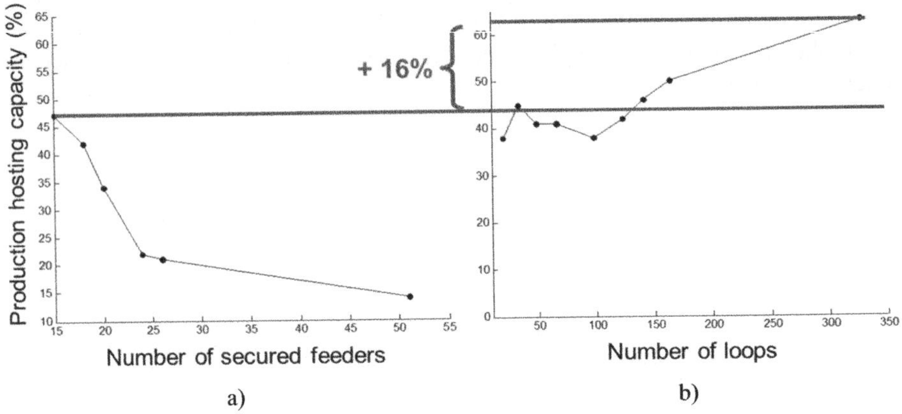

Figure 4.32. *Production hosting capacity as a function of the number of main secured feeders (for the secured feeder structure: a) and as a function of the number of loops (for the hybrid loops: b)*

The limits of variation for the number of secured feeders and the number of loops are set by various constraints:

– conductor sizing constraints for the minimum number of main secured feeders and loops;

– planar graph constraints (no crossing edges) for the maximum number of main secured feeders;

– definition of a loop itself (a loop has at least three loads) for the maximum number of loops.

Analysis of the secured feeder structure

Concerning the secured feeder structure, we can notice the presence of an optimal point at 20 secured feeders where the total discounted cost is minimal. As the number of secured feeders increases, the CAPEX increases, as well as the cost of faults, which is proportional to the conductor lengths. On the other hand, even if the total lengths increase, the currents flowing in each secured feeder decrease, which has a greater impact on the losses (losses are proportional to the square of the current). The SAIDI per feeder varies between 12.5 and 6 min/year/customer, which is the current French order of magnitude. The SAIFI tends to decrease as well, but in a manner that is less pronounced than the SAIDI. The production hosting capacity reaches its maximum when the number of secured feeders is minimal, which is logical since the increase in the number of secured feeders decreases the cross-section of the cables used, and therefore the current constraints due to the evacuation of the power produced and not consumed locally occur more quickly. The capacity of this case study is therefore limited to 47% with the traditional feeder breaker architecture (see Figure 4.32(a)).

Hybrid loop structure analysis

For hybrid loops, the losses are non-monotonic, which is mainly due to the discrete choice of the conductor branch. Nevertheless, we can observe a minimum between 50 and 100 loops. The SAIDI and SAIFI decrease sharply up to 100 loops and then tend to increase if the number of loops becomes too large. We also find here a compromise between the number of loops and the general reliability of the system. Even though the shape is non-monotonic, the production hosting capacity is rather high regardless of the number of loops (always higher than 39%, see Figure 4.32(b)).

Comparison of the secured feeder and the hybrid loop structures

Table 4.6 gives the performance criteria for the secured feeder and hybrid loop structures. If we place ourselves in the context of significant production development, we can see that the secured feeder structure is limited to 47% capacity, while the hybrid loops allow for significant penetration of production. In fact, the division into loops allows for better evacuation of the power produced when it is not consumed locally. The loop protection equipment (circuit breakers at each loop and automated switches) allows for a significant improvement of the SAIDI and the ENS. The secured feeder structure equipment could be improved by remotely controlling all the breaker devices, but this would increase the cost[6]. If we consider the maximum penetration rate achievable with the secured feeder structure, then the cheapest hybrid loop structure is 25 loops, resulting in an architecture that is slightly cheaper than the feeder breaker, by 12%. On the other hand, the SAIDI is divided by a little over half.

Maintenance costs have not been included in the calculation of OPEX for these two types of structures. These could prove to be more significant for hybrid loops, in particular, those linked to the maintenance of loop head substations that have been equipped with remote-controlled switching devices and circuit breakers. Moreover, the risk of failure of these circuit breakers has not been considered in the calculation of reliability indices (SAIDI, SAIFI and ENS). In this case, the resupply of the secondary substations located in the loops could be complex and costly, unless

6 This cost, mainly linked to the motorization of operating devices and the communication chain, could nevertheless decrease in the future. Indeed, the new secondary substations are either pre-equipped with motors, or can be motorized at a negligible cost when compared to the cost of the substation. Moreover, when possible, the DSO could use the communication chain, made up of the communicating smart meters, to remotely operate these switching devices.

isolated micro-grid operations that are based on existing DERs are allowed. Therefore, the theoretical gains described in the previous sections could be less in practice, depending on the type of system (urban or rural). Finally, it would be necessary to study the transition cost from an existing system to a hybrid loop-type system, and in particular, the operation mode of the intermediate systems.

	Secured feeder structure	**Hybrid loops**
Production hosting capacity	(15%; 47%)	(39%; 67%)
SAIDI	(6.7 min; 12.8 min)	(1.8 min; 12 min)
END	(111.53 kWh; 846 kWh)	(90 kWh; 223 kWh)
Total discounted cost	(43.7 M€; 56.5 M€)	(39 M€; 51.4 M€)
For a capacity of 47% (15 primary secured feeder branches and 25 loops)		

Table 4.6. *Comparison of secured feeder and hybrid loop structures*

4.3.1.2.2. Case study 2b: looped secured feeder structure (mesh between two different transformers)

Description

A second possible mesh architecture, without the need for additional conductors, is the looped secured feeder structure that works through the closure of the remote-controlled switches, as shown in Figure 4.33.

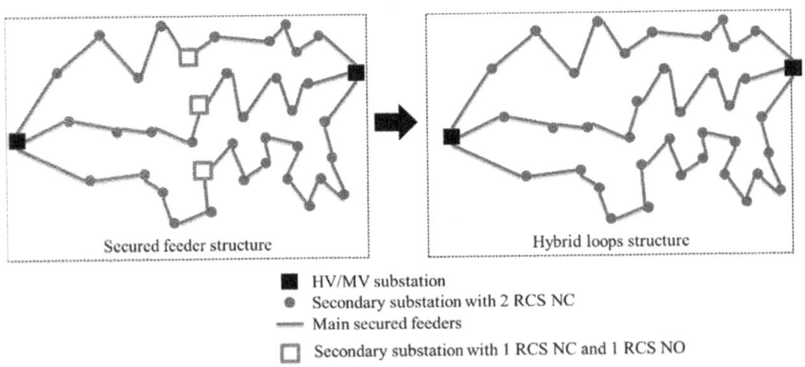

Figure 4.33. *Looped feeder breaker*

However, the looping of the distribution systems requires amending the protection plan, in particular, for the ability to locate the fault, which is then fed by the two primary substations, as well as any connected generators. To maintain the current protection plan, a mixed radial/looped operating mode is proposed. In normal operation, the system would be looped, thus allowing for better distribution of flows and a better voltage plan. On the other hand, in case of a fault, a device, called a quick delooper, rearranges the system into a radial configuration so that the existing protection plan can work. This device must therefore detect the fault before the existing system protection devices. An example of a quick delooper structure is shown in Figure 4.34.

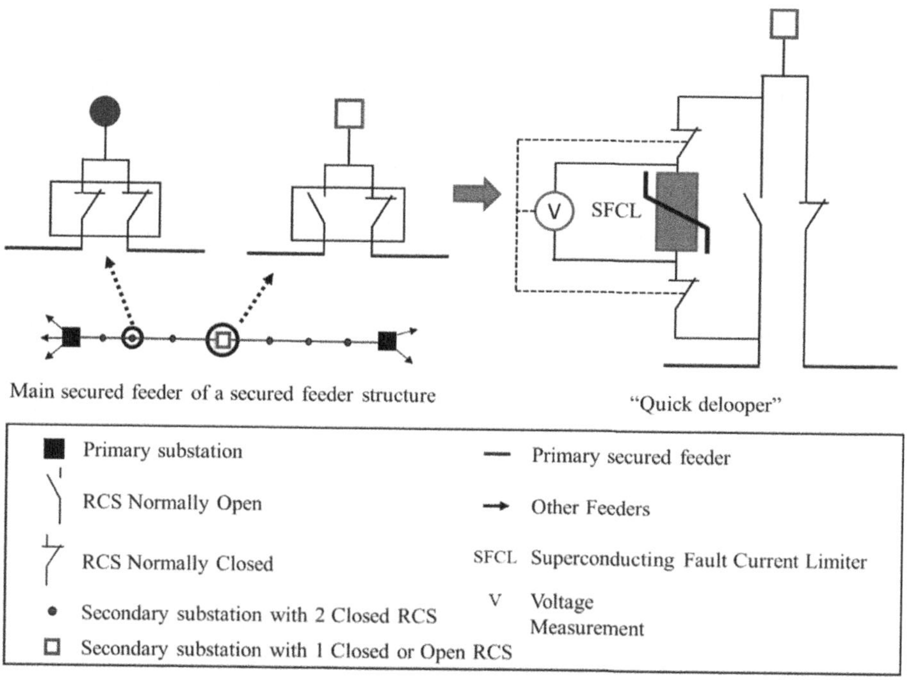

Figure 4.34. *Example of a quick delooper based on superconducting current limiters*

A superconducting fault current limiter (SFCL) would be placed in parallel to the normally open switch via two normally closed switches. In normal operation, the SFCL has an intrinsic resistance of zero, allowing the current to flow without any joule loss. In the event of a fault, the resistance of the SFCL will increase very rapidly and reach very high values so that the distribution system will behave as if it were radial. Moreover, at the looping point, the current decreases strongly and becomes easier to break. However, in the event of a fault, if the SFCL is not disconnected it may deteriorate. For this reason, a voltage measurement at the SFCL can be used to monitor the condition of the SFCL and to detect a fault current. The SFCL is then disconnected by opening the normally closed switches, letting the normally open switch take over. In order to maintain the current protection system, the SFCL must respond more quickly. As seen in Chapter 1, the protection devices do not react until 100 ms after the fault to allow time for the self-extinguishing faults to die out. Therefore, in the event of a fault, the SFCL could be disconnected 80 ms after the fault and then reconnected 15 s later to allow the fugitive and semi-permanent faults to clear. If the fault is still present, then the SFCL is permanently disconnected and human intervention will be required to loop the system back. When the SFCL has changed state[7], it takes about 1 s to recover its superconducting state. This device must therefore be sized on a case-by-case basis and depends on the existing protection system, the type of fault and the type of power system.

Applied example

This solution has been simulated on a French case study composed of seven MV feeders equally distributed on two HV/MV transformers and represented in Figure 4.35. The seven normally open switching devices that can be closed via an SFCL are circled in black-dotted lines and numbered from S1 to S7. When closed, two of them create loops on the same transformer, while five of them create a loop between the two transformers.

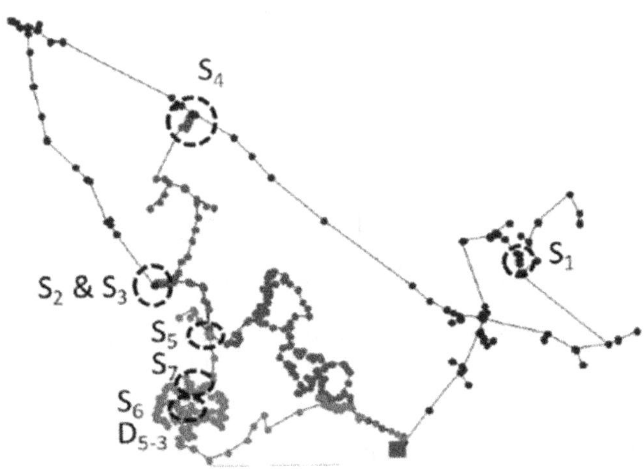

Figure 4.35. *French case study (rural system)*

The renewable generation hosting capacity of this power system, according to the method described in section 4.2.2, is 21.9% and reaches 93.5% if it is operated

7 That is to say, the current has exceeded the critical current for which it was designed.

as a closed system, that is, with all seven switches closed. In this example, it is already possible to increase the capacity by closing only one switch, S2, which allows it to reach 75% of the capacity.

These conclusions are obviously specific to this case study and cannot be generalized. However, it shows that the historical architectures, once considered optimal, can be challenged by other types of architectures that can call into question certain technological choices as soon as we add other performance criteria such as the hosting capacity of DER. Also, a cost–benefit analysis shows the economic significance of the looped secured feeder. If we consider, for example, the Grenelle criteria (27% of RE) and NégaWatt (55% of RE) in 2030, then the cost of the meshed solution is two times less expensive, despite the additional OPEX costs related to the maintenance of the SFCL.

4.3.2. *Case study 3: non-wire alternatives*

In Chapter 3, we classified the upstream and downstream NWAs that planners can use to defer or cancel a system investment. Here, we present possible methodologies and their integration into classical planning tools.

4.3.2.1. *Case study 3a: integrating load curtailment into planning to defer system investments*

The following applied case study was conducted within the framework of the GreenLys project, and we are specifically interested in a secured feeder on the Grenoble system (see Figure 4.36) supplying 29 MV/LV substations with a maximum consumption of 12.47 MW.

We study this secured feeder according to two types of evolution scenario, which present current constraints for part of the cables, and which we solve by using a consumption curtailment strategy as an alternative to classical reinforcement.

– *A deterministic scenario* over a 10-year period with a consumption evolution rate of 1.5% per year that results in the appearance of constraints from the fourth year onward, on a 247 m branch of 150 mm² aluminum line shown in red in Figure 4.36. In this scenario, there are no production units or EVs connected to the system. The discounted cost of reinforcing with a 240 mm² aluminum cable is estimated to be about 25 k€. As an alternative, we study the possibility of curtailing the consumption on the feeder when the 247 m branch starts to show signs of stress. This hypothetical scenario aims to illustrate the rebound and deferral effects, described below, specific to load curtailment, and the importance of anticipating them in the solution of the equivalent mathematical problem.

– *A set of stochastically generated scenarios* over a period of 15 years, for which the parameters for the evolution of consumption, production penetration and the number of EVs connected to the system are simulated randomly, using the projection data from the *Grenelle de l'environnement* and the *NégaWatt* scenarios as terminals (see Table 4.7 and Tables 4.1 and 4.2 in section 4.2.3.3). This approach makes it possible to statistically evaluate the potential of implementing a *demand response strategy* as an alternative to reinforcement.

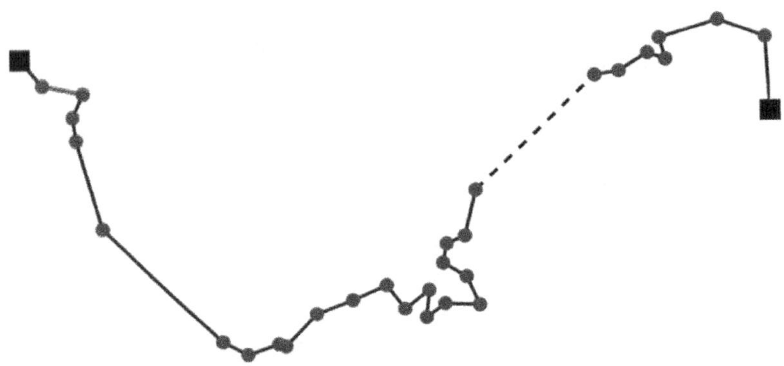

Figure 4.36. *Load shedding study – feeder branch studied*

	Grenelle scenario			NégaWatt scenario		
Periods	2015	2020	2030	2015	2020	2030
Penetration rate of renewable production	17%	24%	27%	17%	33%	55%
Number of EV	0	475	2610	0	63	995
	For the Grenelle and NégaWatt scenarios					
Periods	2015–2020			2020–2030		
Expense growth rates	0.96%/year			0.87%/year		

Table 4.7. *Boundaries for the generation of the parameters of the random scenarios*

4.3.2.1.2. Modeling of the load curtailment strategy

In this case study, we are interested in the partial direct load curtailment program, which consists of reducing the power lost by curtailing some points of consumption for a limited period of time. This type of load curtailment has two side effects:

– *the rebound* is a power peak which corresponds to the recovery in consumption for the loads that were cut, which may be greater than the power curtailed (motor start current, human reactions such as increasing the heating or air-conditioning set point after a period of shutdown). It is defined by the rebound rate $\tau_{rebound}$ expressed as a % of the curtailed power;

– *the deferral* is the energy that should have been consumed and is consumed during a deferred period. It is defined by the deferral time $T_{deferral}$ and the deferral rate $\tau_{deferral}$ expressed as a % of the energy curtailed. These effects are illustrated in Figure 4.37.

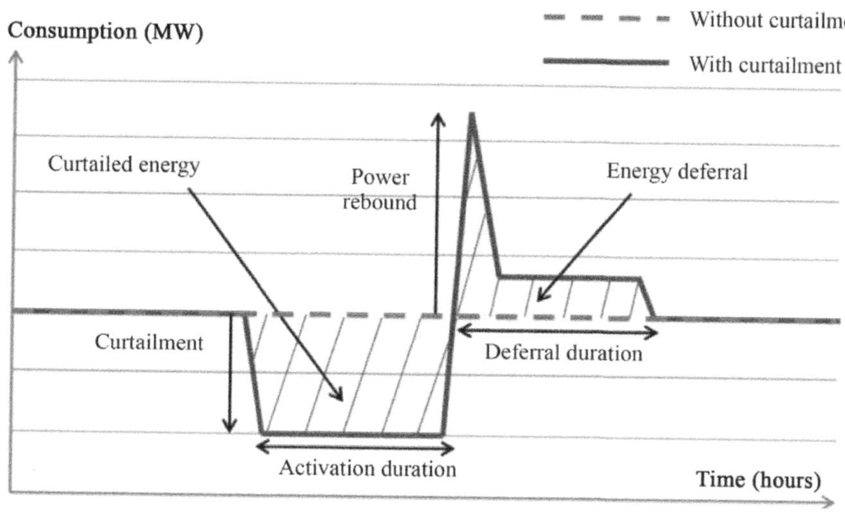

Figure 4.37. *Schematic diagram of curtailment*

The value of the parameters $T_{activation}$, $\tau_{rebound}$, $T_{deferral}$ and $\tau_{deferral}$ depends on the type of customers and electrical appliances cut off. For example, the deferral is 100% of the curtailed energy for hot water, but it does not have the rebound phenomenon. In this case study, we consider that the load curtailment takes place for 1 h, with a rebound of 140% for 30 min and a deferral of 100% of the curtailed energy for 2 h.

The mathematical method proposed to apply a load curtailment strategy is the solution of a linear optimization problem via the simplex method, ensuring the best solution is found without getting stuck by a local optimum. The mathematical form of the problem is described in Chapter 5.

The study is carried out by successive load flow calculations over the study period, with a time step of 30 min. The consumer and production load curves are constructed from the profiles provided by RTE to those responsible for balance, distributed on each MV/LV substation in proportion to the number of customers of each type (residential, professional, business, public lighting, PV generation, wind generation, etc.).

4.3.2.1.3. Ten-year deterministic scenario

Figure 4.36 enables us to visualize the electrical branch which starts to exhibit constraints from the fourth year onward. A first optimization is performed without taking into account the rebound and deferral phenomena in the modeling of the problem to be optimized. Figure 4.38 shows in green the number of hours of constraint per year over the entire study period, and in blue the value of the maximum excess current as a percentage, with and without curtailment. We see that depending on the year, the peak current is either increased or decreased, but the occurrence of constraints is always decreased (from 169 to 127 h in the last year).

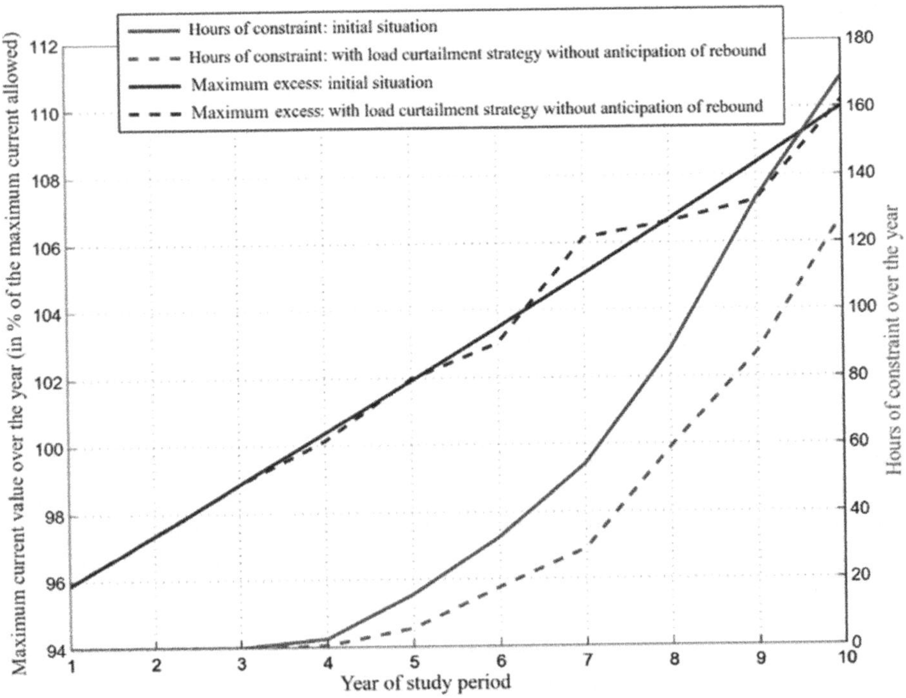

Figure 4.38. *Evolution of the constraint over time with curtailment strategy*

The constraints on the opposite feeder are lower and therefore require less curtailment, in terms of power or energy, as shown in Figure 4.41.

We can also see that the location of the flexibility requirements changes after reconfiguration, as the constraints are then shifted and reduced, which also results in less energy being curtailed. We can also derive maximum flexibility costs below which the load curtailment strategy becomes more attractive than reinforcement.

The cost of reconfiguration is assumed to be negligible because the switching devices are changed only once over their lifetime, estimated at several thousand operations. Similarly, the reduced cost of joule losses associated with curtailment is not considered.

Table 4.8 shows these maximum costs with and without initial reconfiguration, per unit of power and energy. Indeed, the flexibility remuneration mechanisms can integrate both dimensions. We can see that these costs are rather high, leading to the conclusion that the use of flexibility becomes a potentially interesting option.

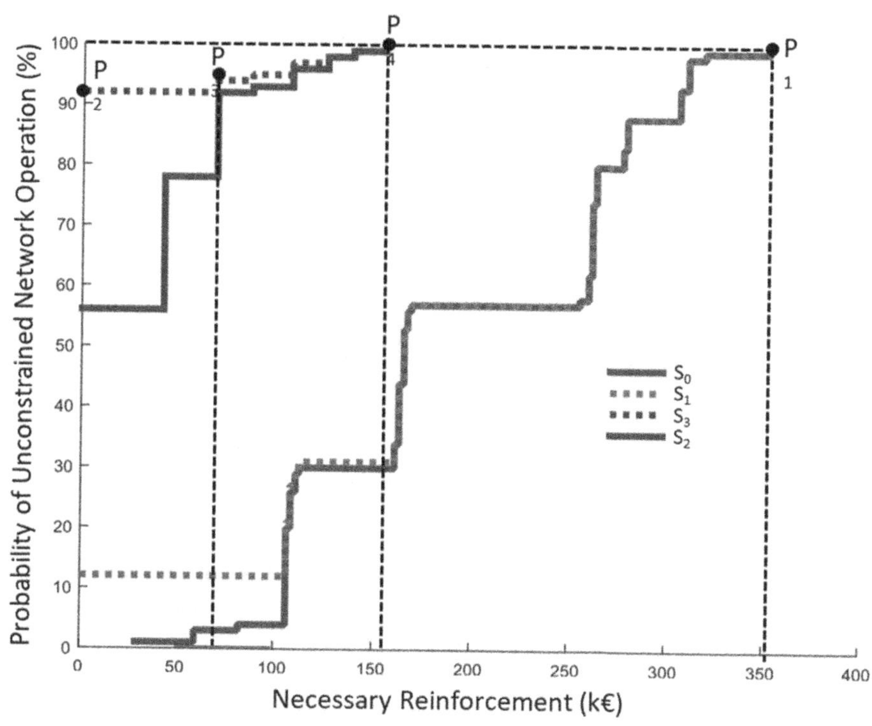

Figure 4.40. *System operating rate depending on the chosen strategy and investment*

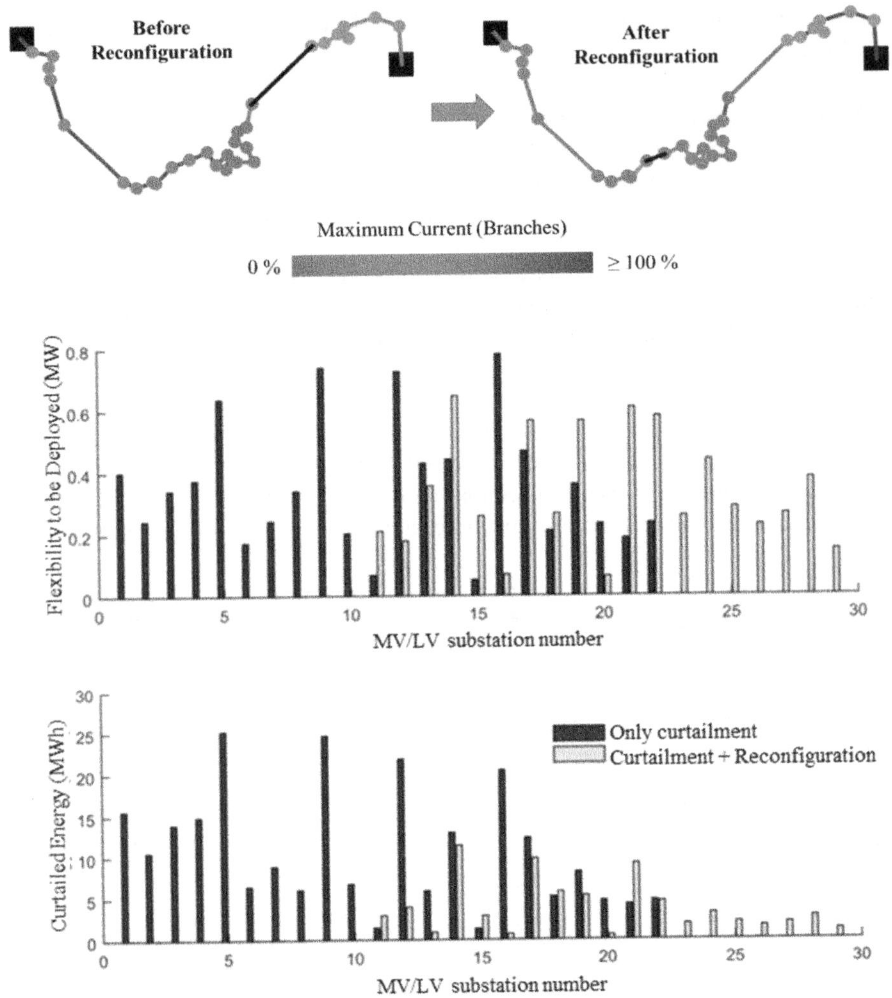

Figure 4.41. *Curtailment strategy + reconfiguration*

	Curtailment without initial reconfiguration	Curtailment with initial reconfiguration
Maximum deployment costs	43 k€/kW 16 k€/kWh	27 k€/kW 10 k€/MWh

Table 4.8. *Maximum flexibility costs below which the curtailment strategy is more attractive than reinforcement*

4.3.2.2. *Case study 3b: integration of generation curtailment to postpone a primary substation investment*

4.3.2.2.1. Case study description[8]

Here, we consider two primary substations whose characteristics are summarized in Table 4.9. For each of the primary substations, thanks to the S3RENR (see Chapter 2), we can define scenarios for the arrival of producers, also defined in Table 4.9. The aim is to calculate whether the primary substation needs to be reinforced and, if so, to determine whether a curtailment contract with these same producers could cancel the need for reinforcement. In these case studies, two different types of generators are considered, which have the advantage of allowing for easy-to-use graphical representation. In the case of three or more generators, the same approach is possible but the graphical representation will be more complex, if not impossible.

Primary substation number 1: maximum capacity of 20 MW							
Maximum connected consumption 15 MW							
Cumulative installed generation capacity per year by type (MW)							
Years	**Year 0**	**Year 1**	**Year 2**	**Year 3**	**Year 4**	**Year 5**	**Year 6**
Photovoltaic	0	0.3	3	7	11	15	19
Wind	0	0	0	0	0	11	21

Primary substation number 2: maximum capacity of 20 MW						
Maximum connected consumption 8 MW						
Cumulative installed generation capacity per year by type (MW)						
Years	**Year 0**	**Year 1**	**Year 2**	**Year 3**	**Year 4**	**Year 5**
PV	1	1	1.5	2.5	3.2	4
Hydroelectric	19	20.5	22	22	22	22

Table 4.9. *Scenarios for the arrival of producers at the two primary substations under consideration*

The decision to reinforce the primary substation according to the traditional method consists of considering the two worst-case scenarios that are most likely to occur on the system: maximum consumption (P_{max}^{cons}) – minimum generation (P_{min}^{prod}) and minimum consumption (P_{min}^{cons}) – maximum generation (P_{max}^{prod}). Since the systems have historically been sized for consumption, the constraint to be checked is the second one. The boundary between the "non-reinforcement" and

8 This work was carried out within the framework of the Enedis Industrial Chair on Smart Grids.

"reinforcement" zones, shown in Figure 4.42, is therefore defined by equation [4.8], where P_{max}^{PS} is the maximum allowable power of the primary substation and P_1, P_2 are the sums of the maximum powers produced by the Type 1 and 2 generators. The objective of this case study is to estimate the intermediate boundaries for "curtailment" and "curtailment + light reinforcement", as illustrated in Figure 4.43, by quantifying the risks when choosing the alternatives for curtailment, with or without reinforcement. It is therefore first necessary to model the behavior of the generators so as to define the 1–2, 2–3 and 3–4 boundaries.

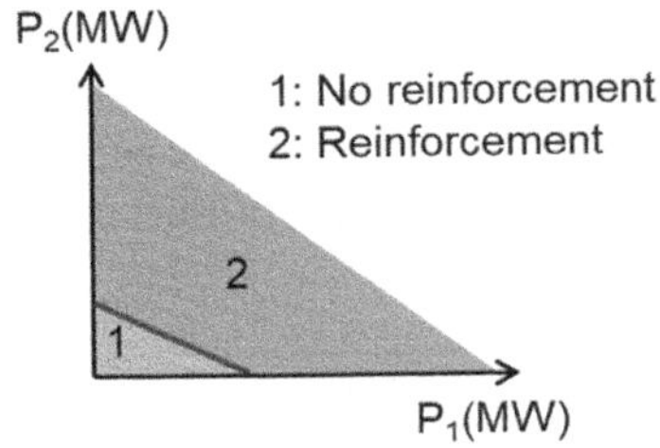

Figure 4.42. *Traditional "reinforcement" – "non-reinforcement" boundary*

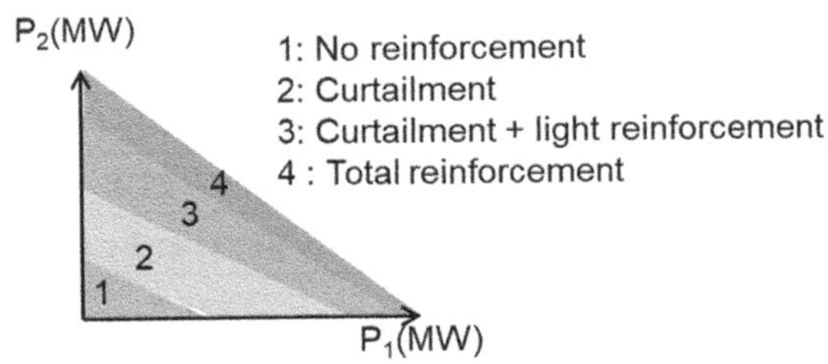

Figure 4.43. *Intermediate borders with curtailment*

$$P_2 = P_{max}^{PS} + P_{min}^{cons} - P_1. \quad [4.8]$$

4.3.2.2.2. Probabilistic modeling of producers

The different distributed generators (wind, PV, hydroelectric, and cogeneration) do not have the same behavior or correlation, that is, the degree of dependence on each other and with regard to the load.

In this case study, 4,254 generation readings (514 cogeneration, 1,202 hydroelectric, 806 wind, and 676 PV), at 10-minute time intervals for 2 years (2012 and 2013), and six primary substation load curves made it possible to identify:

– the seasonality effects for each type of generation, as summarized in Table 4.10. In the example of primary substation 1, where photovoltaic and wind generators are planned, a year can be divided into 12 seasons: three seasons for wind, and for each wind season there are four photovoltaic seasons. Therefore, we can consider that each season is represented by a monotone average production;

	Winter (from November to March included)	**Spring (from April to June included)**	**Summer (from July to October included)**
Hydroelectric	Undetermined	Undetermined	Weak
Wind	Large	Large	Weak
Cogeneration	Active	Inactive	
Photovoltaic	Four solar seasons on a daily scale: $\begin{cases} \vec{R}\cdot\vec{u} \in [-1;0] \text{ (night)} \\ \vec{R}\cdot\vec{u} \in [0;0.5]\ (low - angled\ sun) \\ \vec{R}\cdot\vec{u} \in [0.5;0.7] \text{ (day)} \\ \vec{R}\cdot\vec{u} \in [0.7;1] \text{ (sun near its zenith)} \end{cases}$ With $\vec{R}\cdot\vec{u}$ = horizontal solar projection		

Table 4.10. *Division into "seasons" of MV production*

– effects of dependencies between certain types of production/production and production/load, as summarized in Table 4.11. We can see that there are three different types of dependency patterns, defined in more detail in section 4.2:

- production of the same type:

  – model 1: model without expansion (wind and photovoltaic cases),

  – model 2: independent model (winter cogeneration case),

  – model 3: Gaussian copula model (hydroelectric case);

- production of different types and production/load:

  – model 2: independent model (except for cogeneration/photovoltaic and cogeneration/hydroelectric),

  – undetermined: cogeneration/photovoltaic; and cogeneration/hydroelectric.

	Wind	Photovoltaic	Hydroelectric	Cogeneration (winters[9])	Load
Wind	Model 1	Model 2	Model 2	Model 2	Model 2
Photovoltaic	Model 2	Model 1	Model 2	Undetermined	Model 2
Hydroelectric	Model 2	Model 2	Gaussian copula model	Undetermined	Model 2
Cogeneration (winters)	Model 2	Undetermined	Undetermined	Model 2	Model 2

Table 4.11. *Dependency between types of production and between production and load*

4.3.2.2.3. Methodology for determining decision boundaries

In order to determine the boundaries defined in Figure 4.43, the methodology applied is as follows:

– Step 1: for each pair (P_1, P_2), we consider the average and at-risk distribution functions arbitrarily fixed at 10% for the two producers in the case study;

– Step 2: calculation of the average and risk distribution function is at 10% of the primary substation;

– Step 3: calculation of the energy required to be curtailed to avoid reinforcement using the last in first out (LIFO) method, which is the current method used by the French DSO (see Chapter 2). In the event of constraints, an order is sent to the generator to reduce its production. The minimum time interval between two orders is 30 min, and the energy erased is remunerated at a cost fixed in this study at 70 €/MWh;

– Step 4: calculation of the economic decision boundaries:

- threshold for the occurrence of constraints (traditional methods): we apply equation [4.8] to our case studies,

- constraint appearance threshold (probabilistic methods): we apply equation [4.8] but instead of considering the minimum consumption and maximum production, we consider the probability density functions of the producers,

- curtailment/reinforcement boundary as seen by the DSO (average): when the cost of curtailment is on average higher than the cost of reinforcing the substation (here the addition of a 10 MW transformer),

9 Winter is the only season when the cogeneration is active. We do not consider the other seasons.

- DSO's view of the load curtailment/reinforcement boundary (10% risk): the probability that the cost of load curtailment is higher than the cost of reinforcing the substation is 10%.

4.3.2.2.4. Results

Figure 4.44 shows the different decision curves obtained.

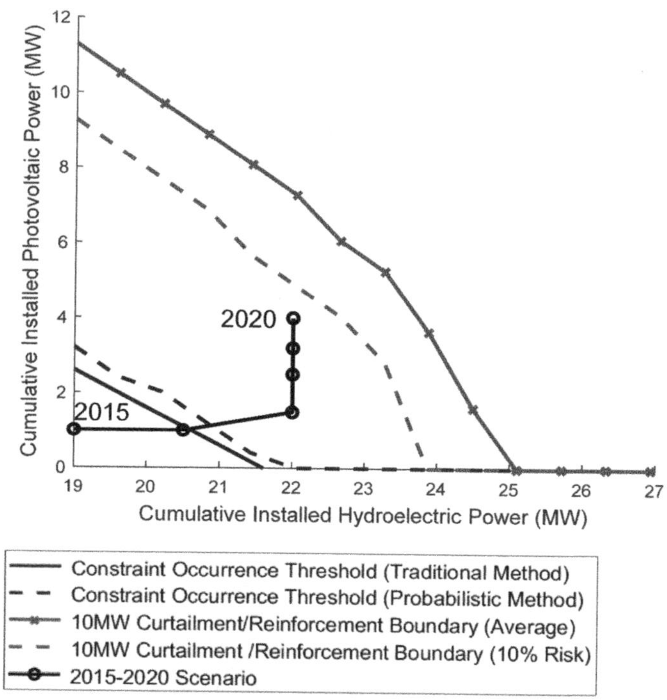

Figure 4.44. *Decision chart between reinforcement and curtailment of primary substation No. 1*

We can see that the threshold for the appearance of constraints through the traditional and probabilistic methods is very close, showing that the hypothesis of considering the minimum consumption–maximum production sizing case is realistic. Then, we can see that as of 2016, the primary substation reaches its limit and either the substation must be reinforced or generation must be curtailed. In case study no. 1, the curtailment solution is the best, with very low risk as of the last scenario in 2020, remaining below the curtailment/reinforcement boundary, set by the DSO with a risk of 10%.

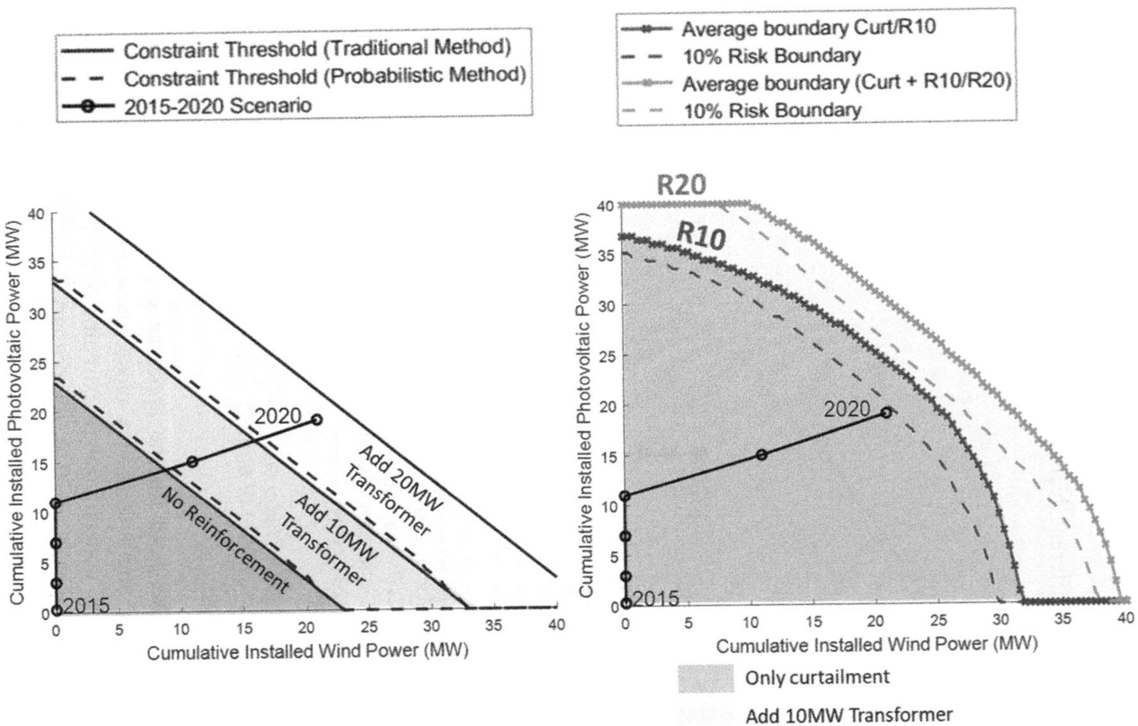

Figure 4.45. *Decision chart between reinforcement of primary substation No. 2 and a curtailment contract with R10 (and respectively, R20) for reinforcement with a 10 MW (respectively, 20 MW) transformer and "Curt" for curtailment contract*

Figure 4.45 shows the decision curves obtained for primary substation No. 2 with the classical method (left) and the probabilistic method (right). The classical method indicates that the primary substation should be reinforced with the addition of a 20 MW transformer. In the figure on the right, two types of boundaries are shown:

– medium boundary (curt/R10): decision boundary between a curtailment contract with the producers, or substation reinforcement with a 10 MW transformer;

– medium boundary (curt + R10)/R20: boundary between reinforcing the substation by adding a 10 MW transformer and setting up a curtailment contract with the producers, or reinforcing the substation with a 20 MW transformer.

The dashed curves represent these same boundaries but considering a 10% risk. We can see that the reinforcement of a 10 MW transformer and a curtailment contract are sufficient. We notice that contrary to primary substation No. 1, the 2020 point is very close to the 10% risk boundary, so the DSO has less margin of certainty. Nevertheless, since the 2019 point is very far from this boundary, the DSO may choose to defer its investment until it has greater visibility and there are more connection requests in future.

4.3.2.2.5. Interest in probabilistic modeling

An alternative method for finding these boundaries is to use the Monte Carlo method presented in section 4.2.3 (see Figure 4.46). This methodology entails:

– randomly simulating a curve for hydroelectric and solar production over 1 year (for case study 1) and a curve for wind and solar production (for case study 2) previously normalized, and then multiplied by the installed capacity of the scenario being considered;

– calculation of the power of the primary substation using the formula in equation [4.7] and determining the necessary curtailment.

Results from Monte Carlo simulations and the probabilistic approach provide similar results. A comparison of the available stochastic approach is discussed in Chapter 5.

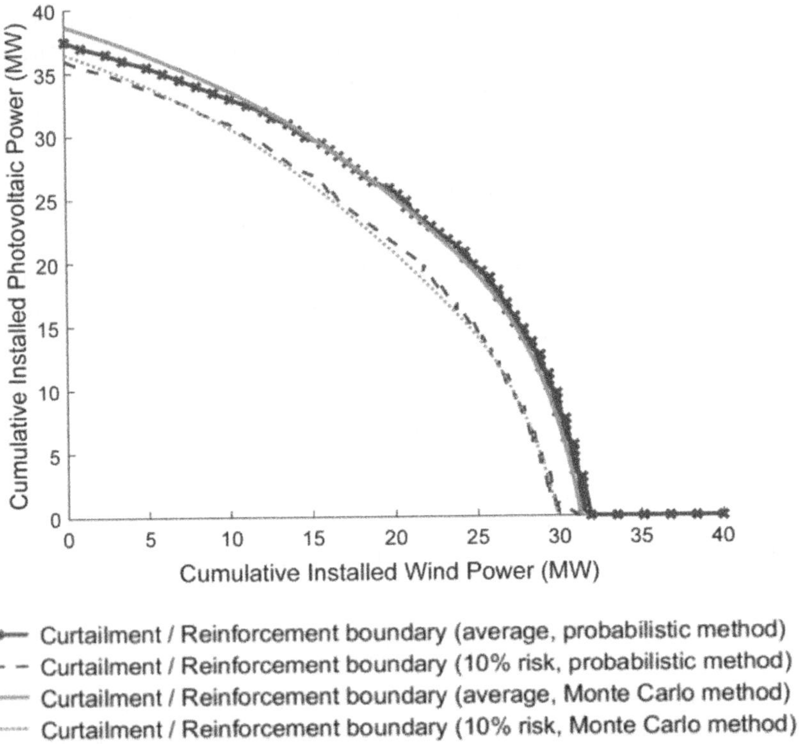

Figure 4.46. *Comparison of curtailment and reinforcement boundaries with a Monte Carlo and probabilistic method*

4.3.2.3. *Case study 3c: impact of EV with and without a curtailment strategy – California*

4.3.2.3.1. Description of case study

The objective of the analysis conducted by Coignard et al. (2019) is to determine whether further developments are necessary to accommodate the insertion of EV onto an adapted system, or whether charging stations can be installed on existing systems without implementing charging control structures. This study evaluates the impact of broad EV penetration on California distribution systems in PG&E's service area. The working assumption is that one electric car is added per housing unit. This analysis includes 39 independent distribution systems with different topologies. Although half of the systems primarily serve residential customers, we also find services for commercial, industrial, and agricultural sites. From a topological point of view, the length of the systems varies from 2 to 64 km, with 68% of the systems under 10 km. On the other hand, the systems are mainly 12 kV. Temporal consumption data from 2016 tell us that 80% of the systems have a peak consumption between 3 and 15 MW. Typically, for systems dominated by residential (where housing accounts for at least 50% of demand), 4,000–8,000 EVs are added as per the number of housing units at each node. The methodology developed in the following section is inspired by the way PG&E conducts its interconnection studies. The impact of EV charging is measured in terms of deviations from a nominal voltage, maximum current in each line and maximum transformer station capacity.

4.3.2.3.2. Modeling method

The modeling of the problem covers the temporal and spatial aspects related to the dispersion of EVs and loads (on the system and over the year). The spatial distribution of EVs is directly obtained from our working hypothesis of one EV per home, the placement of the homes being included in the system topology. However, we note that, as in the other cases, a Monte Carlo approach could have been used to overcome the uncertainty in the spatial distribution of EVs. To determine the temporal consumption profiles from EVs, various scenarios are simulated. In order to be realistic, the simulations are based on traffic data including a typical day of use for 2,952 cars (U.S. Department of Transportation 2009). These routes define when, how many kilometers and to which destinations vehicles travel. This allows us to estimate the energy consumption of the trips and thus the state of battery charge over the course of the day. Once the need for charging has been determined, the charging power (in kW) and the place of charging still need to be determined. EVs can be charged at work, in public places (restaurants, parking lots) or at home with chargers ranging from traditional 1.4 kW plugs to dedicated 120 kW DC chargers. Several scenarios (low, medium, high) are explored, with variations in the number of EVs charging at a workplace, and a variation in the number of EVs having access to a 6.6 kW charger at home (the rest having access to a traditional 1.4 kW charger anyway).

In the most pessimistic scenario (high), EVs only charge at home with a 6.6 kW charger, leading to a peak of 175 kW for 100 EVs. In contrast, in the lowest scenario (low), all EVs have access to a charger during the working day and only charge at 1.4 kW in the evening, leading to a lower residential peak demand of 68 kW (again for 100 EVs and due to the copious amount charging). Finally, the median scenario is constructed so that 50% of the EVs charge during the day, and 50% of the residential chargers can deliver 6.6 kW (the rest at 1.4 kW). This scenario, which is retained for the remainder, leads to a peak residential consumption at 6:30 p.m. of

100 kW per 100 EVs (results obtained from the 2,952 normalized routes per 100 vehicles). We note that we would have expected the "high" scenario to lead to a peak of 660 kW (100 EV × 6.6 kW). In reality, the peak demand is never that dramatic due to the fact that EVs do not all arrive at the same time and only need about 51 min to charge (at 6.6 kW) after a 37 km journey.

For the distribution system, from a temporal perspective, the Supervisory Control and Data Acquisition (SCADA) time series data measured over 365 days at the primary substation for each of the 39 systems are reduced to only 4 days, with the highest peak demands being between 15h00 and 23h00 for each system. This ensures that we simulate days that are representative of power peak coinciding with EV demand, and thus obtain the largest deviations in allowable voltage and current. A load allocation method is then used to vary the power demand of the loads on the system in proportion to their energy consumption over the year, such that the sum of the powers for all the nodes on the system, plus the joule losses, correspond to the power demand recorded at the primary substation at any time. This process allows the power demand at each node to vary over time, from a single time series.

4.3.2.3.3. Results

The impact of EVs is measured using three indicators: the remaining capacity at the primary substation (defined as the maximum power allowed by the transformer minus the maximum power demand including EVs), the lowest voltage recorded at a node and the maximum line current recorded as a percentage of the maximum allowable line current shown in Figure 4.47 as a function of the percentage of residential consumers in the 39 systems studied. We can see that at a minimum, 12 of the 39 systems are affected by current limits being exceeded after adding one EV per house. We note that the impact is all the more important when the system is already close to the limit and when the proportion of homes is high. The systems are not always limited by the same aspect. Regarding the other indicators, 15 systems exceed the maximum capacity of the secondary substation, and three systems have low voltage constraints (Coignard et al. 2019). The systems affected by these three different types of constraints are not necessarily the same. For example, system 16 does not have line current constraints but does have constraints on the capacity of its primary substation. Similarly, system 36 does not exceed the current constraints; however, the maximum voltage drop goes beyond the tolerated 5% (Coignard et al. 2019).

To understand the value of direct or indirect control (through economic incentives), we explore the possibility of shifting EV demand to minimize peak consumption. Previously uncontrolled EVs have their demand clipped between 18h00 and 06h00, as shown in Figure 4.48.

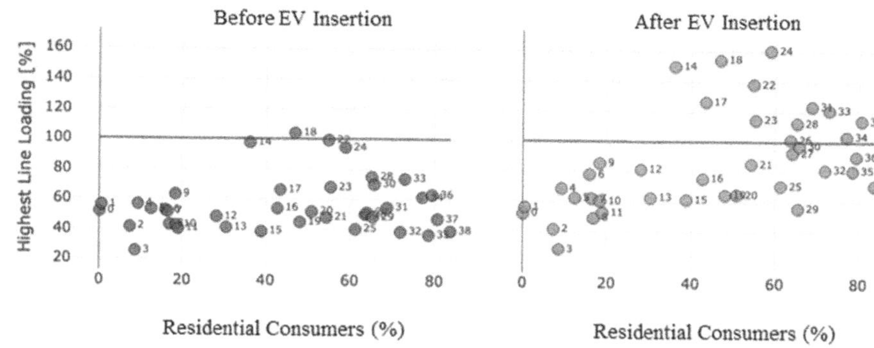

Figure 4.47. *Highest[10] line current as a function of residential customers sharing 39 distribution systems, before (blue) and after (orange) adding one EV per house*

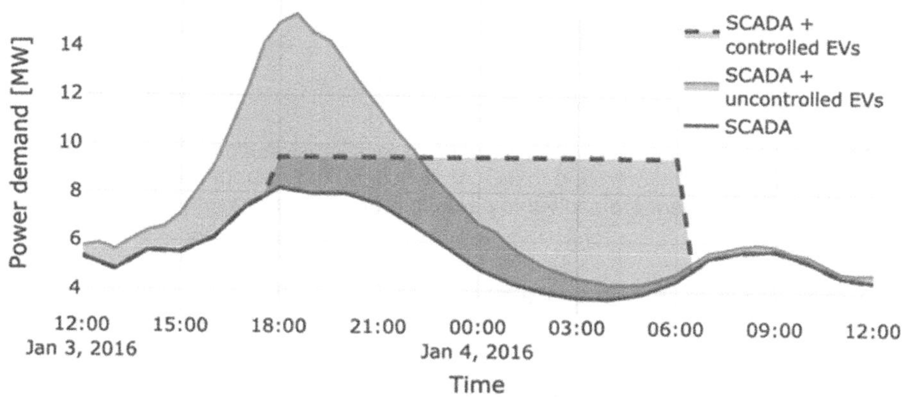

Figure 4.48. *Power demand without EVs (blue), with EVs without control (orange), with EVs controlled to minimize peak power (red)*

This approach assumes perfect knowledge and control of the system. In reality, not all EVs can participate in this control mechanism, consumption forecasts are not perfect and communication systems are subject to delays. In fact, the reduction in peak demand shown in Figure 4.48 is very optimistic. Of the 39 systems available, 19 of them have more than 50% residential consumption (see Figure 4.47). Figure 4.49 shows the increase in peak consumption with EV as a percentage of peak consumption without EV as a function of these 19 systems. We can see that 11 systems are subject to an increase in peak consumption, even in this perfectly

---

10 Expressed as a percentage of the maximum allowable line value.

controlled scenario. However, we note that this increase in peak consumption is limited to 27% of the peak without EVs (the case of system 26) in the worst-case scenario, with an average of only 8%. This is better than the no-control scenario, with a potential 104% increase in peak power (the case of system 36) in the worst-case scenario, with an average of 64% across all residential systems.

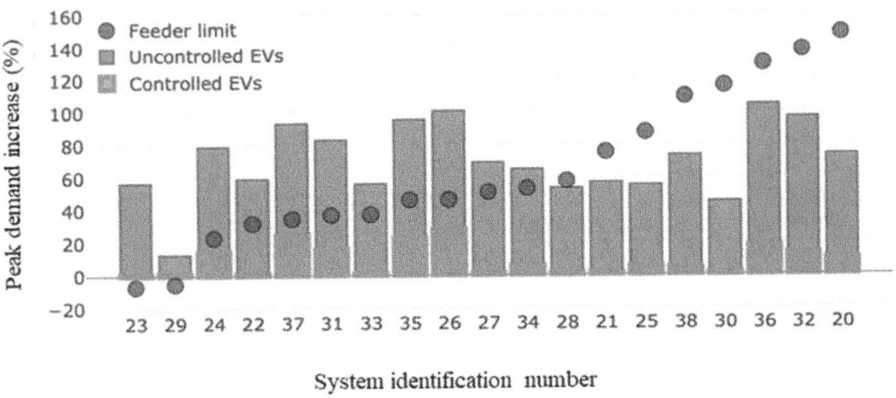

Figure 4.49. *Increase in peak consumption for 19 distribution systems when EVs are left uncontrolled (blue), or are controlled (orange). The red dots mark the allowable peak increase for each system*

4.4. Rural interconnected systems

Rural systems, characterized by low density consumption, have more reduced architectural possibilities and are very expensive, especially for those with lower voltages, where the system is radial and tree-like and has little or no possibility of recovery via normally open switching devices. The application of flexibility here is therefore particularly interesting.

4.4.1. *Case study no. 4: NWA to integrate DERs into LV rural distribution systems*

4.4.1.1. *Applied example: SMAP project*[11]

The SMAP project (CRE 2021), for SMArt grids in natural Parks, aims to facilitate the insertion of PV on the LV system through the experimentation and

11 This work was carried out within the framework of the Enedis Industrial Chair on Smart Grids.

simulation of smart grid "solutions". The study system is rural with rather low consumption and is located in a regional natural park that imposes certain restrictions, such as the prohibition of ground-mounted photovoltaic plants. It is also located in a TEPCV (*Territoire à énergie positive pour la croissance verte* – Positive energy territory for green growth), which bestows strong incentives to promote the insertion of PV on the territory. This case study is particularly interesting because the CVRC (*Centrales villageoises de la région de Condrieu*), a company regrouping village power plants, had to decrease the installed power of its PV installations because of the connection costs which became prohibitive following the resulting system reinforcements. Considering the scenarios of consumption and production evolution established by the project partners, the question of the cost–benefit analysis of flexibilities in comparison with reinforcement to increase the share of RE in the system was raised.

4.4.1.2. *Flexibilities studied*

In this case study, four types of flexibilities are compared to traditional system reinforcement: the addition of an on-load tap changer at the secondary substation, fixed and dynamic reactive control of PV panels, and limiting PV generation.

4.4.1.2.1. Addition of an on-load tap changer at the MV/LV transformer

As seen in Chapter 1, MV/LV secondary substations are equipped with off-load tap changers, that is, the transformer must be disconnected from the electrical system for a technician to manually change the tap. In France, there are three positions +2.5%, 0% and −2.5% (sometimes five, with the positions −5% and +5% in addition), but the taps are rarely changed because of the resulting service interruption. In this case study, the voltage constraints posed by PV generation justify our interest in installing an on-load tap changer in place of the off-load tap changer. The proposed on-load tap changer has nine taps and varies the voltage by 1.5% intervals.

4.4.1.2.2. Reactive control of PV inverters

In MV, reactive control of generation resources is already used and tested (Braun 2008; Demirok et al. 2009; Turitsyn and Šulc 2010; Delille et al. 2013; Minaud et al. 2013). For LV systems, which are by nature mainly resistive, many studies suggest the use of active power curtailment instead (Lopes 2008; Gaudin et al. 2012; Minaud et al. 2013). Two types of reactive controls are studied here:

– *a control with* $\tan\varphi$ *fixed at −0.4*: the producer permanently absorbs 0.4 times the power they produce. This solution has the advantage of being simple to implement and without additional cost, but generates additional losses in the LV system and encourages the producer to oversize their installation in order to ensure they are able to inject the maximum power into the system;

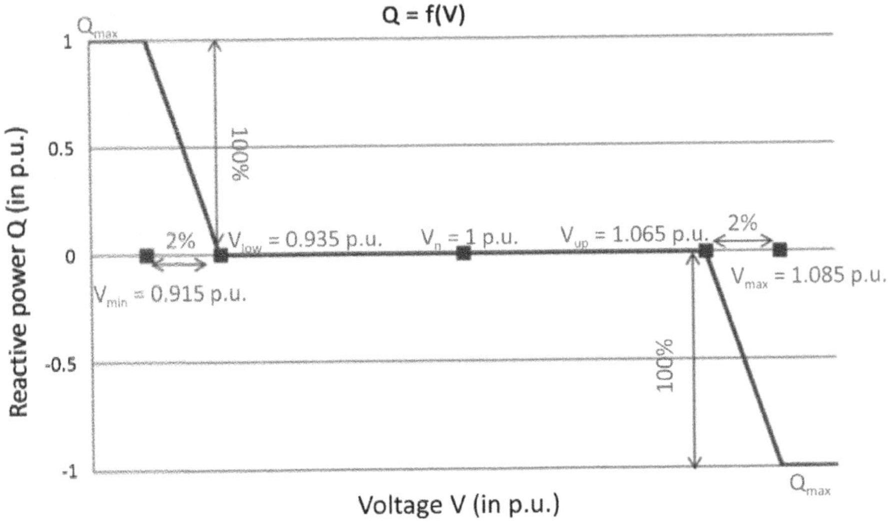

Figure 4.50. *Reactive control used*

– *a dynamic control called "Q(U)", with priority to the reactive power*: that is, if the reactive power set point given by the control causes the maximum apparent power of the inverter to be exceeded, then the active power of the inverter is reduced. The control characteristic used is given in Figure 4.50. If the voltage measured at the output of the inverter is between V_{low} and V_{up}, then the reactive power of the PV inverter remains at 0. On the other hand, if the voltage leaves this range, then:

$$Q(V_{measured}) = \frac{-Q_{max}}{V_{low}-V_{min}} \times (V_{measured} - V_{low})\ si\ V_{measured} < V_{low},$$

$$Q(V_{measured}) = \frac{Q_{min}}{V_{up}-V_{max}} \times \left(V_{measured} - V_{up}\right)\ si\ V_{measured} > V_{up}.$$

4.4.1.2.3. Limitation of PV production

It is possible to limit the PV power produced by clamping the inverter, in our case to 70% of the nominal power produced by the PV.

This 30% power reduction may seem large, but in reality it only occurs a limited number of times in a year and over a short period of time.

To illustrate this, Figure 4.51 plots the normalized monotone of estimated PV generation from 2007 to 2017 for the village of Les Haies at 15-min intervals. Over each 15-min period, the maximum power output of each photovoltaic panel is normalized by the theoretical power rating of the panel. With the available data, we obtain 160,512 values, which are then ranked in descending order so as to obtain a power monotone.

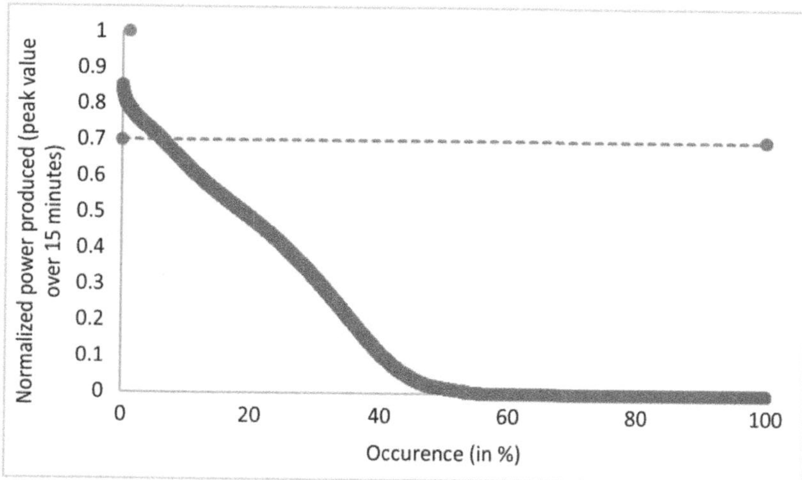

Figure 4.51. *Monotone (in blue) of production peaks over 15-min intervals from 2007 to 2017 inclusive, Les Haies and limited at 70% (orange)*

We can see that only 6.5% of the time, production is higher than 70% of the nominal power. Moreover, according to Solargis (2020), the average solar production in the north of France varies from 1,022 to 1,168 kWh/kWp, while in the South it varies from 1,461 to 1,607 kWh/kWp, which yields an average north/south ratio of 71.4%. By clipping the maximum power to 70%, we inject 98% of the initial energy, which is still well below a PV installation located in the north of France.

4.4.1.2.4. Methodology

In order to evaluate the technical and economic benefit of the four flexibilities solutions considered as an alternative to reinforcement, it is necessary to reflect on time-stamped consumption and production curves, and not only with the two most constrained sizing cases (maximum consumption/minimum production and minimum consumption/maximum production), in order to be able to evaluate the operating costs of these solutions. A large number of uncertainties require the adoption of a stochastic rather than a deterministic method. Indeed, in our case, the LV power system data are known (which is not the case for all the case studies); however, sources for uncertainties are of two kinds:

– *uncertainties related to production*: respect of an annual objective between 50 and 100%, location, connection mode (three-phase or single-phase with choice of the connection phase), production curves and orientation;

– *uncertainties related to consumption*: connection phase (when not in the database), consumption curves, electrical equipment of each consumer.

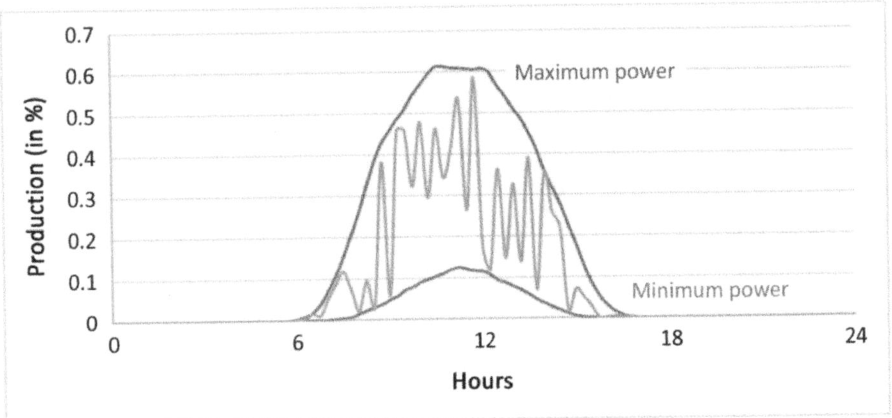

Figure 4.52. *Example of maximum and minimum production curve for the fall season and example of a random simulation curve between these two borders (in green)*

Therefore, for a given system, the uncertainties on generation and consumption must be modeled, then a *load flow* calculation must be simulated in order to evaluate the steady state current and voltage constraints and determine the flexibilities and/or degree of reinforcement needed. Since the load flow equations are nonlinear, a Monte Carlo approach, as presented in section 4.2.3, was chosen. A "high" scenario (high generation, low consumption) and a "low" scenario (low generation, high consumption) were defined in the SMAP project. These scenarios set targets for 2050, in terms of increased and/or decreased consumption and production, compared to 2015. Therefore, for each of these two scenarios, a Monte Carlo simulation has been carried out. A Monte Carlo test consists of the following:

– Step 1: assignment of producers and consumers in the LV system. For each system node, production and consumption are randomly simulated as well as the connection phases until the objectives set by the studied scenario ("high" or "low") are respected.

– Step 2: assignment of generation and consumption curves. To reduce the simulation time, rather than considering annual curves of a 30-min interval, representative daily curves were used:

- producers: a database provided by HESPUL (11 years with several orientations) has been used to define four PV seasons (spring, summer, autumn and winter) with a minimum production curve (corresponding to overcast weather) and a maximum production curve (corresponding to sunny weather) for each. A random curve is simulated between these two boundaries, as shown in Figure 4.52;

- consumers: for each season, two types of days were identified: work days and weekend days. From a database of eight day–night curves of 2016, provided by Enedis, and a table of probabilities for the evolution of electrical appliances, defined by AURAEE, equivalent consumption curves have been defined for 2050.

Therefore, one year will be represented by four PV production curves (one per solar season) and two consumption curves (work days and weekend days). This makes it possible to go from 8760 × 2 = 17,520 time intervals (30 min scale) to 24 × 2 × 8 = 384 for each Monte Carlo simulation (eight combinations of load and generation curves over 24 h at 30-min time intervals).

– Step 3: calculation of load distribution over the 35 years studied at a 30-min time interval.

– Step 4: simulation of possible actions by the DSO. In case of constraints, activation of the flexibility mechanism, or reinforcement (if the flexibility mechanism alone is not sufficient, a reinforcement is performed in addition).

– Step 5: evaluation of the solutions' performance indicators. For each solution, the following criteria are calculated:

- system infrastructure development rate, system reinforcement ratio (SDR), given by equation [4.9]:

$$SDR = \frac{\sum_{k\in\Omega_{L_r}} L_k}{L_{tot}}, \qquad [4.9]$$

with:

– Ω_{L_r}: all the lines that have been reinforced or added;

– L_k: the length of line L_k such that $k \in \Omega_{L_r}$;

– L_{tot}: the total length of the studied system;

- calculation for the benefit of the studied solution compared to the reference solution (equation [4.11] through to equation [4.12]).

$$Profit(n) = \frac{C_s(n) - C_0(n)}{C_0(n)}, for\ s \in [0;4], \qquad [4.10]$$

with: $C_s(n)$: the total discounted cost of solution No. s in year n back to year 0 ($s = 0$ for the reinforcement reference solution, $s = 1$ for the dynamic reactive control of the PV inverters, $s = 2$ for the generation curtailment, $s = 3$ for the on-load tap changer, and $s = 4$ for the fixed reactive control of the PV inverters)

$$C_s(n) = \frac{1}{1+i^{(n-2016)}}\left(C_{CAPEX}(n) + C_{OPEX}(n)\right), \quad [4.11]$$

$$C_{CAPEX}(n) = \left(C_{reinforcement}(n) + C_{flexibility}(n)\right) \times \frac{2050-n+1}{T_{AM}}, \quad [4.12]$$

$$C_{OPEX}(n) = C_{losses} \times E_{losses}(n) + C_m(n) + C_{ni}(n) \times E_{ni}(n), \quad [4.13]$$

- total discounted cost of the solutions given by equation [4.14].

$$C_s = \sum_{n=2016}^{2050} C_s(n), \quad [4.14]$$

with:

- n: the current year (year);
- i: the discount rate (in %);
- $C_{reinforcement}(n)$: the cost of changing/adding a line (k€);
- $C(n)$: the investment related to the implementation of flexibility (k€);
- T_{AM}: the amortization period (years);
- C_{losses}: the cost of losses (k€/kWh), see Chapter 2;
- $E_{losses}(n)$: total energy lost in year n (kWh);
- $C_m(n)$: the maintenance cost (k€);
- $C_{ni}(n)$: the cost of energy not produced by a producer (k€/kWh);
- $E_{ni}(n)$: total energy not produced in year n (kWh).

4.4.1.2.5. Results for the village of Les Haies

Description of the systems and simulations studied

The algorithm presented previously has been applied at two different scales to the village of Les Haies, presented in Figure 4.53. It is composed of six LV distribution systems belonging to the same feeder of an MV/LV substation. Table 4.12 shows the characteristics of this system in which 70% of the consumers have a subscribed power between 6 and 9 kW, 18% between 12 and 15 kW, 7% above 18 kW and 5% between 0 and 3 kW. The installed capacity of PV generators as well as the possible deposit are given as a percentage of all six systems.

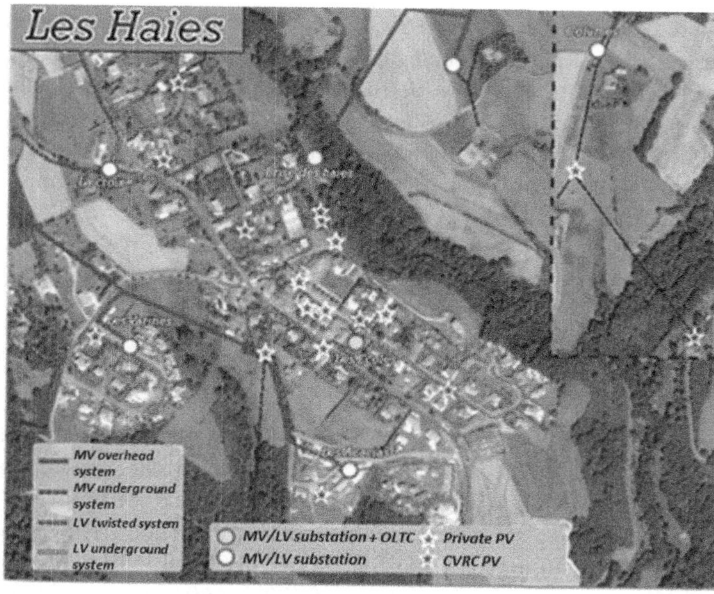

Figure 4.53. *The village of Les Haies*

No.	MV/LV substations (kVA)	Maximum extended length (m)	Number of customers	PV producers			Consumption
				Number	Installed capacity (kW)	Potential output (kWp)	Peak power (2016) (kW)
1	50	500	6	2	16.4%	8%	14.2
2	250	770	61	11	61.2	44%	191.1
3	50	200	12	1	6.1%	5%	33.6
4	250	270	36	1	4.1%	18%	80.1
5	160	300	20	1	4.1%	4%	57.1
6	160	360	35	2	8.1%	21%	78.5

Table 4.12. *Characteristics of the Les Haies system*

For the high and low scenarios, the number of Monte Carlo simulations performed resulted from a trade-off between computation time and the accuracy of results. For the Les Haies perimeter, a Monte Carlo simulation without any reinforcement or flexibility action consists of generating the random variables presented above (Steps 1 and 2, see section 4.4.1.2.4), then simulating (Step 3) eight typical days at 48 time intervals of two scenarios (high and low) over 35 years and for six different systems, that is, 161,280 load distribution calculations. When reinforcement or flexibility mechanisms are tested, other load flows must be performed to recalculate the new performance indicators (Steps 4 and 5). By choosing the minimum number of Monte Carlo iterations at 50 (corresponding to a 95% probability that the result obtained is within 13.86% of the actual result, see Chapter 5), the minimum number of simulations for the five solutions is about 40.3 million (5 × 50 × 161,280).

Need for reinforcement/development of LV system

To simulate the reference case "reinforcement only", or when current and/or voltage constraints are not solved by the flexibility mechanisms and it is necessary to reinforce the system, the possible actions are as follows:

– *in the case of a constraint on lines*: the feeders to which the lines belong are identified. We start by increasing the branch of the feeder line closest to the secondary substation. If this is not enough, we also increase the branch of the next line, and so on, until there are no more constraints on the feeder. The cross-sections used are 70, 95, 150, 240 mm², two to four lines in parallel with a cross-section of 240 mm² representing virtual cross-sections of 480 to 960 mm². So, if the system is to be reinforced and several parallel lines are used, then the SDR can exceed 100%, which might seem counterintuitive;

– *in the case of a constraint on a secondary substation*: the transformer is replaced by a transformer with an immediately higher power.

Table 4.13 shows the reinforcement rate obtained as a function of the mechanisms used for systems 1, 2, 5 and 6. Systems 4 and 5 are shorter and slacker than the others and are built with larger conductor cross-sections. In addition, the initial PV penetration rate and the potential deposit are very low. Therefore, there is no risk of a constraint occurrence in any scenario. Table 4.13 shows that the results are quite variable from one system to another and two groups of systems stand out: systems 1 and 2 and systems 3 and 6. Systems 1 and 2 have a significant reinforcement requirement, which can be explained by Table 4.12, where we can see that the maximum length developed from the secondary substation is two to three times greater than the other systems. System 2 also has a higher penetration of PV (initially installed and in the field) than the other systems. We can also notice in Table 4.13 that the two technical solutions that stand out are the centralized control Q(U) and the on-load tap changer, which drastically reduce the required reinforcement rate. On the other hand, generation curtailment does not seem to be very effective, and fixed reactive control is far less effective than centralized control Q(U).

	High scenario				Low scenario			
	R_1	R_2	R_3	R_6	R_1	R_2	R_3	R_6
S_0: reinforcement	197	140	84	65	92	25	17	9
S_1: centralized control Q(U)	47	5	6	1	16	1	12	0
S_2: fixed production curtailment (70%)	189	123	54	50	84	22	14	7
S_3: on-load tap changer	21	4	1	0	0	0	23	0
S_4: fixed reactive control ($tan\ \varphi = -0.4$)	119	40	7	20	52	9	11	6

≤ 50	≤ 100	≤ 150	≤ 200

Table 4.13. *Rate of reinforcement, SDR (system infrastructure development rate), (in %) required according to the mechanisms used, the scenarios and the systems (R1–R6)*

Cost–benefit analysis of different mechanisms

The cost–benefit analysis presented above was conducted on Les Haies for the five mechanisms according to the assumptions laid out in Table 4.14. Since the system topology is unchanged, we assume that the costs related to quality of supply (undelivered energy and power outages) are constant, regardless of the solution being considered here.

Mechanism	CAPEX	OPEX
S_0: reinforcement	– Change of line branch – Change or transfer of MV/LV substation – Addition of new LV feeders	Losses
S_1: centralized control Q(U)	Infrastructure (considered negligible in comparison to a reinforcement)	Losses
S_2: fixed production curtailment (70%)	None	Unproduced energy Losses
S_3: load controller	Mutation of the MV/LV substation	Maintenance of on-load tap changer Losses
S_4: fixed reactive control ($tan\ \varphi = -0.4$)	None	Losses

Table 4.14. *Cost assumptions for the cost–benefit analysis*

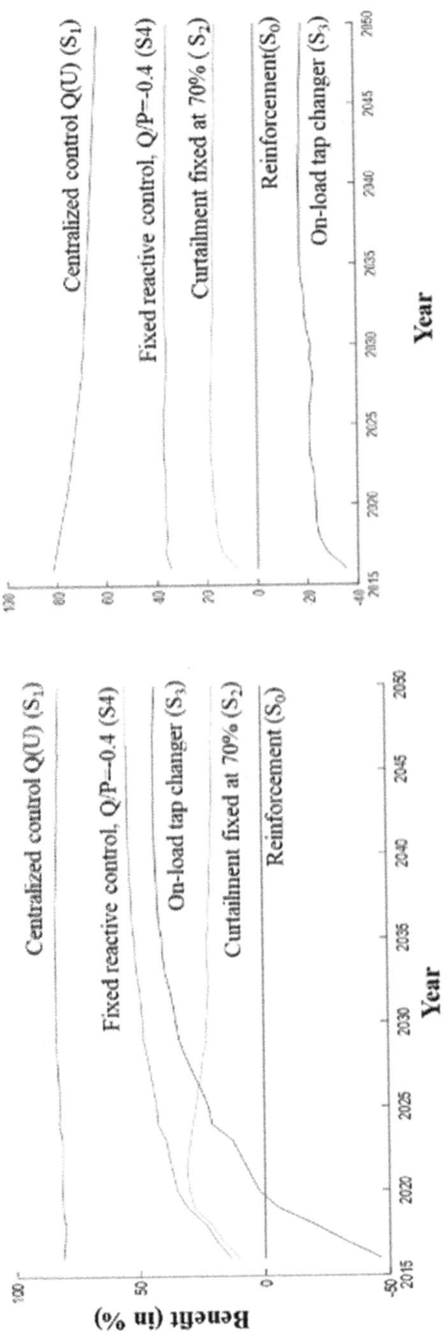

Figure 4.54. *Benefit of each solution: a) high scenario, b) low scenario) – Les Haies*

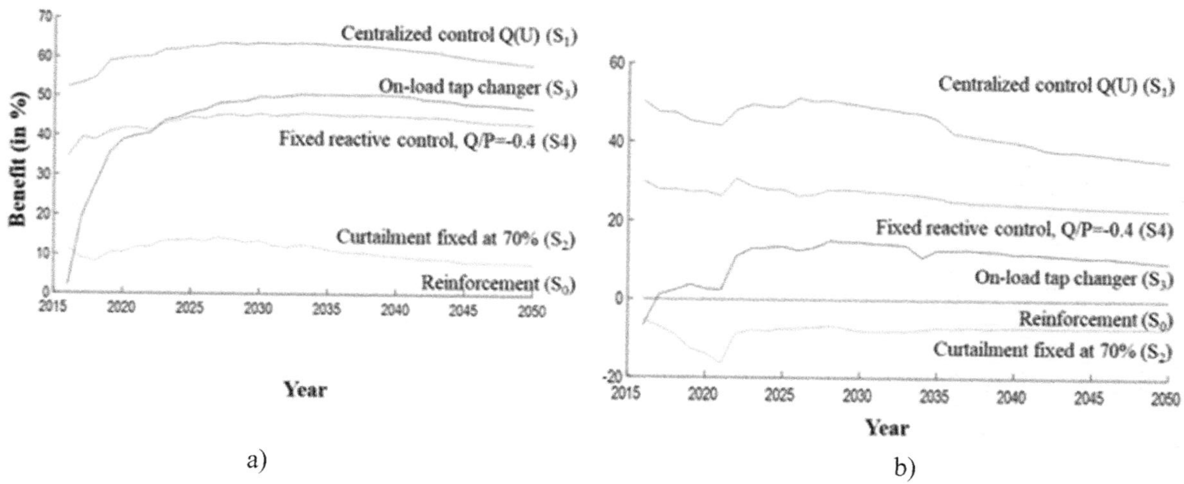

Figure 4.55. *Benefit of each alternative: a) high scenario, b) low scenario – CCRC*

Figure 4.54 shows the calculation of the benefits to the DSO for each solution tested in each year until 2050. We can see that the on-load tap changer solution is not at all robust to scenario uncertainties given how we can observe opposite effects, in terms of benefit, depending on whether the scenario is high or low in relation to reinforcement. The centralized control in Q(U) is the best solution, both in terms of benefit for the DSO, amounting to 80% compared to reinforcement, and in terms of robustness to scenario uncertainties. Fixed reactive control comes in second place, allowing for gains to be reduced by half, followed by fixed generation control, which reduces gains by a factor of 4 when compared to the centralized control in Q(U). These results are obviously specific to the Les Haies system and therefore their generalization requires additional analyses to study the variability of the system parameters (topology, load density, R/X ratio).

	Benefits for the DSO in 2050 (low scenario; high scenario)		**Variability each year**	**Variability depending on the scenario**
	Les Haies	**CCRC**		
Centralized control in Q(U)	(62%; 81%)	(35%; 58%)	Fairly stable	Fairly stable
Fixed control	(34%; 53%)	(23%; 43%)	Variable for the first 15 years for Les Haies, stable for the rest of the time	Fairly stable
On-load tap changer	(–19%; 42%)	(10%; 47%)	Highly variable	Highly variable
Curtailment at 70%	(16%; 20%)	(–7%; 8%)	Les Haies: variable for the first 5 years and otherwise stable CCRC: stable enough	Les Haies: fairly stable Highly variable for the CCRC

Table 4.15. *Comparison of results between Les Haies and the CCRC*

The impact for the structure of the rural systems, in terms of line lengths and branches, has been carried out by enlarging the perimeter to that of the CCRC (*Communauté de communes de la région de Condrieu*). It is composed of 195 LV distribution systems distributed on three different, mostly rural primary substations 57% of the systems are very rural with a load density lower than 10 kW/m², 38% of the systems are rural with a load density lower than 100 kW/m² and 5% are semi-urban with a load density lower than 5 MW/km² (Gain 1993). In this context, a single simulation was performed on two typical days (winter and summer) on the 195 CCRC systems, the results of which are shown in Figure 4.55. The conclusions remain broadly the same even though the values for the maximum benefits achieved by the flexibilities are lower than those of Les Haies, as shown in Table 4.15. The conclusions regarding the 70% PV with set point curtailment are far more mixed in the CCRC case, making this solution not very robust.

In order to refine the results obtained with regard to the centralized control in Q(U), a CAPEX of 4 k€ is added to represent the potential extra cost of the inverter. This rather large value is on the order of the price of a 36 kVA three-phase PV inverter. For the CCRC perimeter, the benefits obtained in the year 2050 go from 35% to 22% for the low scenario, and from 58% to 48%, thereby approaching the benefits of the fixed tanφ solution. Therefore, in a context of uncertainties for the CAPEX required to deploy centralized control of PV inverters, the fixed tanφ solution may be less risky.

4.4.2. Case study 5: using storage to defer investments in LV systems

Storage systems can provide several services to the power system, including the possibility of postponing or canceling an investment in LV due to voltage or current constraints from PV generation. Up until now, the high cost of storage systems has made it economically unattractive to consider this solution. However, the rapid decrease of these costs makes a cost–benefit analysis for its use in the distribution system relevant. We therefore consider an LV distribution system, such that at year 0, its capacity to receive PV generation is maximal. Each year, PV production is connected to the system imposing constraints. The first reference strategy is the reinforcement of the system by increasing the line branches. The second strategy is to use the storage systems already present in the system to remove the constraints. As DSOs are not systematically allowed to own storage systems (*exemption* is possible in the absence of storage players in some countries, or even allowed in Australia), we consider that these are owned by third parties who provide services to the system. Since the location and power of PVs are not predictable in advance, a Monte Carlo simulation is used to calculate the average benefit of using storage.

4.4.2.1. *Description of the implemented algorithm: general algorithm*

For each year of the study period and for each Monte Carlo simulation:

– Step 1: load sharing calculation to evaluate the magnitude and duration of system constraints;

– Step 2a: strategy 0 – system reinforcement (reference strategy):

- determination of the lines to be reinforced,

- calculation of the new system's load distribution to evaluate the losses,

- calculation of the CAPEX and OPEX of the current year;

– Step 2b: strategy 1 – use of storage devices:

- determination of the total storage energy used,

- calculation of the storage manager's remuneration;

– Step 3: calculation of the cost–benefit ratio (CBR) between strategy 0 and strategy 1 according to the formulas described in Chapter 2.

For each year of the study period, the average profit from the Monte Carlo simulations is calculated. As with the previous case study, we consider that there is no significant variation in the costs of energy not supplied and power cut between the two strategies. They are therefore not included in the calculation of OPEX. In the case of reinforcement, the CAPEX will be the cost of new lines, and the OPEX will be the cost of losses calculated, as in Chapter 2. For the strategy using storage, the CAPEX will be zero and the OPEX will consist of the cost of losses and the remuneration of the storage service. However, as this service is not yet commercialized, there is no economic model for remunerating storage which must therefore be defined in advance.

4.4.2.1.1. Algorithm for determining the storage fee

Definition

The cost of remunerating storage depends on the CAPEX it avoids, the load and production curves and their growth rates. For it to be worthwhile to use storage, its remuneration cost must be both lower than the cost of reinforcement and higher than the gain that the storage manager would have had if it had been used initially. The upper bound for the storage remuneration cost can therefore be determined using the CBR presented in Chapter 2. The "do nothing" solution corresponds here to using storage.

As explained in Chapter 2, reinforcement will be justified if the CBR of equation [4.15] is greater than or equal to $\frac{i}{1+i}$. Conversely, storage will be preferred if it respects equation [4.16].

$$CBR(n) = \frac{C_{OPEX}^{strategy\,2}(n) - C_{OPEX}^{strategy\,1}(n)}{C_{OPEX}^{strategy\,1}(n)}, \qquad [4.15]$$

$$\frac{C_{OPEX}^{strategy\,2}(n) - C_{OPEX}^{strategy\,1}(n)}{C_{OPEX}^{strategy\,1}(n)} \leq \frac{i}{1+i}, \qquad [4.16]$$

$$-\Delta C_{losses}(n) + C_{\lim} \times E_{S2}(n) \leq C_{OPEX}^{strategy\,1}(n) \times \frac{i}{1+i}, \qquad [4.17]$$

with:

$$\Delta C_{losses}(n) = C_{losses}^{strategy\,1}(n) - C_{losses}^{strategy\,2}(t),$$

$$C_{\lim} \leq \frac{\Delta C_{losses} + CAPEX_{strategy\,1}(n) \times \frac{i}{1+i}}{E_{S2}(n)}, \qquad [4.18]$$

with:

– n: the year studied;

– i: the discount rate;

– $C_{\lim}$: the cost limit of storage fee (in k€/kWh/year);

– $E_{S2}(n)$: the amount of storage used in year n, both in charge and discharge in absolute value (in kWh).

Algorithm

We consider a simplified system consisting of a secondary substation, a line of variable length with a 95 mm² cross-section and a node to which a load, a PV and a storage system are connected. A Monte Carlo-type study is used to vary the growth rate of the loads and the lengths of the lines, thereby varying the number and magnitude of the constraints. Each time there is a constraint, reinforcement is applied to estimate the CAPEX required to cancel the constraint and, in parallel, the storage service is simulated to estimate the energy required to cancel the constraint. Finally, it is possible to define the average boundary cost of paying for storage as a function of the number of constraints on the system. We used the same classification as per E-CUBE (2016) to define three system categories:

– low constrained system (total duration of constraints less than 50 h/year or 10 consecutive h/year);

– moderately constrained system: total duration of constraints between 50 h/year and 500 h/year or between 10 h and 20 consecutive h/year;

– highly constrained system: total duration of constraints greater than 500 h/year or 20 consecutive h/year.

Finally, the boundary costs of storage as a function of system type are defined in Table 4.16. Logically, the more the DSO will need to use storage, the lower the compensation cost will need to be in order to remain competitive with reinforcement.

In Obi et al. (2017), the LCOE of storage technologies that can be deployed in distribution systems ranges from €0.03/kWh to €0.27/kWh, which is between 12 and 116 times smaller than the borderline compensation costs found. Under these conditions, the storage system operator has an interest in offering services to the DSO.

System type	C_{lim} (in €/kWh/year)
Weakly constrained system	4,616 €/kWh
Moderately constrained system	75.24 €/kWh
Highly constrained system	3.48 €/kWh

Table 4.16. *Cost limit for storage remuneration according to the number of annual constraints*

4.4.2.2. *Description of the case study*

The 12-node study system in Figure 4.56 consists of a 50 kVA MV/LV transformer supplying 10 consumers, six connected in three-phase and four connected in single-phase, representing a total maximum consumption of 42 kW. This system is long (total length of 2,220 m with a developed length between the secondary substation and the most distant load of 1,050 meters), lightly loaded and composed of conductors of 70 mm² cross-section (except line 7–8 which is 35 mm²).

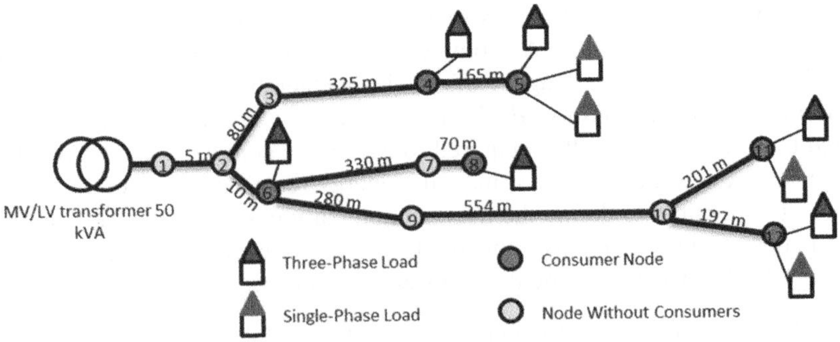

Figure 4.56. *Actual LV distribution system studied*

4.4.2.2.1. Results

The PV generation hosting capacity for this system is estimated to average 21.7% using the methodology presented in section 4.2.2. We assume here that PV panels can only be installed on the roofs of existing houses and thus model an aggregated PV generation capacity at consumer nodes only (green in Figure 4.56). The location and power of the PV generators vary with each Monte Carlo iteration to account for uncertainty. We assume that the installed PV capacity increases linearly every year until it reaches 47.5% in year 30.

– Strategy 0 – system reinforcement (reference strategy): the occurrence of voltage constraints requires the reinforcement of 1,650 m of lines, which represents 187 k€ of investment discounted to year 0;

– Strategy 1 – use of storage devices: the maximum number of constraints appears at year 7 and reaches 626 h, which classifies the system as “highly constrained”. The remuneration cost of the storage chosen is therefore 3.48/kWh, as shown in Table 4.16. We assume that storage is available at all consumer nodes (green in Figure 4.56);

– calculation of the CBR between strategies 0 and 1: Figure 4.57 shows the benefit of the “storage“ solution (in red) for each year of the study period.

The benefit is positive over the entire study period, which shows that the storage solution is to be preferred here over reinforcement. This benefit is highest in the initial year and then decreases until year 8, as the penetration of PV increases, the OPEX associated with the use of storage also increases, which reduces the benefits of storage compared with reinforcement. After year 8, the benefit increases slightly again. Indeed, beyond a certain installed PV power, the connection is made in three-phase, which reduces the imbalance and the constraints and thus the use of storage. An uncertainty of ±30% on the remuneration price of the storage has also been studied. The greater the number of constraints to be solved, the greater the impact of this uncertainty, as shown in Figure 4.57. In year 8, a variation of −30% change in the storage remuneration price increases the benefit by 33%. The benefit of storage will therefore strongly depend on the characteristic of the system under study.

Another case study consisting of a shorter (total length divided by 2.7) and more heavily loaded (maximum power multiplied by 4.33) system was analyzed. The PV potential is twice as high in the initial year and 1.4 times higher in the final year. The maximum number of constraints is twice as high as for the previous system, which increases the OPEX related to the use of storage. On the other hand, since the system is much shorter, the reinforcement cost is halved. In this example, the benefit of storage is negative since the OPEX required to resolve the constraints exceeds the additional cost of reinforcement.

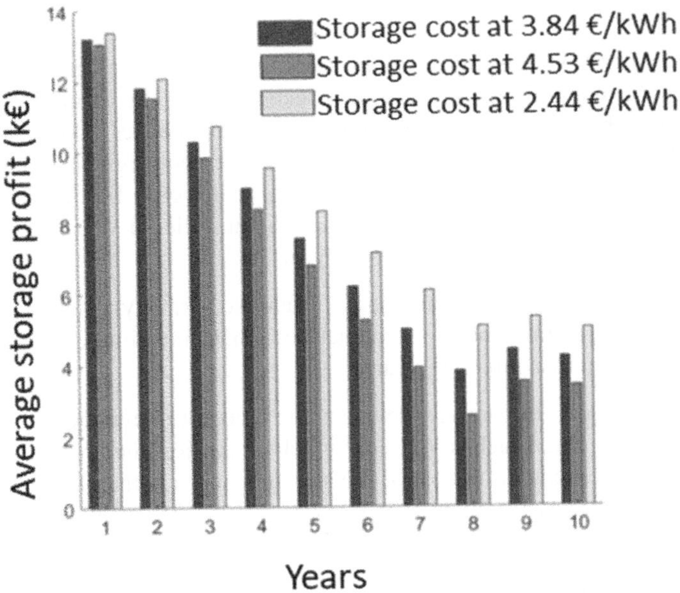

Figure 4.57. *Benefit of the "storage" strategy (six storage units connected to the six consumer nodes) ±30% variation around the compensation cost for the "highly constrained system" case*

Initial studies have shown that the use of storage as an alternative to system reinforcement is not always economically viable. It is necessary to consider multiservice use associated with better defined economic models.

4.5. Off-grid systems

4.5.1. *Case study 6: rural electrification – Cambodia*

4.5.1.1. *Description of case study*

The electrification of developing countries has very different objectives to those of industrialized countries, the challenge being to provide access to electricity for at least part of the day without any particular quality objective. Since HV and MV systems are often far from non-electrified or weak areas, DERs must be planned to provide all or part of the power locally. Given that 80% of Cambodians live in rural areas with 75% of them without access to electricity, the case study considered is the village of Bavel, shown in Figure 4.58, located in Battambang Province. The maximum consumption for this area is estimated at 23.6 kW with 68 consumers to be connected.

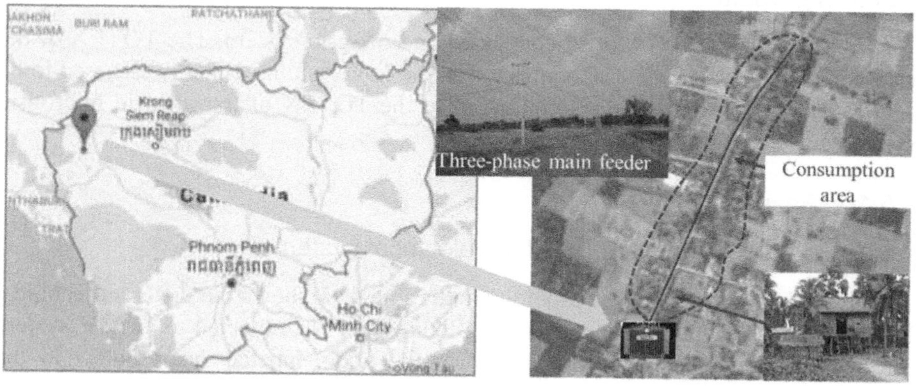

Figure 4.58. *Case study – Bavel Village, Battambang Province, Cambodia*

4.5.1.2. *Planning stage*

The planning steps for the implementation of a self-sufficient electrical system are as follows:

– Step 1: construction of the electrical system. The consumers are connected to each other and to a future secondary substation with the methods typical of the country, that is, single-phase with a three-phase and radial structure. The system topology is determined with consideration to loads at maximum power;

- Step 1a: load balancing. The loads are assigned *a priori* to each phase in order to minimize the power imbalance using a mixed integer linear programming (MILP, see Chapter 5) resolution method,

- Step 1b: connecting the loads. The loads of each phase are then connected to the MV/LV secondary substation in a tree-like manner in order to minimize the total length of the electrical system;

– Step 2: sizing and placement of photovoltaic panels, batteries and diesel generator. For the Cambodian LV, the best production potential is photovoltaic. The objective is to jointly size the production and the associated storage system in order to minimize the total cost of the system and ensure a certain percentage of electricity autonomy in case they are not connected to the MV, or if the MV system is not robust enough to supply the total power of the LV system. This is a fairly standard problem in the literature that can be solved in different ways. In our case study, a genetic algorithm was used (see Chapter 5). In this step, daily and yearly load and generation curves are considered in order to determine the charge and discharge profiles of the batteries, but also to quantify the losses and the energy produced by the diesel generator;

– Step 3: comparison of grid-connected and off-grid system. The TOTEX of the grid-connected system, that is, connected to the MV system, which is assumed to be able to supply all the power, is compared to the TOTEX of the off-grid microgrid system.

4.5.1.3. *Results*

4.5.1.3.1. Step 1: construction of the electrical system

Figure 4.59 shows the tree topology and Table 4.17 shows the different technical and economic results obtained, assuming the system is connected to an MV system through a secondary substation located at the node, labeled "start node" in the figure. We can find the characteristics of a rural system, that is, not very loaded (the most loaded line is at 26.5% of its maximum capacity) but with voltage constraints that appear quickly (minimum voltage of 0.94 pu). The TOTEX is mostly explained by the cost of losses (95.7%) since the conductors have small branches.

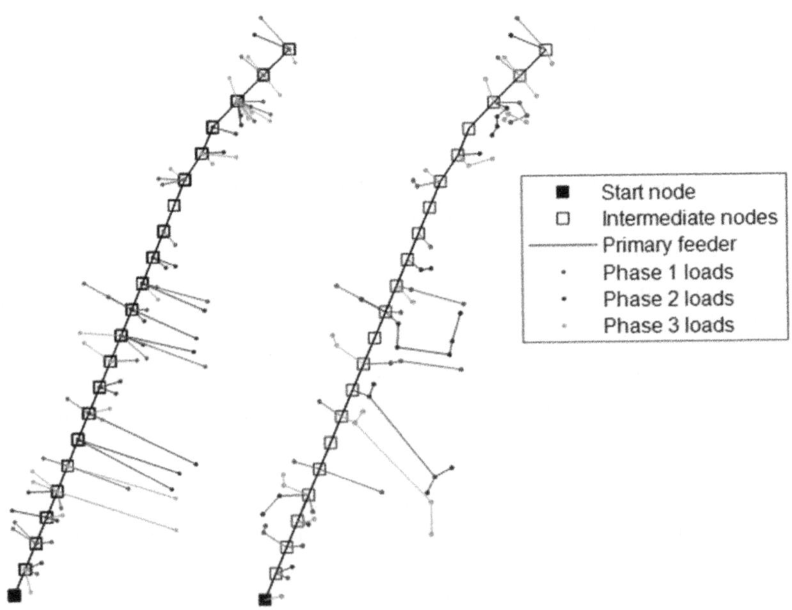

Figure 4.59. *Direct connection topology on the left and tree topology on the right*

By defining the power imbalance rate with equation [4.19], in our case study we obtain a power balanced topology ($\Delta P_i = [0.0023256; 0.0023256; 0.0046512]\%$). The voltage imbalance, ΔV_p, is given by equation [4.20].

$$\Delta P_p = \frac{P_p - P_{avg}}{P_{avg}} \; for \; p \in [1;3], \tag{4.19}$$

$$\Delta V_p = max\left(\left|\frac{V_{ip}}{V_{dp}}\right|\right) \; for \; p \in [1;3]. \tag{4.20}$$

with:

– p: the phase number;

– P_p: total power consumed on phase No. p;

– P_{avg}: the average power consumed on the three phases;

– V_{ip}: the vector of the inverse voltages of each load connected to the phase No. p;

– V_{dp}: the vector of direct voltages of each load connected to the phase No. p.

Figure 4.60 shows the detailed voltage map at each node of the system and for each phase. The squares represent the power poles[12] where the loads represented by the dots are connected. Since the energy comes only from the "feeder node", the voltage map decreases as we move away from that node.

Sizes	Values
Total length (conductors) Longest phase of the primary feeder (single wire) Total length of secondary feeders (single wire)	8.9 km 955.43 m 2.54 km
Loading of the least loaded line[13] at maximum load	0.2%
Loading of the most loaded line[14] at maximum load	26.5%
Minimum voltage (in pu) at maximum load	0.94 pu
Maximum voltage (in pu) at maximum load	1 pu
Maximum voltage imbalance at maximum load	0.89%

Table 4.17. *Technical results, grid-connected mode*

12 The location of the poles is usually defined in advance and follows the routes.

13 As a percentage of the maximum allowable line current.

14 As a percentage of the maximum allowable line current.

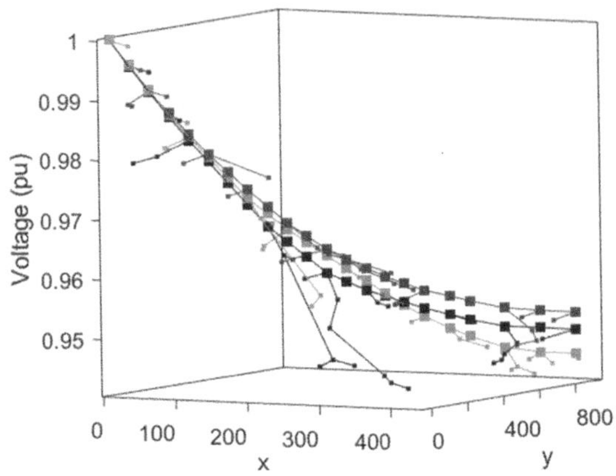

Figure 4.60. *System voltage plan, full load, grid-connected mode*

4.5.1.3.2. Step 2: sizing and placement of PV panels and storage

The proposed structure for supplying the loads of this microgrid is shown in Figure 4.61.

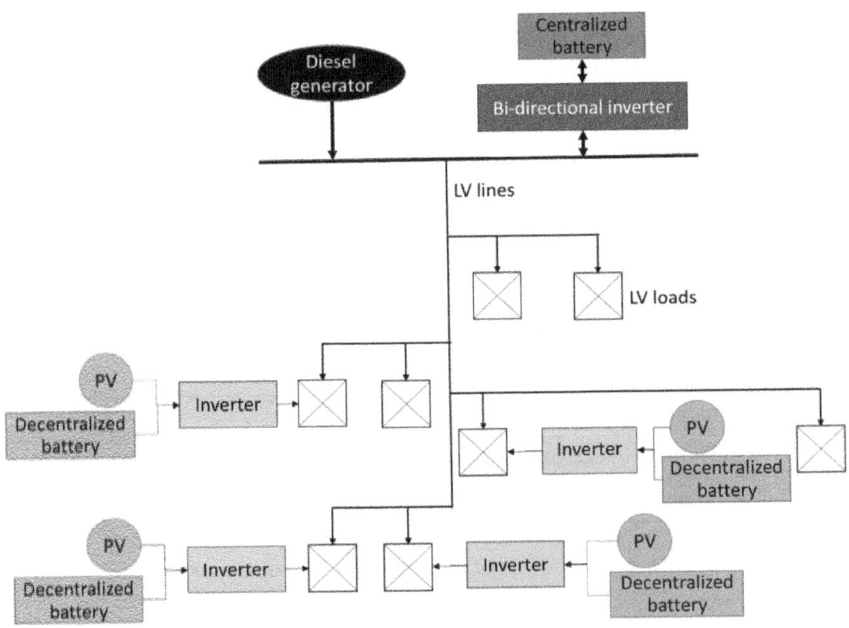

Figure 4.61. *Schematic diagram of the placement of PV, batteries and diesel generator*

In this structure, a photovoltaic panel is automatically associated with a decentralized storage system whose objective is to maximize local consumption. These decentralized PV-storage systems are identical in terms of the maximum power produced, maximum charging, discharging power and storage capacity, to facilitate their placement and reduce purchase costs. To impose and maintain the voltage at the microgrid level and ensure the production–consumption balance, a battery and a diesel generator are located at the future location of the substation. The genetic algorithm used converges to the placement of 29 PV + battery systems shown in Figure 4.62. The sizing of the PV, batteries and diesel generator is given in Table 4.18, and Figure 4.63 shows the daily load curve used, based on real measurements.

Figure 4.64 shows the variation in the voltages on the different phases of the system as a function of the time of day using a box plot: the red bars represent the median, the blue box the quantiles, at 25% and 75%, and the bars the extreme values. We can see that in the grid-connected mode, the voltage varies significantly, unlike the off-grid mode where the local production/consumption balance minimizes the variations of the voltage plane.

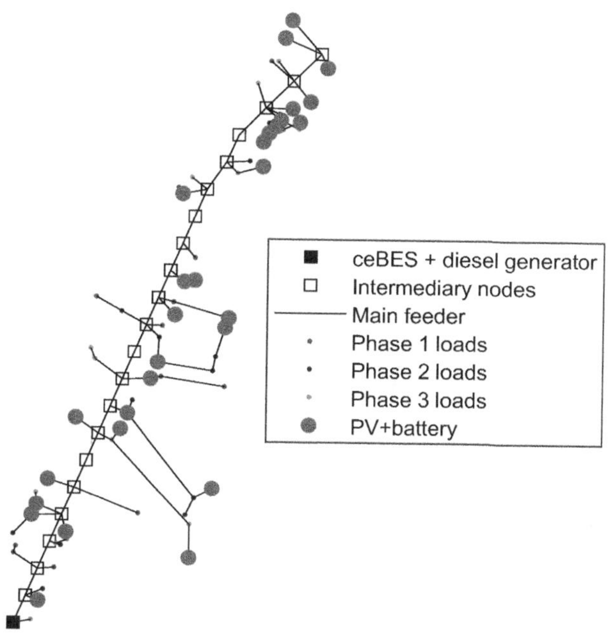

Figure 4.62. *Placement of "PV + battery" systems – off-grid mode*

PV + battery systems	Value	Battery system + diesel generator	Value
Maximum power PV (kW)	0.5	**Maximum diesel generator power**	11.8 kW
Maximum battery power (kW)	0.77	**Maximum battery power (kW)**	1.16
Battery capacity (kWh)	3.06	**Battery capacity (kWh)**	3.22

Table 4.18. *Sizing of PV + battery and battery + diesel generator systems*

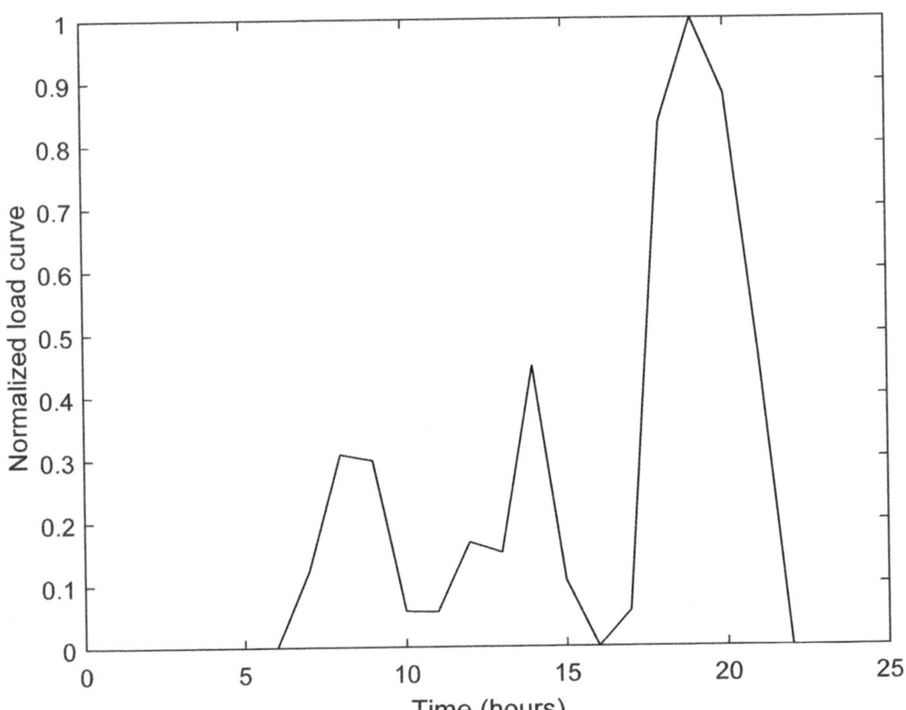

Figure 4.63. *Normalized rural load curve from actual measurements*

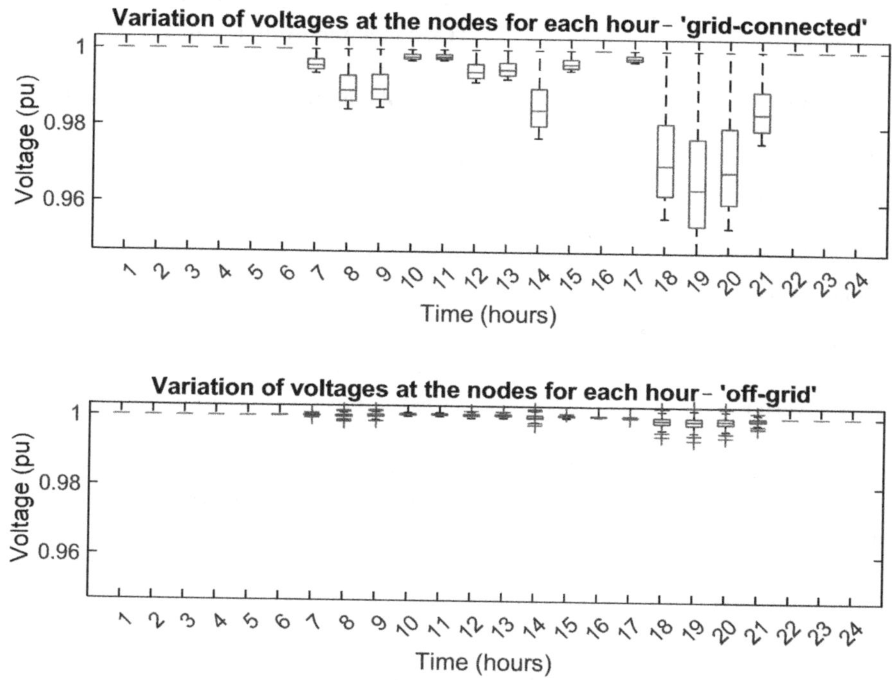

Figure 4.64. *Variation of voltages in the system for each hour of the day*

4.5.1.3.3. Step 3: comparison between the connected and the off-grid system

Table 4.19 shows the cost values used to calculate the TOTEX for the grid-connected and off-grid solutions. The battery sizing and loss calculations require the use of a daily load curve. This curve is repeated in order to build an annual profile as there is no noticeable seasonal effect. The method for calculating losses is different from the one presented in Chapter 2 because it consists of calculating the total energy lost valorized by the cost of electricity.

Table 4.20 gives the results obtained for the system in grid-connected and off-grid mode. We can see that consumption is indeed more local because annual losses are reduced by 83.7% and 90.4% of total energy is produced locally. The TOTEX is therefore almost halved thanks to an 80.4% reduction in OPEX, which compensates for the increase in CAPEX due to PV, batteries and diesel generators.

Discount rate	6%	Cost of a photovoltaic panel	0.6 k$/kW
Cost of cable Section 4 mm² Section 35 mm² Section 50 mm² Section 70 mm² Section 95 mm²	 0.3 k$/km 0.5 k$/km 0.8 k$/km 1 k$/km 1.5 k$/km	Cost of a battery	0.105 k$/kW
Cost of inverter	0.3 k$/kW	Cost of a two-way inverter	1.155 k$/kW
Cost relating to the maintenance of PV and batteries	0.0115 k$/kW	Cost related to the maintenance of inverters	0.023 k$/kW
Cost of electricity	0.27 $/kWh		

Table 4.19. *Costs used for TOTEX calculations*

Sizes	Grid-connected mode	Off-grid mode
Annual losses (MWh)	0.41	0.067
Annual energy from either the MV system upstream (grid-connected case) or the diesel generator and the battery located at the "start node" (MWh)	43.6	4.18
TOTEX Including CAPEX Including OPEX	106.97 k€ 4.6% 95.7%	56.08 k€ 64.2% 35.8%

Table 4.20. *Comparison of grid-connected and off-grid mode*

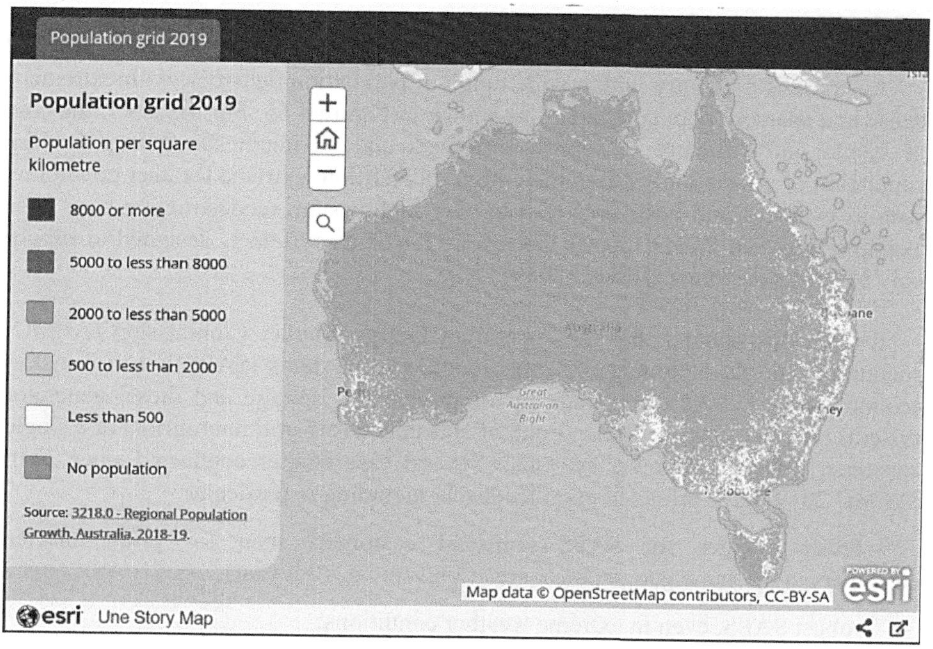

a)

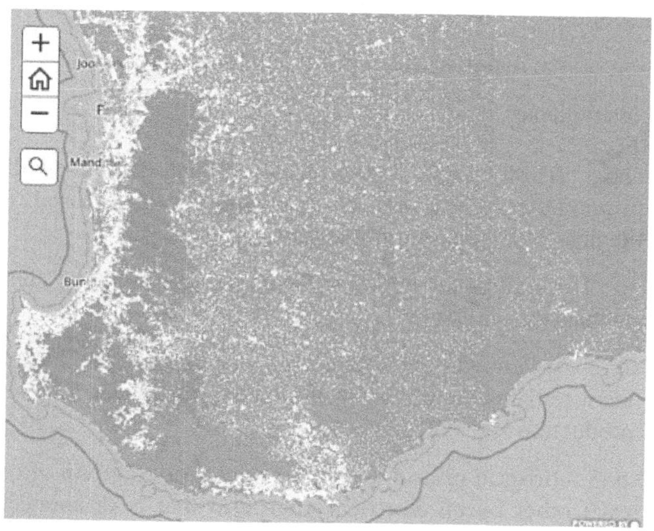

b)

Figure 4.65. *Population density in Australia, a) with a focus on the southwestern area, b) (2018–2019 data) (Australian Bureau of Statistics 2020)*

4.5.2. Case study 7: high cost, difficult access areas – Australia

Australia is a country with a wide range of population densities, with extremely dense and sparsely populated areas, as shown in Figure 4.65. For the latter, the cost of system infrastructure to supply customers and its maintenance (e.g. for the simulation of vegetation) is extremely high. In addition, extreme weather conditions such as cyclones and bush fires lead to very high system reconstruction costs. For example, Western Power reports that over 52% of its system is designed to supply only 3% of its customers (Power 2019).

It is in this context that the Australian Energy Market Commission (AEMC) encourages the development of stand-alone power systems (SAPS) when it makes economic sense. SAPS are made up of production, storage and diesel generator systems that can supply either a group of consumers (off-grid microgrids) or a single consumer (individual power systems). Several case studies conducted since 2016 (AEMC 2019) have provided good feedback, including in particular:

– fewer outages for SAPS-connected customers than for grid-connected customers in the same geographical area (5 h/year vs. 70 h/year);

– robust SAPS, even in extreme weather conditions;

– power supply mainly based on solar energy (more than 90%), the rest coming from diesel generators;

– elimination of line maintenance and simulation of expenses;

– reduced overall costs.

More recently, AEMC (2020) proposes a regulatory framework for SAPS deployment. In particular, it presents the four supply models based on population densities and the possible regulatory framework:

– Model No. 1 – traditional power system: the consumer is supplied by the power system managed by a DSO;

– Model No. 2 – private system: this is a system owned and managed by a third party. This system remains connected to the traditional power system and may or may not have production and storage units;

– Model No. 3 – off-grid or grid-connected microgrid: a group of consumers are connected to each other by an electrical microgrid that is disconnected from the traditional power system. Their power supply is ensured by production units, storage units and diesel generators. These microgrids can be owned and managed by DSOs, local organizations, ESCOs, local communities or other actors. The logic behind off-grid or grid-connected microgrids has not yet been determined and varies according to the actors;

– Model No. 4 – individual power systems: each consumer has their own generation-storage-diesel plant and is autonomous. Similarly, these facilities can be owned and managed by DSOs, local organizations, local communities or other actors.

Some of AEMC's recommendations are shared by DSOs like Western Power.

For the 10-year development plan (Western Power 2019), the plan is to evolve its secondary system toward a more "modular" and decentralized system (see Figure 4.66), where the traditional system (in dark gray) coexists with the autonomous system (in light gray) made up of microgrids connected, or not, to the system, and SAPS not connected to the system.

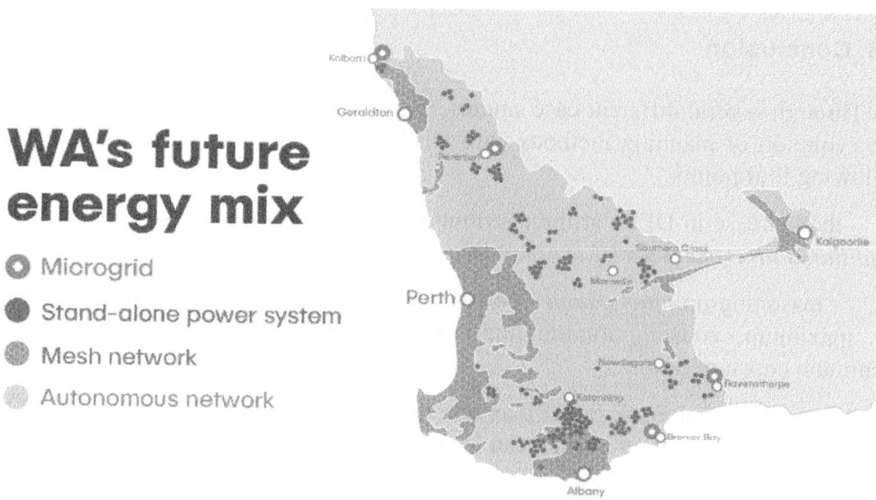

Figure 4.66. *Western Power's distribution system evolution in the coming years (Power 2019)*

Table 4.21 summarizes the potential for SAPS deployment over a 5- to 20-year time horizon as assessed by Australian DSOs.

DSO	Number of identified sites
Western Power	15,000
SAPN	Some sites
AusNet Services	300–400 in fire risk areas
Citipower/Powercor/United Energy	None
Energy Queensland	1,000–2,000
Essential Energy	2,000
Endeacour Energy and Ausgrid	Some sites
TasNetworks	5

Table 4.21. *Potential for SAPS deployment over a 5- to 20-year horizon (AEMC 2020)*

4.6. Conclusion

Through several different case studies, this chapter has provided an indication of the evolution of planning methods with DER, which we have synthesized into the following four points:

– the increase in DERs at the distribution level makes it necessary to *evolve the traditional investment decision procedures* used in planning:

- reasoning on *time-stamped power curves* rather than on extreme cases (such as maximum consumption/minimum production and maximum production/minimum consumption),

- use of *probabilistic* rather than deterministic methods to take into account the large number of uncertainties that can lead to erroneous choices,

- systematically integrate new technological solutions and flexibilities (NWA) as an alternative to reinforcement and quantify the risks associated with them;

– make *detailed hosting capacity maps public* allowing all actors to have a vision of the sites favorable to DER;

– testing *solutions that break* with historical solutions, in certain specific cases, as shown in the last case study in which large systems coexist with a multitude of independent local systems. This raises technical, economic and governance issues;

– *anticipate regulations* and *generalize exemptions* to allow the development of full-scale demonstrators to test new planning solutions.

4.7. References

AEMC (2019). New, cheaper options to power remote communities [Online]. Available at: https://www.aemc.gov.au/news-centre/media-releases/new-cheaper-options-power-remote-communities [Accessed 2020].

AEMC (2020). Updating the regulatory frameworks for distributor-led stand-alone power systems, Report [Online]. Available at: https://www.aemc.gov.au/market-reviews-advice/updating-regulatory-frameworks-distributor-led-stand-alone-power-systems [Accessed 2021].

Alvarez-Hérault, M.-C., Picault, D., Caire, R., Raison, B., Hadjsaid, N., Bienia, W. (2010). A novel hybrid network architecture to increase DG insertion in electrical distribution systems. *IEEE Trans. Power Syst.*, 26(2), 905–914.

Alvarez-Hérault, M.-C., N'doye, N., Gandioli, C., Hadjsaid, N., Tixador, P. (2015). Meshed distribution network vs reinforcement to increase the distributed generation connection. *Sustain. Energy, Grids Netw.*, 1, 20–26.

Applegate, D.L. (2006). *The Traveling Salesman Problem: A Computational Study*. Princeton University Press, Princeton.

Australian Bureau of Statistics (2020). Regional population change – Interactive map [Online]. Available at: https://www.abs.gov.au/AUSSTATS/abs@.nsf/Latestproducts/3218.0Main%20Features152018-19?opendocument&tabname=Summary&prodno=3218.0&issue=2018-19&num=&view= [Accessed 2020].

Bollen, M. and Rönnberg, S. (2017). Hosting capacity of the power grid for renewable electricity production and new large consumption equipment. *MDPI Ener.*, 10(9), 1325.

Braun, M. (2008). Reactive power supply by distributed generators. *Power and Energy Society General Meeting*, Pittsburgh, USA.

Coignard, J., MacDougall, P., Franz, S., Vrettos, E. (2019). Will electric vehicles drive distribution grid upgrades – The case of California. *IEEE Electrification Magazine*.

CRE (2021). Smart grids [Online]. Available at: http://www.smartgrids-cre.fr/index.php?p=smap [Accessed 2021].

De Martini, P. and Kristov, L. (2015). Distribution systems in a high distriubted energy resources future. Rapport, Berkeley Lab, Berkeley.

Delille, G., Malarange, G., Gaudin, C. (2013). Analysis of the options to reduce the integration costs of renewable generation in the distribution networks. Part 2: A step towards advanced connection studies taking into account the alternatives to grid reinforcement. *Proceedings of the 22nd International Conference and Exhibition on Electricity Distribution (CIRED)*, Stockholm, Sweden.

Demirok, E., Sera, D., Teodorescu, R., Rodriguez, P. (2009). Clustered PV inverters in LV networks: An overview of impacts and comparison of voltage control strategies. *Proceedings of the IEEE Electrical Power Energy Conference (EPEC),* Quebec, Canada.

Dijkstra, E.W. (1971). EWD316: A short introduction to the art of programing [Online]. Available at: https://www.cs.utexas.edu/users/EWD/transcriptions/EWD03xx/EWD316.html.

E-CUBE (2016). Étude sur la valeur des flexibilités pour la gestion et le dimensionnement des réseaux de distribution.

Gain, E. (1993). Réseaux de distribution : conception et dimensionnement. *Techniques de l'ingénieur*.

Gaudin, C., Ballanti, A., Lejay, E. (2012). Evaluation of PV curtailment option to optimize PV integration in distribution network. *CIRED Workshop*.

Lopes, R.T. (2008). Voltage regulation in radial distribution feeders with high penetration of photovoltaic. *Proceedings of IEEE Energy 2030 Conference,* Atlanta, Georgia.

Minaud, A., Gaudin, C., Karsenti, L. (2013). Analysis of the options to reduce the integration costs of renewable generation in the distribution networks. Part 1: Impact of PV development in France and global analysis of considered alternatives to reinforcement. *Proceedings of 22nd International Conference and Exhibition on Electricity Distribution (CIRED),* Stockholm, Sweden.

NégaWatt (2013). Scénario négawatt 2011.

Obi, M., Jensen, S., Ferris, J., Bass, R. (2017). Calculation of levelized costs of electricity for various electrical energy storage systems. *Renew. Sustain. Energ. Rev.,* 67, 908–920.

SEPA (2020). Proposing a new distribution system planning model [Online]. Available at: https://sepapower.org/knowledge/proposing-a-new-distribution-system-planning-model/ [Accessed 2021].

SOLARGIS (2020). Solar resource maps of France [Online] Available at: https://solargis.com/maps-and-gis-data/download/france [Accessed 2020].

Turitsyn, K., Šulc, P., Backhauss, S., Chertkov, M. (2010). Distributed control of reactive power flow in a radial distribution circuit with high photovoltaic penetration. *Proceedings of the IEEE PES General Meeting,* Minneapolis.

U.S. Department of Transportation (2009). National household travel survey. California.

Western Power (2019). The future of the grid: What is it made up of? [Online]. Available at: https:// westernpower.com.au/community/news-opinion/the-future-of-the-grid-what-is-it-made-up-of/?utm_primary=linkedin&utm_medium=organic&utm_campaign=news&utm_content=article-future-of-the-grid-120520 [Accessed 2020].

World Bank Group (2014). More power to India. The challenge of electricity distribution [Online]. Available at: http://documents.worldbank.org/curated/en/815021468042283537/More-power-to-India-the-challenge-of-electricity-distribution [Accessed 2021].

5

Mathematical Tools for Planning

5.1. Introduction

Chapter 4 presented applied case studies demonstrating how planning methods might evolve in the presence of DERs, and how the associated growing uncertainties could be modeled. These case studies were described in a pedagogical way, that is, by presenting the main methodological guidelines as well as the results obtained. The objective of this chapter is to explore the mathematical modeling of the planning problem in a generalized way, so as to give the reader mathematical tools for application to their own case studies, as per the objectives and constraints being encountered. Therefore, this chapter focuses on the modeling of the technical and economic variables of the electrical distribution system, and then, on the mathematical formulation of the objectives and constraints. This first step will identify the type of optimization problem to be solved, in order to identify the best resolution from the possible list of methods. These methods are illustrated in Chapter 6 using some of the applied examples and case studies covered in Chapter 4.

5.2. Inputting data for the planning problem

5.2.1. *Preliminary definitions*

5.2.1.1. *Notion of graph and source, customer and intermediate vertices*

Historically, the DSO only planned the MV system, given that at the LV system level usage was reduced to consumption, and the technological choices in terms of

---

This chapter was written by Oana IONESCU RIFFAUD.

LV architectures and equipment were therefore very simple. With the rise of DERs and flexibilities, in general at the MV level, but especially at the LV level, it becomes necessary to think about the planning of LV systems and their interaction with MV systems, or even the planning of LV and/or MV microgrids, interconnected or not (see the example of Australia in section 4.5.2 of Chapter 4). The architectures of electrical distribution systems can be assimilated to graphs where LV and MV end-users, public or private MV/LV primary and secondary substations can be assimilated to vertices (or nodes) and the electrical conductors connecting them to arcs (or edges). An undirected graph $G = (V_G, E_G)$ is defined by the finite sets $V_G = \{v_1, \dots, v_n\}$[1] of $\mathrm{N_{vertices}}$ and $E_G = \{e_1, \dots, e_m\}$ of $\mathrm{N_{arcs}}$ arcs. An arc is defined by its two ends v_k and $v_{k'}$, with $k, k' \in \{1 \dots \mathrm{N_{vertices}}\}^2$, and denoted $e_l = (v_k - v_{k'})$ with $l \in \{1 \dots \mathrm{N}_{arcs}\}$. A vertex v_k can be of three different natures, depending on the domain under study (see Table 5.1).

– Source vertex: this type of vertex represents the upstream system with a very high power level, as compared to that of the scope under study. In general, we speak of an infinite power vertex. In the case of a microgrid, it is the vertex that has the function of voltage reference (thanks to a battery or to a controllable production generator for example).

– Customer vertex: this is a vertex that consumes and/or produces power and can offer flexibility services.

In Table 5.1, we can see that depending on the scope under consideration by the DSO, the secondary substation can be modeled in the following three ways:

– as a voltage source: in the case of LV planning, the scope of the planner starts from the MV/LV secondary substation and goes to the LV customer. The MV/LV substation represents the connection to the main system that has infinite power. It is thus modeled as a voltage source;

– as an aggregated customer: in the case of MV planning, the MV/LV substation is a combination of public and private substations and MV customers. The MV/LV substation is therefore modeled as an aggregation of LV customers;

– as an intermediate node: in the case of MV and LV planning, the MV/LV substation is seen as an intermediate node only allowing for a change of the voltage level, and thus representing neither a customer nor a source.

1 The notation V_G to denote the vertices of a graph should not be confused with the notation used for voltage levels at power system nodes, namely V_k where k is the vertex index. Hereafter, to denote sets of vertices, we use the symbol Ω instead.

	Source vertices	Customer vertices
LV scope	MV/LV substations	LV customers (residential, DER, producers, etc.) Clusters of LV customers (in the case of grid-connected microgrids)
MV scope	HV/MV substations	MV/LV substations (aggregation of LV systems) MV customers (directly supplied with medium voltage or with their own MV/LV transformer, manufacturers, producers, DER) MV/LV substation cluster (in the case of grid-connected microgrids) and/or MV customers
LV+MV scope	HV/MV substations	LV and MV customers Clusters of LV and MV customers (in the case of grid-connected microgrids)

Table 5.1. *Examples of source and customer vertices according to the planning scope*

Graph G can be completed through a set of intermediate vertices for which no extraction or injection of power takes place. In urban or rural areas, this can be MV/LV substations (if the scope is for an MV and LV system), street/road/path intersections, pre-located crossing points to optimize a route between natural obstacles or system access points (location of the bypass accessory) for a customer connection.

5.2.1.2. *Architectural elements*

The architecture can be divided successively into the sub-elements shown in Figure 5.1 and outlined below.

– Lines: these, in medium or low voltage, are supplied by one or more sources and serve one or more customers (MV, MV/LV substations and LV) or no customers (backup). There are several types of lines according to the type of architecture (see Chapter 2): secured feeder, double branch, mesh, tree structure, etc.

– Feeders: a line is divided into several feeders that are supplied by a single source and thus supply customers in a radial manner. The division into several feeders is done via the open/closed configuration of the switching devices installed along the line. For example, for source-to-source secured feeder disconnection, the line is split into two feeders by opening a remotely controlled switching device located on the line. For an MV tree structure or an LV feeder, the line consists of only one feeder.

– Branches: a branch connects two important elements of the system, namely a source, a customer, or a junction. An outgoing feeder consists of several branches.

– Sections: a single branch may consist of several sections, each of which is characterized by its overhead/underground nature, the material used, the conductor, the cross-section and the insulation technology. For example, the division of a branch into several sections is justified when passing an obstacle that requires an overhead-to-underground-to-overhead transition (train tracks), or an underground-to-overhead-to-underground transition (river). There can also be several parallel sections to connect two vertices.

– Arcs: a section uses a path that is constrained by its environment (streets, buildings, natural obstacles). The possible paths are modeled in the form of a graph (see next section): to go from one point to another, a section will follow one or more arcs of this graph. Similarly, the same arc can be traversed by several sections belonging to the same line or to different lines.

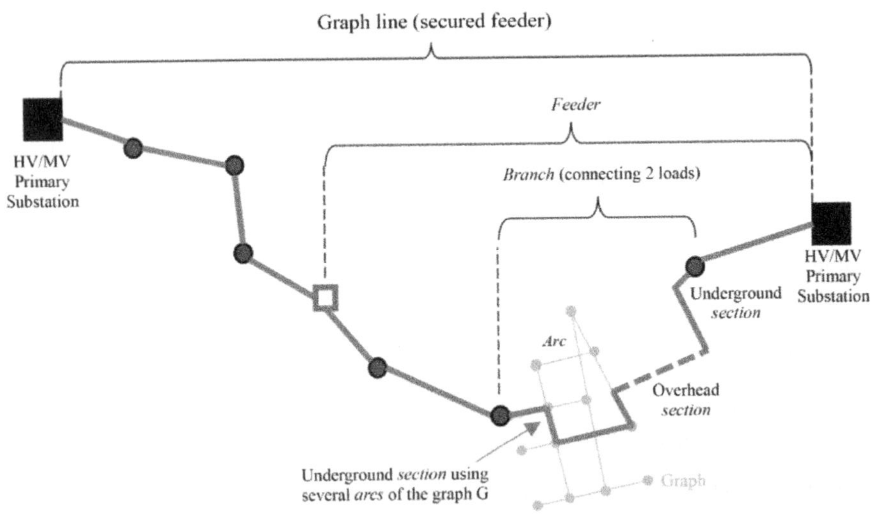

Figure 5.1. *Definition of the elements constituting the architecture. For a color version of this figure and all subsequent figures in this chapter, see www.iste.co.uk/alvarezherault/distribution.zip*

5.2.1.3. *Illustrative example*

Figure 5.2 shows a planning example where five customers are connected to two sources. This is an illustrative example that does not correspond to any particular traditional topology.

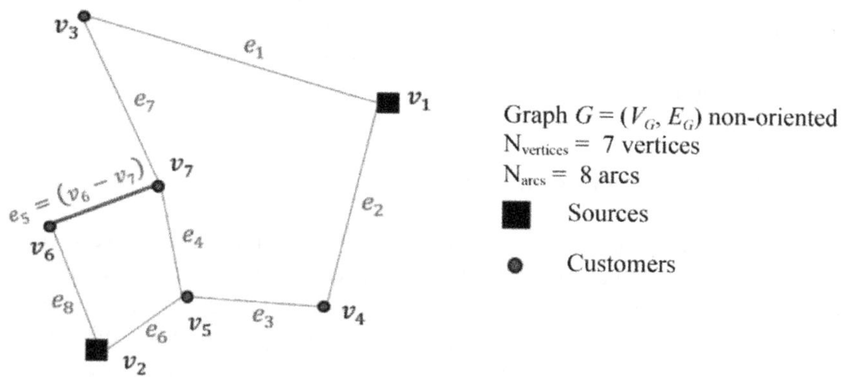

Figure 5.2. *Example graph*

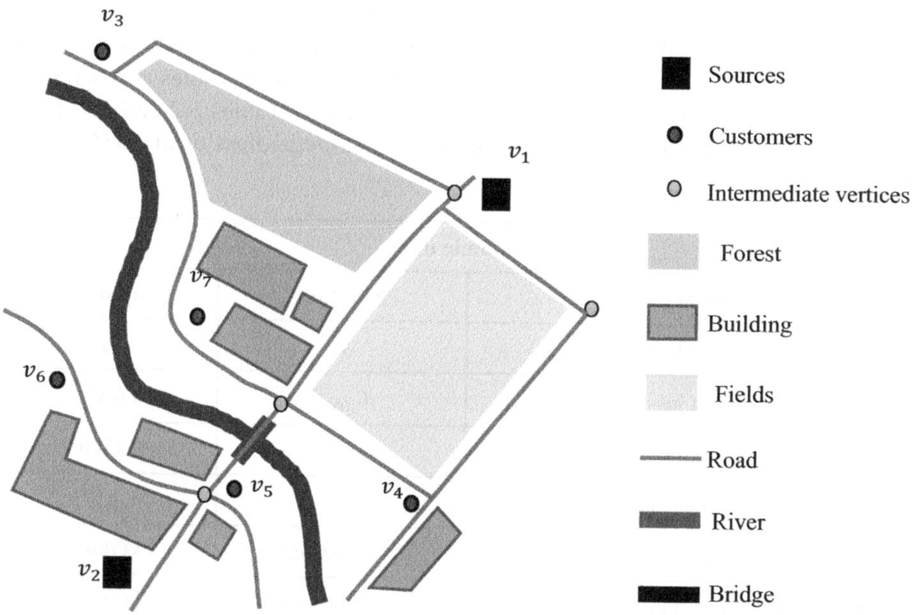

Figure 5.3. *Defining intermediate vertices*

Figure 5.3 shows the example illustrated in Figure 5.2 without the arcs, in which we find several types of obstacles: buildings, forests, fields and a river. A connection as per the shortest path is not possible, so four intermediate vertices have been added. This is an illustrative example in which some intersections close to a customer have not been assigned to an intermediate vertex in order to simplify the illustration; it is up to the planner to consider, or not, each intersection as a vertex to be added to the model, or not.

5.2.2. *Technical and economic data*

The technical and economic parameters are defined for the entire electricity system being planned. These include the reference costs used for the calculation of the economic indicators, whether operational (Joule effect losses, energy not supplied and power outages) or related to investments (laying/removal of lines, costs according to the type of conductor, switching devices), and the technical characteristics of the elements to be sized, such as the conductors of the medium and low voltage lines. This information depends mainly on the distributor (own policy, size, type of system operation), the legislation in force and the catalog of materials used. We also define the total period of the planning study, as well as the different sub-periods that constitute it, according to the information we know at the start. This segmentation of the study period allows for a more refined techno-economic study by defining the different indicators for each period, such as the rates of change in consumption and production of distribution system usage. Tables 5.2 and 5.3 detail the common technical and macroeconomic input data, regardless of the DSO scope under study.

Economic data		
Data	**Notation**	**Comments**
Discount rate	i	%
Cost of Joule effect losses at the peak	C_{peak}	€/kW
Cost of energy not supplied	C_{ENS}	€/kWh
Cost of interruptions due to failures	C_{PC}	€/kW
Linear cost of laying an overhead line	$C_t(overhead)$	k€/km
Linear cost of laying an underground cable	$C_t(underground)$	k€/km
Cost of a remote-controlled switching device (RCS)	C_{OMT}	k€

Table 5.2. *Common economic data*

Technical data			
Data		**Notation**	**Comments**
Total study period		N_{years}	Years
Number of types of conductors		N_{cond}	No unit
By type of conductor $\forall b \in \{1..N_{cond}\}$	Linear cost	$C_c(b)$	k€/km
	Linear resistance	$r(b)$	Ω/km
	Linear reactance	$x(b)$	Ω/km
	Linear capacity	$c(b)$	F/km
	Permissible thermal limit	$I_{th}(b)$	A
Overhead line failure rate		$\tau_{fail}(overhead)$	Incidents/year/km
Underground line failure rate		$\tau_{fail}(underground)$	Incidents/year/km
Fault isolation time		T_{iso}	Hours
Fault repair time		T_{rep}	Hours

Table 5.3. *Common technical data*

Historically, the traditional users of the distribution system are predominantly end-users (and in some cases producers). The DSO therefore plans its system on the basis of consumption and production growth forecasts in power or energy. These parameters are defined over different study periods. There are several methods for estimating them, such as the "in-depth" method, which relies on studies carried out at the national or regional level by different actors, according to "multi-criteria" studies carried out at the local level and then scaled up to ensure the consistency of the assumptions (Coiffier 2013). The methods for estimating consumption and production are not integrated into the scope of the planning problem, as presented in this chapter; only the results of these studies are used as input data.

The development of DERs modifies not only the evolution of future peak consumption and generation but also the shape of the power curves.

If flexibility is not integrated into the planning studies, then it is possible to reason with the growth rates of the different uses, as defined in Table 5.4 (conventional consumption, generation and electric vehicles), and by considering limited scenarios (high consumption, low generation and high generation and low consumption).

However, in the case of flexibility services via uses such as storage and electric vehicles, or other means such as load curtailment (see the flexibility measures listed in Chapter 3), it is necessary to reason with power curves in order to evaluate the flexibility needs and their cost. We can see that the level of uncertainty is an important key to the effective integration of DERs in planning studies, since we go from relatively well-controlled uncertainties with regards to consumption to:

– uncertainties pertaining to location and usage power curves;

– uncertainties regarding the development of flexibilities and their economic model.

We return to the treatment of these uncertainties in section 5.4.4. Table 5.4 details the DSO's local data based on the information known at year 0.

Data on the evolution of usage		
Data	**Notation**	**Comments**
Number of sub-periods in the study	$N_{periods}$	No unit
Duration of sub-periods	$T_p, \forall p \epsilon \{1..N_{periods}\}$	Years
Annual rate of change in average u usage, $u \epsilon$ {conventional consumption, generation, storage, electric vehicles, flexibility, etc.}	$\tau_{p,u}^{avg}, \forall p \epsilon \{1..N_{periods}\}$	%
Annual rate of change for usage u at peak, $u \epsilon$ {conventional consumption, generation, storage, electric vehicles, flexibility, etc.}	$\tau_{p,u}^{peak}, \forall p \epsilon \{1..N_{periods}\}$	%

Table 5.4. *Data for forecasting the evolution of usage connected to the distribution system*

5.2.3. *Structure of the initial electrical system*

We report in Tables 5.5–5.7 the necessary input data for the planning problem, related to sources and loads.

HV/MV primary substations		
Data	**Notation**	**Comments**
Number of primary substations	N_{PS}	
Rated secondary voltage of HV/MV transformers	U_{MV}	Between phases, in kV
Number of HV/MV transformers	$N_s^{HV/MV\ transfo}$ $\forall s \in \{1..N_{PS}\}$	Example: from 1 to 3
Geographic coordinate systems		EPSG, WGS, UMT[2] systems, etc.
Range of variation of the on-load tap changer	$\Omega_{s,u}^{taps\ on}\ \forall u \in 1..N_s^{HV/MV\ transfo}$, $\forall s \in \{1..N_{PS}\}$	Depends on the minimum and maximum limits and the discrete adjustment step of the on-load tap changer
Number of taps of the on-load tap changer	$N_{s,u}^{taps\ on} \forall u \in 1..N_s^{HV/MV\ transfo}$, $\forall s \in \{1..N_{PS}\}$	Typical value: 33 (one to baseline, 16 to increase ratio and 16 to decrease ratio) (Siemens Energy Sector 2016)
Maximum power per transformer	$S_{s,u}^{HV/MV\ transfo} \forall u \in 1..N_s^{HV/MV\ transfo}$, $\forall s \in \{1..N_{PS}\}$	Example: 20, 36 or 70 MVA (Coiffier 2013) in MV
Maximum number of MV feeders per transformer	$N_{s,u}^{MV\ feeders} \forall u \in 1..N_s^{HV/MV\ transfo}$, $\forall s \in \{1..N_{PS}\}$	Example: from 1 to 50 (Coiffier 2013)
Maximum power per MV feeder in normal operation	$S_{s,u,f}^{max,MV\ feeders}$ $\forall f \in \{1..N_{s,t}^{MV\ feeder}\}$, $\forall u \in 1..N_s^{HV/MV\ transfo}$, $\forall s \in \{1..N_{PS}\}$	Example: 5 or 6 MVA at 20 kV (Coiffier 2013)
Type of connection of MV/LV substations		Depending on the country (see Chapter 2)

Table 5.5. *Input data on the composition of the distribution system: primary substations*

2 EPSG: European Petroleum Survey Group, WGS: World Geodetic System, UTM: Universal Transverse Mercator.

MV/LV secondary substations		
Data	**Notation**	**Comments**
Number of secondary substations	N_{SS}	
Nominal voltage at the secondary of the MV/LV transformer	U_{LV}	Between phases, in kV
Geographic coordinate systems		EPSG, WGS, UMT systems, etc.
Range of variation of the off-load tap changer	$\Omega_d^{taps\ off}$ $\forall d \in \{1..N_{SS}\}$	Depends on the minimum and maximum limits and the discrete adjustment of the on-load tap changer
Number of taps of the off-load tap changer	$N_d^{taps\ off}$ $\forall d \in \{1..N_{PD}\}$	Three taps (0%, +2.5% and +5%) or five taps (We add the -5% and the -2.5%) (see Chapter 1)
Maximum permissible power of the transformer	$S_d^{MV/LV\ transfo}$ $\forall d \in \{1..N_{SS}\}$	Example: from 50 kVA to 1 MVA in LV (SéQuélec 2016)
Maximum number of LV feeders per transformer	$N_d^{LV\ feeders}$ $\forall d \in \{1..N_{SS}\}$	Example: from 1 to 8 depending on the type and power of the MV/LV substation (SéQuélec 2016)
Maximum power per LV feeder in normal operation	$S_{d,g}^{max,\ LV\ feeders}$ $\forall g \in \{1..N_d^{LV\ feeder}\}$ $\forall d \in \{1..N_{SS}\}$	Example: 120 kVA at 400 V (excluding housing estates) according to the Enedis technical department
Type of connection of MV/LV substations and MV and LV customers		Depending on the country (see Chapter 2)

Table 5.6. *Input data related to the composition of the distribution system: MV/LV secondary substations*

All customers		
Data	**Notation**	**Comments**
Number of customers	N_{cust}	
Rated power (or "connection power" for LV customers)	$S_i^{rated}, \forall i \in \{1..N_{cust}\}$	Unit: kVA
Maximum power output	$S_i^{cons}, \forall i \in \{1..N_{cust}\}$	Unit: kVA
Maximum injected power	$S_i^{prod}, \forall i \in \{1..N_{cust}\}$	Unit: kVA
Extraction power factor	$FP_i^{cons}, \forall i \in \{1..N_{cust}\}$	
Power factor of injection	$FP_i^{prod}, \forall i \in \{1..N_{cust}\}$	
Economic data	The data presented in Tables 5.1, 5.3 and 5.4 can be broken down by the customer	
MV customers and MV/LV substations (MV planning)		
Information on LV customers	Position, number, type (residential, professional, tertiary, industrial), subscribed powers, types of customer subscription. These data are described in a macro way, unlike the description for each LV customer (see below)	
LV system characteristics	Number of feeders, length and type of lines, the maximum authorized voltage drop at the end of the feeder	
LV customers (LV planning and attached MV-LV planning)		
MV/LV substation supply	$d \in \{1..N_{SS}\}$	

Table 5.7. *Input data on the composition of the distribution system: customers*

5.2.4. *Topological data*

The first objective of planning is to connect customers to the sources of the electrical distribution system. From the perspective of graph theory, this consists of connecting *customer vertices* to *source vertices* (see prior definitions in section 5.2.1.1), that is, choosing a set E'_G of arcs from among the arcs E_G of the complete graph G, that is, the graph where each vertex is connected to all others by an arc. In this case, we do not discriminate against any solution: it is possible to connect any pair of vertices. The number of arcs in a complete graph with $N_{vertices}$ is defined by equation [5.1].

$$\sum_{k=1}^{N_{vertices}} (N_{vertices} - k) = \sum_{k=0}^{N_{vertices}-1} k = \frac{N_{vertices}(N_{vertices} - 1)}{2} \quad [5.1]$$

In the example with 11 vertices, shown in Figure 5.3, the complete graph has 55 arcs ($\frac{11(11-1)}{2}$), as shown in Figure 5.4.

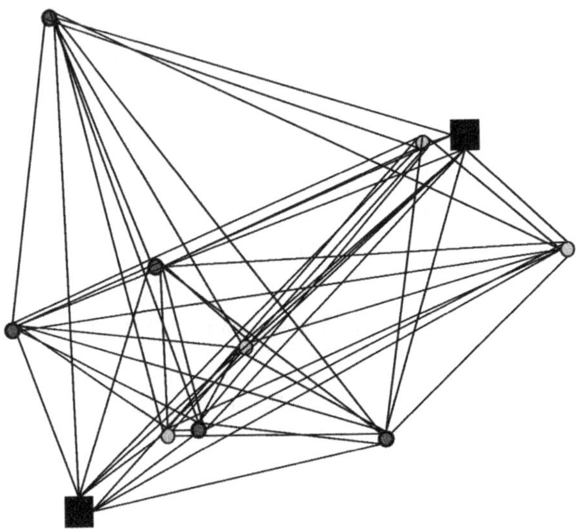

Figure 5.4. *Complete graph, 11 vertices, 55 arcs*

It is possible to reduce the number of arcs in the set E_G to a subset E_r, such that $E_r \subset E_G$. In this case, we decide that it is not possible to directly connect some pairs of vertices. The number of solutions is small, but this can greatly simplify the problem and reduce the duration of future technical and economic calculations. Figure 5.5 gives an example of a set E_r of candidate arcs for the planning problem.

If graph G is *weighted* (or labeled), each arc is assigned a positive real number, called the *arc weight*. We denote $p(e_l)$ the weight of the arc e_l. The weight of a chain, that is, a set of arcs, is equal to the sum of the weights of the arcs composing it. In the planning problem, graph G is weighted: the weight, which we can also call cost, will not be the same depending on the pair of vertices to be connected.

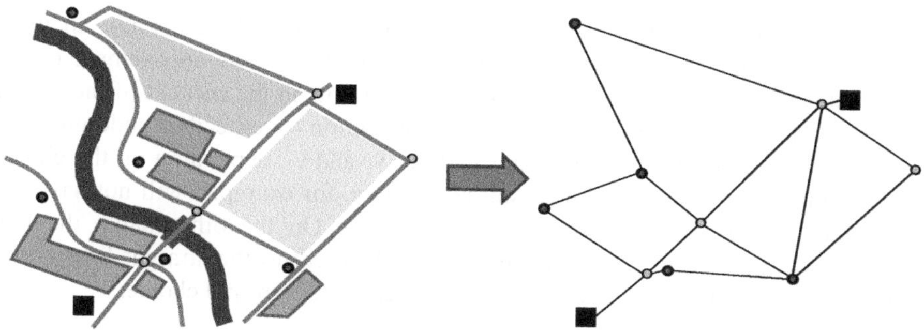

Figure 5.5. *Example of reduction of possible arcs: 13 arcs instead of 55*

As we can see in Table 5.5, sources and customers are geographically located by their coordinates. The first possible weighting of our graph G is to assign to each arc the geographical distance, "as the crow flies", between the two vertices it connects. This first way of doing things requires very little pre-processing of the input data to model the optimization problem but is not very realistic. In the field, a power line cannot pass through obstacles, whether they are buildings, street layouts, bridges, natural features or areas restricted for administrative or legislative reasons (protected area, unobtained permits, etc.). Mapping is necessary to determine the geographic path to best connect two vertices. This can be according to the street layout in urban areas, or the path between natural obstacles in rural areas. Each arc of graph G is then weighted by the distance traveled to connect its two vertices using the route that is the most feasible on the ground.

This second type of weighting is more realistic but it still does not fully reflect the reality of the situation on the ground. For example, the route may be underground or overhead, with different costs (Doulet and Horson 2008). The cost of burying a line is also variable, depending on the nature of the soil, or the presence of other buried systems (electricity, gas and telecommunications). The cost of burying a line can be extremely prohibitive (for example, in a historic street with heritage cobblestone paving requiring specialist work) or, on the contrary, can be reduced by roadwork opportunities (work sites already planned for a tramway line or work on another system). These cost differences can be integrated with the weight of the arcs by means of improving (<1) or depreciating (>1) coefficients. A particularly costly connection between two vertices can also be removed from the list of candidate arcs for planning. Finally, the cost of laying an underground cable in a trench is degressive if the trench accommodates several conductors.

The first element of complexity is thus identified: this information is only available after the optimization phase of the architecture that determines the layout of each line. The number of conductors in each trench is therefore not an input data; we come back to this in section 5.3.1. Figure 5.6 shows an example of the graph obtained by integrating the topological constraints in the context of the reality on the ground, as well as the possible type of connection (overhead or underground). For example, the power line connecting vertices v_9 and v_3, depending on the choice of underground and overhead due to soil constraints, for example, will not have the same route (hence the two possible route options). On the other hand, the line connecting vertices v_7 and v_3 has a route that is different to the straight line due to topological constraints, independent of the overhead/underground choice.

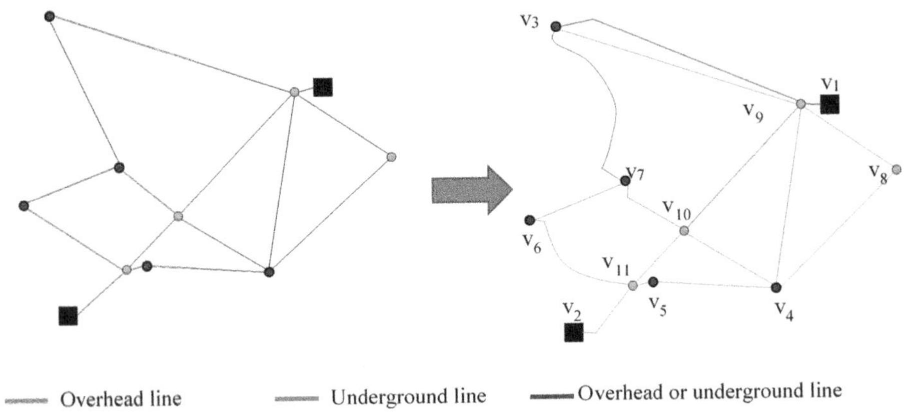

Figure 5.6. *Example of weighting the candidate arcs of the graph G*

We therefore define as topological inputs, summarized in Tables 5.8 and 5.9, the graph G that contains the set E_r of candidate arcs to receive an electrical section and the set Ω_G of vertices consisting of sources, customers, and possibly intermediate vertices. Each candidate arc has an associated weight that can be defined in different ways (length "as the crow flies" or actual length, cost depending on the technology used, that is, overhead or underground). We can also assign to each arc a limited number of sections that can be buried in the same trench.

HV/MV primary substations		
Data	**Notation**	**Comments**
System graph	$G = (\Omega_G, E_G)$ with $N_{vertices} = \lvert\Omega_G\rvert$ and $N_{arcs} = \lvert E_G\rvert$	

Candidate arcs	$E_r \subset E$	
Set of sources	$\Omega_S \subset \Omega_G$	$\Omega_S + \Omega_{Cl} + \Omega_{other} = \Omega_G$
Set of customers	$\Omega_{Cl} \subset \Omega_G$	
Set of intermediate vertices	$\Omega_{other} \subset \Omega_G$	
Arc weights	$p(e_l), \forall e_l \in E_r$	In meters or in k€
Maximum number of sections on a single arc	$N_{\text{max sections}}(e_l), \forall e_l \in E_r$	
Type of possible sections	$\Omega_{\text{sections types}}(e_l), \forall e_l \in \mathrm{E}_r$	Subset of {overhead, underground}

Table 5.8. *Topological input data*

Method	**Arc weights**	**Unit**
No. 1	$p_1(e_l) = 1$	
No. 2	$p_2(e_l)$ = distance as the crow flies between the vertices v_k and $v_{k'}$	km
No. 3	$p_3(e_l)$ = distance between the vertices v_k and $v_{k'}$ respecting a realistic route	km
No. 4	$p_4(e_l) = \big(p_2(e_l) \; or \; p_3(e_l)\big) \times Cost \; (line \; type)$	k€
No. 5	$p_4(e_l) = \big(p_2(e_l) \; or \; p_3(e_l)\big) \times Cost \; (line \; type) \times Coefficient \; (route \; characteristics)$	k€ corrected
$\forall e_l = (v_k - v_{k'}) \in \mathrm{E}_r$ $\forall k, k' \in {\Omega_G}^2$		

Table 5.9. *Possibilities of weighting the topological graph*

The choice of the MV distribution system (with or without a distributed neutral) as well as the system architecture (see Chapters 1 and 2) modifies the constraints applied to the planning problem, as we will see in section 5.3.3.1. It also has a very important influence on the modeling of the electrotechnical elements of the electrical distribution system, and thus on the solution methods that can be used, in particular, for the load flow calculation.

5.2.5. Definition of sizing situations

For the evaluation of the objective function and the constraints of the planning problem, two operation situations of the power system must be considered.

– *Normal operating regime*: this is the situation wherein the system operates normally, that is, without incident. All the operating devices are in their normal operating state (be they open or closed).

– *Emergency mode*: this is the situation in the event of an incident on the system resulting in the loss of a major element, such as the HV arrival of a primary substation, an HV/MV transformer, or part of an MV line. The state of the switching devices has been modified to isolate the faulty zone and continue to supply a maximum number of customers while guaranteeing their safety, as well as that of the equipment (see Chapter 2). There are several emergency backup situations depending on the nature and location of the lost element, the structure of the primary substations (number of HV feeders, HV busbars, number and power of transformers, MV busbars), the architecture, the location of the remote-controlled switching devices and the envisaged emergency plans. The emergency situation will differ depending on the element considered. Figure 5.8 shows a typical MV urban system main feeder on a secured feeder (see Chapter 2), on which the conductor being studied (in green) supplies loads 2, 3 and 4 in normal operation. This conductor is endangered by several situations during an emergency backup mode:

- loss of HV/MV transformer no. 1: the conductor feeds load 1;

- loss of HV/MV transformer no. 2: the conductor feed loads 2–8, as shown in Figure 5.8;

- loss of a conductor along a secured feeder: depending on the location of the lost conductor, the investigated conductor feeds load 2, or one or more loads.

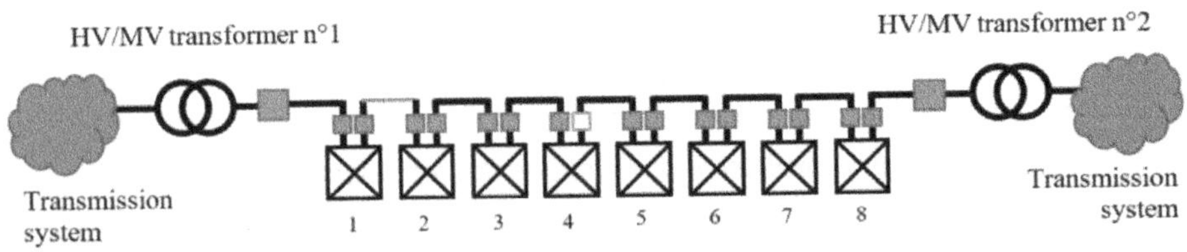

Figure 5.7. *Example of a typical main feeder of an urban MV secured feeder*

We define in Table 5.10, that for each element, the average and maximum voltages, powers and currents transmitting in the normal and emergency backup states, are necessary to evaluate the objective function and the constraints.

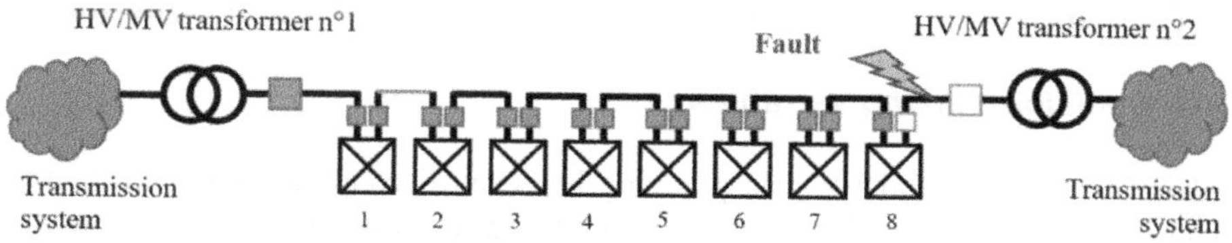

Figure 5.8. *Example of a fault condition for a conductor (fault downstream of HV/MV transformer no. 2, situation 2)*

For example, we use the average apparent power (or maximum apparent power, or maximum current, depending on the calculation method used) in normal operation to evaluate the technical losses along a section, while we use the maximum current in a backup emergency operation to determine the minimum cross-section area of the section's conductor(s) in order to meet the current constraints.

Element	Size and granularity	Normal regime	Emergency plan
Customers $\forall \boldsymbol{i} \in \{\mathbf{1}..N_{cust}\}$	Voltage at any time	V_i	V_i^{N-1}
MV sections $\forall \boldsymbol{j} \in \{\mathbf{1}..\boldsymbol{N}_{\boldsymbol{MV\ sections}}\}$	Apparent power, average value	S_j^{avg}	Not applicable
	Apparent power and current, maximum value	S_j^{max} I_j^{max}	$I_j^{max,N-1}$
LV sections $\forall \boldsymbol{j} \in \{\mathbf{1}..\boldsymbol{N}_{\boldsymbol{LV\ sections}}\}$	Apparent power and current, maximum value	S_j^{max} I_j^{max}	No recovery in general
HV/MV transformers $\forall \boldsymbol{u} \in \mathbf{1}..\boldsymbol{N}_{\boldsymbol{s}}^{\boldsymbol{HV/MV\ transfo}}$ $\forall \boldsymbol{s} \in \{\mathbf{1}..\boldsymbol{N}_{\boldsymbol{PS}}\}$	Apparent power, maximum value	$S_{s,u}^{HV/MV\ transfo}$	$S_{s,u}^{HV/MV\ transfo,N-1}$
MV/LV transformers $\forall \boldsymbol{d} \in \{\mathbf{1}..\boldsymbol{N}_{\boldsymbol{SS}}\}$	Apparent power, maximum value	$S_d^{LV/MV\ transfo}$	Not carried out

Table 5.10. *Definition of the powers according to the operating regime*

5.3. Planning: a multi-objective optimization problem under constraints

5.3.1. *Decision-making variables*

Historically, MV and LV grid planning has been decoupled for simplicity and because there were no particular issues at the LV grid level. However, with the rise of DERs and flexibilities in general, the interaction between LV and MV systems becomes necessary to model. The planner is therefore looking for the location of MV and LV lines, their type, the location of the RCSs, the HV/MV on-load tap changers, the powers drawn and injected by conventional loads and the DERs. In this section, we give the most general formulation of the planning problem so that everyone can adapt it to their own situation.

5.3.1.1. *Decision vector*

The vector containing the variables of the problem is noted $X_{\delta t}$:

$$X_{\delta t} = \begin{bmatrix} A \\ B \\ C \\ D \\ \underline{S} \\ \underline{J} \\ \underline{V} \end{bmatrix}$$

with:

– $\delta t \in \{1..N_{\delta t}\}$ with $N_{\delta t}$: the number of sizing time steps in the year (see section 5.3.1.3);

– A, B, C, D, $\underline{S}$, $\underline{J}$ and $\underline{V}$ the solution sub-vectors described in Table 5.11;

– each solution sub-vector has as many components as there are MV and LV lines used to supply the distribution substations and the MV and LV customers, of index a. We note N_{lines}, the number of MV and LV lines.

Sub-vector	Description	Formulation	Type and size
$A = \begin{bmatrix} A_1 \\ A_2 \\ \vdots \\ A_a \\ \vdots \\ A_{N_{lines}} \end{bmatrix}$	Location of the sections	$A_a[l] = \begin{cases} 1 \text{ if the arc } E_l \text{ is used} \\ 0 \text{ if not} \end{cases}$ $\forall l \in \{1..N_{arcs}\}, \forall a \in \{1..N_{lines}\}$ [5.2]	*Boolean* $N_{lines} \times N_{arcs}$

Sub-vector	Description	Formulation		Type and size
$B = \begin{bmatrix} B_1 \\ B_2 \\ \vdots \\ B_a \\ \vdots \\ B_{N_{lines}} \end{bmatrix}$	Type of conductor	$B_a[l] = \begin{cases} b \in \{1..N_{cond}\} \\ 0 \text{ if not} \end{cases}$ $\forall l \in \{1..N_{arcs}\}, \forall a \in \{1..N_{lines}\}$	[5.3]	*Discrete* $N_{lines} \times N_{arcs}$
$C = \begin{bmatrix} C_1 \\ C_2 \\ \vdots \\ C_a \\ \vdots \\ C_{N_{lines}} \end{bmatrix}$	Location of open sections during normal operation	$C_a[l] = \begin{cases} 1 \text{ if the section is open} \\ 0 \text{ if not} \end{cases}$ $\forall l \in \{1..N_{arcs}\}, \forall a \in \{1..N_{lines}\}$	[5.4]	*Boolean* $N_{lines} \times N_{arcs}$
$D = \begin{bmatrix} D_1 \\ D_2 \\ \vdots \\ D_a \\ \vdots \\ D_{N_{lines}} \end{bmatrix}$	Location of RCSs	$D_a[k]$ $=$ number of RCS on the vertex v_k $\forall k \in \{1..N_{vertices}\}, \forall a \in \{1..N_{lines}\}$	[5.5]	*Discrete* $N_{lines} \times N_{vertices}$
$\underline{S} = \begin{bmatrix} \underline{S_1} \\ \underline{S_2} \\ \vdots \\ \underline{S_a} \\ \vdots \\ \underline{S_{N_{lines}}} \end{bmatrix}$	Active and reactive powers of the customers' peaks	$\underline{S_a}[k] = P_a[k] + j.\,Q_a[k]$ $P_a[k]$ and $Q_a[k]$ are active and reactive powers of vertex k, positive if drawn and negative if injected $\forall k \in \{1..N_{vertices}\}, \forall a \in \{1..N_{lines}\}$	[5.6]	*Complex number* $N_{lines} \times N_{vertices}$
$\underline{J} = \begin{bmatrix} \underline{J_1} \\ \underline{J_2} \\ \vdots \\ \underline{J_a} \\ \vdots \\ \underline{J_{N_{lines}}} \end{bmatrix}$	Current levels in the lines	$\underline{J_a}[l] = \underline{J_l} = J_l \times e^{j\gamma_l}$ $l \in \{1 \ldots N_{arcs}\}, \forall a \in \{1..N_{lines}\}$	[5.7]	*Complex number* $N_{lines} \times N_{arcs}$
$\underline{V} = \begin{bmatrix} \underline{V_1} \\ \underline{V_2} \\ \vdots \\ \underline{V_a} \\ \vdots \\ \underline{V_{N_{lines}}} \end{bmatrix}$	Distribution substation voltage levels	$\underline{V_a}[k] = \underline{V_k} = V_k \times e^{j\theta_k}$ $k \in \{1 \ldots N_{vertices}\}, \forall a \in \{1..N_{lines}\}$	[5.8]	*Complex number* $N_{lines} \times N_{vertices}$

Table 5.11. *Definition of the variable sub-vectors of the planning problem*

with:

– A_a, B_a, C_a and $\underline{J_a}$ the sub-vector solution of size $1 \times N_{arcs}$;

– D_a, S_a and $\underline{V_a}$ the sub-vector solution of size $1 \times N_{vertices}$;

– N_{arcs} and $N_{vertices}$ respectively the number of candidate arcs and vertices for the planning problem;

– θ_k the angle of the voltages with respect to a reference vertex;

– γ_l the angle of the currents with respect to a reference vertex;

– $b \in \{1..N_{cond}\}$ the type of line, which depends on whether it is overhead or underground, the material used, the cross-section and type of insulation and the number of sections used in parallel. Table 5.12 provides a non-exhaustive example of a list of conductor types. Each type of conductor is associated with a linear cost and the electrotechnical characteristics that allow for the calculation of the load distribution (see Table 5.3).

b	Type	Material	Cross-section (mm²)	Number of conductors in parallel
1	Overhead	Aluminum	150	1
2	Overhead	Aluminum	240	1
3	Underground	Aluminum	150	1
4	Underground	Aluminum	240	1
5	Underground	Copper	240	1
6	Underground	Aluminum	150	2
7	Underground	Aluminum	240	2

Table 5.12. *Example of a list of conductor types*

5.3.1.2. *Applied example*

The purpose of defining the decision vector X in this way is to show as exhaustively as possible the set of decision variables that the planner must define. On the other hand, not all the variables contained in the vector X can be determined simultaneously, for reasons of mathematical complexity, as we will see later on in this chapter (see section 5.3.4). Furthermore, they do not always need to be determined together, as this depends on the type of planning: short-/long-term, with or without DER, "green field"/system expansion. Table 5.13 shows some planning examples in which decision sub-vectors need to be determined.

Applied cases	Decision sub-vectors
Construction of the architecture: drawing of the lines	A
Construction of the architecture: line layout + conductor sizing	A, B
Construction of the architecture: line layout + conductor sizing + determination of the normal operating scheme	A, B, C, D
Reinforcement (excluding the creation of new lines) to resolve constraints due to changes in loads for example	B
Reconfiguration	C
Addition of new RCSs to improve continuity of service	C, D
Load balancing calculation	$\underline{S}, \underline{I}, \underline{V}$
Optimization of the voltage plan	$\underline{S}, \underline{I}, \underline{V}$
DER optimization: location and power levels injected/supplied by generation, storage, curtailment	$\underline{S}, \underline{I}, \underline{V}$
Connection of a customer-generator	A, B
Connection of a customer-generator via a "smart connection offer" (ORI): we size both the feeder or the antenna dedicated to the connection and the production curtailment necessary to avoid constraints on the system	A, B, $\underline{S}, \underline{I}, \underline{V}$

Table 5.13. *Decision variables used, depending on the application – examples*

The list of applied cases given in Table 5.13 is not exhaustive since there are plenty of other case studies for which the combinations are numerous. For each study case, we seek to define the variables of the sub-vectors corresponding to the problem, however, the other sub-vectors are also important. The sub-vectors other than those to be solved can be:

– *indeterminate*: for the construction of the architecture, sub-vectors A, B, C and D can be determined simultaneously according to the choice that has been made for the modeling and optimization, or successively; we can, for example, determine the layout of the lines via sub-vector A without the location of the RCSs being decided (sub-vector D) or without the conductors being sized (sub-vector B);

– *determined*: for a load flow calculation, the variables to be solved are the extracted and injected powers (sub-vector S), currents (sub-vector $\underline{I}$) and voltages (sub-vector $\underline{V}$) on the system, but the sub-vectors A, B, C and D must be fixed beforehand in order to be able to calculate, among other things, the line parameters needed for the calculation and to reconstruct the power flows.

The relationships between the decision sub-vectors depend on the strategy of the planner. For example, it may be necessary to determine the structure of the system (sub-vectors A, C and D) in order to size the conductors (sub-vector B). This sizing can be done:

– only with knowledge of the loads supplied by each section (sub-vector $\underline{S}$) to estimate the technical constraints;

– by calculating precisely the level of these constraints via a load distribution calculation (sub-vectors $\underline{S}$, $\underline{I}$, $\underline{V}$);

– totally independently of the other sub-vectors: For example, in dense urban environments, the planner may decide to use a single cross-section of 240 mm^2 aluminum underground cables in order to avoid future costly works (trenches).

5.3.1.3. *Temporal aspect*

We add to vector X the index δt to take into account the temporal aspect of planning. This temporal aspect can correspond to time horizons (short-, medium- and long-term), sizing situations (consumption/production, normal/backup emergency regime) or time periods for operational planning that depend on the mechanism(s) under study. We consider four time horizons here:

– Definition of the master plan, "historical" method ($N_{\delta t} = N_{normal} + N_{backup}$, with N_{normal} = number of normal regimes and N_{backup} = number of emergency backup regimes): the sizing case is the peak consumption since the 10-year generation is considered marginal. We consider that the occurrence of peak consumption takes place in the last year of the study period of the system, at 30 years for example. We also need to evaluate the state of the system for the considered emergency backup regime. Therefore, we have two δt instants that may increase if the DSO defines several normal and emergency backup regimes.

– Definition of the master plan, taking into account the generation ($N_{\delta t} = 2 \times (N_{normal} + N_{backup})$): if we take into account distributed generation, we have to integrate the sizing case for which the consumption is minimum and the generation is maximum. We therefore have twice as many δt time instants through which to study the system, compared to the first case.

– Definition of the master plan and intermediate targets ($N_{\delta t}' = N_{\delta t} \times N_{ci}$ with N_{ci} the number of intermediate targets): If the planner's goal is to equally define intermediate targets to move from the initial target to the long-term target, δt will need to be multiplied by the number of intermediate targets. For both normal and emergency backup schemes, the decision sub-vector $C_{\delta t}$ must be determined for each δt time, but not the vectors A, B and D due to the fact that the system topology does not change when switching from a normal to an emergency backup situation. On the other hand, for different targets, vectors A through D can be modified for each δt to model the installation, reinforcement or removal of lines, substations or switching devices over time so as to reach the ultimate target.

– Integration of operational constraints in planning: as planning moves towards a more operational approach, that is, by taking into account flexibilities such as curtailment, reconfiguration, etc., the study of a limited number of sizing cases is no longer possible. For example, automatic reconfiguration to minimize technical losses on the system must take into account a number of sizing cases, at least equal to the number of desired reconfigurations: strictly equal, if we already know to determine that the sizing situation for each period corresponds to an RCS configuration, or higher if the best configuration must be averaged across all time steps of the periods considered (the number of δt then depends on the desired accuracy). In order to study flexibility solutions such as load curtailment, an approach limited to a sizing situation (which could work for the load curtailment of a generation facility) does not allow modeling phenomena such as rebound and deferral, which can generate constraints on the system if they are not anticipated. The study of the system based on load curves of the different users of the system is thus necessary.

Table 5.14 gives some examples of how the time aspect is taken into account in the decision vector.

5.3.1.4. *Consideration of the existing system*

To take into account the existing system, a part of the solution vector can be fixed in advance. The layout of the existing lines must be added to graph G of the planning problem and selected in the solution sub-vectors, according to the type of line and the type of conductor. Depending on the optimization method used, the existing system can remain fixed (the corresponding variables cannot be modified) or can evolve (the corresponding variables can be modified, either by removing some existing lines or by changing their type). We can also not select the lines used in the existing system for the initial solution but instead assign to the corresponding arcs a zero cost $p(E_l)$ so as to "encourage" the solution method to use the existing one (see section 5.2.4).

Applied case	Decision-making vector	Number of δt ($N_{\delta t}$)
"Historical" planning of electrical systems, without integration of DERs	$X_{\delta t} = \begin{bmatrix} A \\ B \\ C_{\delta t} \\ D \\ S_{\delta t} \\ J_{\delta t} \\ V_{\delta t} \end{bmatrix}$	(One normal scheme + N backup schemes) × (One sizing situation in consumption)
Power system planning with integration of distributed generation	$X_{\delta t} = \begin{bmatrix} A \\ B \\ C_{\delta t} \\ D \\ S_{\delta t} \\ J_{\delta t} \\ V_{\delta t} \end{bmatrix}$	(One normal scheme + N backup schemes) × (One sizing situation in consumption + one production sizing situation)
"Historical" power system planning, without integration of DERs with the construction of intermediate plans (5-year target, 10-year target, etc.)	$X_{\delta t} = \begin{bmatrix} A_{\delta t} \\ B_{\delta t} \\ C_{\delta t} \\ D_{\delta t} \\ S_{\delta t} \\ J_{\delta t} \\ V_{\delta t} \end{bmatrix}$	(One normal scheme + N backup schemes) × (One sizing situation in consumption) × (N targets)
Automatic reconfiguration of RCSs to reduce technical losses	$X_{\delta t} = \begin{bmatrix} A \\ B \\ C_{\delta t} \\ D_{\delta t} \\ S_{\delta t} \\ J_{\delta t} \\ V_{\delta t} \end{bmatrix}$	Depends on the desired reconfiguration periodicity: hourly, daily, weekly, monthly, etc., and on the study period considered. For example, for a weekly reconfiguration over a year, the planner will need to determine the corresponding 52 δt sizing.
Implementation of flexibility strategies such as power modulation (e.g. consumption or generation curtailment), voltage regulation (VVC for Volt-Var Control, on-load tap changer) as an alternative to reinforcement	$X_{\delta t} = \begin{bmatrix} A \\ B \\ C \\ D \\ S_{\delta t} \\ J_{\delta t} \\ V_{\delta t} \end{bmatrix}$	For example, over a typical year with a time step of 30 min, we will have 17,520 δt instances to study.

Table 5.14. *Temporal aspect of the decision-making vector according to the applied cases – examples*

5.3.2. *Definition of the multi-objective function to be optimized*

5.3.2.1. *Minimization of the overall discounted cost*

The first objective of the planning problem is to minimize the overall discounted cost of the system across the study period.

The cost objective function f_{costs} is divided into:

– *CAPEX* (*CAPital EXpenditures*), f_{CAPEX}, which includes the laying of lines, conductors and switching devices;

– operating costs called *OPEX* (*OPerational EXpenditures*), f_{OPEX}, which include losses, energy not supplied, the cut-off of power in case of failure, and the flexibility used.

Chapter 2 describes the CAPEX and OPEX for the traditional planning case in detail, to which must be added the action of the flexibility provided by DERs.

The following equations are adapted to the decision vector formulation presented in the previous section and differ slightly from the techno-economic equations in Chapter 2.

$$f_{costs} = \sum_{n=1}^{N_{years}} \frac{1}{(1+i)^n} [f_{CAPEX}(n) + f_{OPEX}(n)] \qquad [5.9]$$

$$f_{CAPEX}(n) = C_{CAPEX}^{conductors}(n) + C_{CAPEX}^{equipment}(n) + C_{CAPEX}^{DER}(n)$$

$$f_{OPEX}(n) = C_{OPEX}^{losses}(n) + C_{OPEX}^{reliability}(n) + C_{OPEX}^{DER}(n)$$

$$\begin{aligned} f_{CAPEX}(n) = & \sum_{a=1}^{N_{lines}} \sum_{l=1}^{N_{arcs}} A_{a,n}[l] \times p(E_l) \\ & \times \left(C_t\left(B_{a,n}[l], N_{sections,t}[l]\right) + C_c\left(B_{a,n}[l]\right)\right) \\ & + \sum_{a=1}^{N_{lines}} \sum_{k=1}^{N_{vertices}} D_{a,n}(k) \times C_{RCS} + C_{CAPEX}^{DER}(n) \end{aligned} \qquad [5.10]$$

$$f_{OPEX}(n) = \sum_{a=1}^{N_{lines}} \sum_{l=1}^{N_{arcs}} A_{a,n}[l] \times \big(f_{losses}(a,l,n) + f_{ENS}(a,l,n) + f_{PC}(a,l,n)\big) + C_{OPEX}^{DER}(n) \quad [5.11]$$

$$f_{losses}(a,l,n) = C_{peak} \times 3 \times r(B_{a,n}[l]) \times p(E_l) \times (I^{max}(a,l,n))^2 \quad [5.12]$$

$$f_{ENS}(a,l,n) = C_{ENS} \times \tau_{fail}(B_{a,n}[l]) \times p(E_l) \times \left(T_{iso} \times P_{avg}^{feeder}(a,l,n) + T_{rep} \times P_{avg}^{zone}(a,l,n)\right) \quad [5.13]$$

$$f_{PC}(a,l,n) = C_{PC} \times \tau_{fail}(B_{a,n}[l]) \times p(E_l) \times P_{avg}^{feeder}(a,l,n) \quad [5.14]$$

with:

– $I^{max}(a,l,n)$: the maximum current flowing through the section of the line *a* along the arc *l* in year *n*;

– T_{iso}: the fault isolation time (in hours);

– T_{rep}: the time to repair the fault (in hours);

– $P_{avg}^{feeder}(a,l,n)$: the average power of the feeder supplying the *l* arc used by line *a* in year *n*;

– $P_{avg}^{zone}(a,l,n)$: the average power of the area of which arc *l* is a part, in year *n*, delimited by the switching devices (see section 5.3.3.5);

– $p(E_l)$: the weighting coefficient expressed in units of length, corresponding to the distance between vertices *k* and *k'* connected by E_l either as the crow flies or via a realistic route (see Table 5.9).

The C_{CAPEX}^{DER}(n) corresponds to the investment made by the DSO for DER in year *n*. This can be, for example, a storage asset, if the DSO obtains the right on the exemption to install it on its system in order to cope with congestion, for example. In most cases, the costs related to DERs will be OPEX since the DSO will request flexibility from them to solve system constraints. Estimating the OPEX related to the use of flexibility therefore requires knowing the total annual energy corresponding to the service rendered by DERs (e.g. total energy curtailed, not produced, provided by storage, etc.) as well as the remuneration mode (fixed, variable, etc.). In general, we can write that:

$$C_{OPEX}^{DER}(n) = C_{FF} + C_{VF} \times E_F(n) \qquad [5.15]$$

with:

– C_{FF}: the fixed remuneration cost of flexibility (k€);

– C_{VF}: the variable remuneration cost of flexibility (k€/kWh);

– $E_F(n)$: total energy of flexibility supplied in year n (kWh).

5.3.2.2. *Balancing of PL products*

The balancing of PL products, defined in Chapter 2, is a purely French criterion that adds an objective of equity between end-users with respect to continuity of supply. It can be expressed in different ways, such as in equation [5.16].

$$f_{PL} = \sum_{z=1}^{N_Z} (PL(z) - \overline{PL})^2 \qquad [5.16]$$

with:

– N_z: the number of areas to have the same PL;

– PL(z): the PL product of the z-zone;

– $\overline{PL}$: the average PL of the structure.

The concept of a z-zone is defined in section 5.3.3.5.

5.3.2.3. *Weighting the sub-objectives*

To use several objective sub-functions simultaneously, we assign a weight w_h to each of them. Each objective sub-function can also be subdivided into weighted objectives, like the cost of the objective sub-function. Below we propose a solution for weighting the objective sub-functions. The overall objective function f must be minimized.

$$f = w_1 \times f_{costs} + w_2 \times f_{PL} \qquad [5.17]$$

$$f_{costs} = \sum_{n=1}^{N_{years}} \frac{1}{(1+i)^n} [w_3 \times f_{CAPEX} + w_4 \times f_{OPEX}] \qquad [5.18]$$

$$f_{CAPEX} = w_{3,1} \times \sum_{a=1}^{N_{lines}} \sum_{l=1}^{N_{arcs}} A_a[l] \times p(E_l) \times \big(C_t(B_a[l], N_{sections}[l]) + C_c(B_a[l])\big) + w_{3,2} \times \sum_{k=1}^{N_{vertices}} D(k) \times C_{RCS} + w_{3,3} \times C_{CAPEX}^{DER} \quad [5.19]$$

$$f_{OPEX} = \sum_{a=1}^{N_{lines}} \sum_{l=1}^{N_{arcs}} A_a[l] \times \big(w_{4,1} \times f_{losses} + w_{4,2} \times f_{ENS} + w_{4,3} \times f_{PO}\big) + w_{4,4} \times C_{OPEX}^{DER} \quad [5.20]$$

Each system operator may have a different internal strategy leading to an imbalance in the weights assigned to each objective.

For example, if the primary objective is service quality, then $w_{4,2}$ and $w_{4,3}$ may appear more prominently than the others. On the other hand, if cost is the primary objective, then $w_{3,1}$ to $w_{3,3}$ will be used. There is thus a real issue in adjusting these weightings.

5.3.3. *Defining constraints*

5.3.3.1. *Topological constraints*

Table 5.8 defines the constraints for each arc of graph G that define the possible routes for building the system's architecture, namely the type of sections and the maximum number of lines.

$$N_{sections}[l] \leq N_{\max sections}(E_l), \begin{cases} \forall l \in \{1..N_{arcs}\} \\ \forall a \in \{1..N_{lines}\} \end{cases} \quad [5.21]$$

$$A_a[l] \neq 0 \Rightarrow \mathrm{Type}(B_a[l]) \in \Omega_{\text{types sections}}(E_l) \begin{cases} \forall l \in \{1..N_{arcs}\} \\ \forall a \in \{1..N_{lines}\} \end{cases} \quad [5.22]$$

$$N_{sections}[l] = \sum_{a=1}^{N_{lines}} n_{sections}(B_a[l]), \forall l \in \{1..N_{arcs}\} \quad [5.23]$$

We recall the distinction between an arc, which connects two vertices in a graph, and *a line* which is composed of a set of arcs. A line can also be subdivided into branches, themselves subdivided into sections (see section 5.2.1.2). In the illustrative example shown in Figure 5.9, there are three arcs, (v_1; v_4), (v_4; v_2) and (v_4; v_3), and two lines, line no. 1 is composed of arcs (v1; v4), (v4; v2) and line no. 2 is composed of arcs (v_1; v_4), (v_4; v_3).

If the closed arcs were modeled by a single sub-vector A, the two lines in the example below would be connected and represent a single connected system (case no. 1 in Figure 5.9). The notation A_a, B_a and C_a, with $\forall a \in \{1..N_{lines}\}$, is used to distinguish between the lines. Therefore, for case no. 2 in Figure 5.9, where arc E_1 is used by line nos. 1 and 2, we will therefore have $A_1[1] = 1$ and $A_2[1] = 1$.

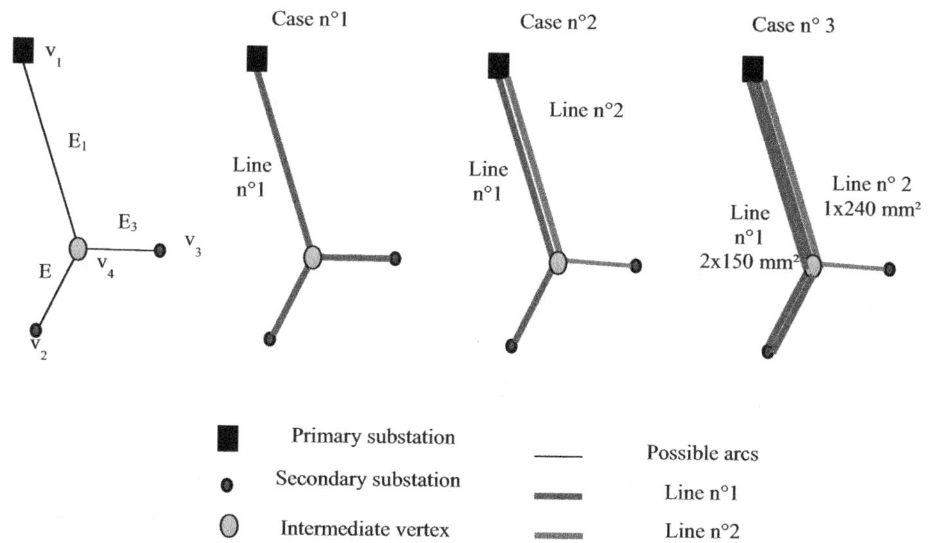

Figure 5.9. *Example of reducing a graph G to a pre-selected set of candidate arcs*

The number of electrical sections along the same route, in addition to being limited for each arc, has a direct impact on the installation cost, which can be reduced (e.g. for underground, the cost of trenching work can be shared) and on the permissible thermal limits (several underground cables in parallel have their thermal limit reduced (Afnor 2011).

If we take the example of Figure 5.9, and consider case no. 3 where line no. 1 uses two aluminum 150 mm² conductors in parallel (type β = 6 from Table 5.12) and line no. 2 uses a copper 240 mm² conductor (type β = 5 from Table 5.12), then arcs 1–3 will respectively have a number of sections equal to $N_{sections}[1] = 3$, $N_{sections}[2] = 2$ and $N_{sections}[3] = 1$.

	Case no. 1	Case no. 2	Case no. 3
Line layout	$A[1] = 1$ $A[2] = 1$ $A[3] = 1$	$A_1[1] = 1$ and $A_2[1] = 1$ $A_1[2] = 1$ and $A_2[2] = 0$ $A_1[3] = 0$ and $A_2[3] = 1$	$A_1[1] = 1$ and $A_2[1] = 1$ $A_1[2] = 1$ and $A_2[2] = 0$ $A_1[3] = 0$ and $A_2[3] = 1$
Number of arc sections	$N_{sections}[1] = 1$ $N_{sections}[2] = 1$ $N_{sections}[3] = 1$	$N_{sections}[1] = 2$ $N_{sections}[2] = 1$ $N_{sections}[3] = 1$	$N_{sections}[1] = 3$ $N_{sections}[2] = 2$ $N_{sections}[3] = 1$

Table 5.15. *Calculation of the vectors A and $N_{sections}$ for the example in Figure 5.9*

5.3.3.2. *Structural constraints*

Equation [5.24] reflects the fact that a type of overhead line (or underground cable) and a conductor type (section and technology) must be assigned to each line selected from the candidate arcs.

$$A_a[l] \neq 0 \Rightarrow B_a[l] \neq 0, \begin{cases} \forall l \in \{1..N_{arcs}\} \\ \forall a \in \{1..N_{lines}\} \end{cases} \quad [5.24]$$

The following equations reflect the fact that we can only open a line under normal operation if it is selected from among the candidate arcs (equation [5.25]), and if at least one of the vertices to which it is connected has an RCS (see equation [5.26] which reflects the fact that if the vertices v_k and $v_{k'}$ do not have an RCS, then the $v_k - v_{k'}$ section is closed)

$$A_a[l] = 0 \Rightarrow C_a[l] = 0, \begin{cases} \forall l \in \{1..N_{arcs}\} \\ \forall a \in \{1..N_{lines}\} \end{cases} \quad [5.25]$$

$$D_a[v_k] + D_a[v_{k'}] = 0 \Rightarrow C_a[l] = 0\,, \begin{cases} \forall l \in \{1..N_{arcs}\} \\ \forall a \in \{1..N_{lines}\} \end{cases} \text{with } e_l \quad = (v_k - v_{k'}) \qquad [5.26]$$

Equation [5.27] reflects the fact that the lines must be placed so as to supply all customers. We denote $\mu[v_k, v_{k'}]$ any path with the origin v_k and end $v_{k'}$, defined by a finite sequence of consecutive arcs belonging to E_k, connecting v_k to $v_{k'}$.

$$\forall v_k \in \Omega_{Cl}, \exists \mu[v_s, v_k] \text{ with } s \in \Omega_S \qquad [5.27]$$

The type of customer connection is also expressed as a constraint. We use as a parameter the degree (or valence) for a vertex of a graph, which is equal to the number of arcs connecting this vertex, and is noted $deg(k)$. The degree of intermediate vertices (which are neither source nor end-user nodes) must be greater than or equal to 2; the opposite means that such a vertex is an antenna without any load and is thus useless. A source node can have as many feeders as needed, and if its degree is zero, it does not feed the system under study.

$$degree(k) = \begin{cases} 1 \text{ for an antenna connection} \\ 2 \text{ for a secured feeder connection} \end{cases}, \forall k \in \Omega_{Cl}$$

$$deg(k) \geq 2, \forall k \in \Omega_{other} \qquad [5.28]$$

$$deg(k) \geq 1, \forall k \in \Omega_S$$

Finally, the radiality constraint must be applied.

– The case where the system is tree-like and therefore totally radial (each end-user is supplied by a single source): for each line a, consisting of $N_{vertices}$, the number of closed sections retained for the final solution must be equal to $N_{vertices}$-1. The supply of a load by a source is provided by equation [5.28].

$$\sum_{l=1}^{N_{arcs}} A_a[l] - C_a[l] = N_{vertices} - 1 \;\; with \begin{cases} \forall l \in \{1..N_{arcs}\} \\ \forall a \in \{1..N_{lines}\} \end{cases} \qquad [5.29]$$

– The case where the system is meshed (a customer can be supplied by one or more sources via several paths) but operated in a radial manner: each customer is supplied by a single source under normal operating conditions due to the presence of open RCSs. Radiality then applies to the feeders constituting each line a: for each feeder f of line a, consisting of $N_{vertices}$, the number of closed sections retained for the normal operating regime of the final solution must be equal to $N_{vertices}$-1. The supply of a load by a source is provided by equation [5.28].

$$\sum_{l=1}^{N_{arcs}} A_{a,f}[l] - C_{a,f}[l] = N_{vertices} - 1 \text{ with } \begin{cases} \forall a \in \{1..N_{lines}\} \\ \forall f \in \{1..N_{feeder}(a)\} \end{cases} \quad [5.30]$$

with:

– $N_{feeder}(a)$: the number of departures of line a (for example, 2 for the main feeder of a secured feeder architecture);

– $A_{a,f}$ and $C_{a,f}$ the sub-vectors of A and C corresponding to the feeder f from line a;

– when the system is operated as a mesh: there is no radiality constraint to apply.

5.3.3.3. *Power constraints*

Customers must be connected to sources within the power limits of the transformers to which the feeders are connected. The constraint is expressed for the sizing case of the emergency situation for HV/MV transformers at the primary substation and the normal operating situation for MV/LV transformers at secondary substations.

$$\sum_{i \in \Omega_{s,u,N-1}} S_i^{cons,MV} \leq S_{s,u}^{HV/MV\ transfo,N-1}, \quad \forall u \in \{1..N_s^{transfo}\}, \forall s \in \{1..N_{PS}\} \quad [5.31]$$

$$\sum_{i \in \Omega_d} S_i^{cons,LV} \leq S_d^{MV/LV\ transfo}, \quad \forall u \in \{1..N_d^{LV\ transfo}\}, \forall d \in \{1..N_{SS}\} \quad [5.32]$$

with:

– $S_i^{cons\ MV}$: the maximum power drawn from the MV load i (MV customer substation or MV/LV secondary substation);

– $\Omega_{s,u,N-1}$: all MV loads (MV customer substation or MV/LV distribution substation) supplied by the primary substation transformer u in emergency backup mode for this transformer's most restrictive situation;

– $S_i^{cons\ LV}$: the maximum power drawn by the LV customer i; all LV customers supplied by the MV/LV transformer of the MV/LV secondary substation d under normal operating conditions;

– Ω_d: all LV customers supplied by the MV/LV transformer of the MV/LV secondary substation *d* under normal operating conditions.

A power limit may also be imposed for each MV or LV feeder under normal operating conditions.

$$\sum_{i\in\Omega_{s,u,f}} S_i^{cons\,MV} \leq S_{s,u,f}^{max,feeders}, \forall f \in \{1..N_{s,u}^{feeders\,MV}\}$$
$$\forall u \in \{1..N_s^{transfo\,MV}\}, \forall s \in \{1..N_{PS}\} \quad [5.33]$$

$$\sum_{i\in\Omega_{d,g}} S_i^{cons\,LV} \leq S_{d,g}^{max,feeders}, \forall g \in \{1..N_{p,t}^{feeders\,LV}\}, \forall d \in \{1..N_{SS}\} \quad [5.34]$$

with:

– $S_{s,u,f}^{max,feeders}$: the maximum power of the MV feeder *f* supplied by the HV/MV transformer *u* of the primary substation *s*;

– $\Omega_{s,u,f}$: all the secondary substations supplied in normal operation by the MV feeder *f* supplied by the transformer *u* of the primary substation *s*;

– $S_{d,g}^{max,feeders}$: the maximum power of the LV feeder *g* supplied by the MV/LV transformer of the secondary substation *d*;

– $\Omega_{d,g}$: all LV customers supplied in normal operation by the LV feeder *g* supplied by the MV/LV transformer of the secondary substation *d*.

5.3.3.4. *Constraints on voltage and current levels*

As explained in Chapters 1 and 2, the voltages of any electrical peak in the system must remain within a range defined by the distribution system operators (equation [5.35]) and the currents in the conductors must remain below the allowable thermal limits (equation [5.36]). The voltage limits at the connection point of each MV customer, distribution substation or LV customer are imposed by legislation. The limits may vary, for example, between an MV generator substation and an MV/LV distribution substation because a generation facility does not have the same characteristics and uses as a consumer facility. The admissible voltage limits are respected via the voltage plan set up by the DSO: set point of the on-load tap changer, of the off-load tap changer of the secondary substation transformer, and other means of voltage regulation on the system (capacitor bank, distributed generation with local control of the voltage by injecting and/or absorbing active and/or reactive power, autotransformer in the middle of the system, etc.) and the characteristics of the MV or LV system conductors (larger cross-section, less resistive material and optimized routing)

$$V_{min}^{normal} \leq V_a[k] \leq V_{max}^{normal} \begin{cases} \forall k \in \{1 \dots N_{vertices}\} \\ \forall \mathrm{a} \in \{1..\mathrm{N}_{lines}\} \end{cases} \qquad [5.35]$$

$$V_{min}^{backup} \leq V_a^{N-1}[k] \leq V_{max}^{backup} \begin{cases} \forall k \in \{1 \dots N_{vertices}\} \\ \forall \mathrm{a} \in \{1..\mathrm{N}_{lines}\} \end{cases}$$

$$J_a^{max,backup}[l] \leq I_{th}(B[l], N_{sections}[\mathrm{l}]) \ if \ A[l] = 1, \begin{cases} \forall \mathrm{l} \in \{1..N_{arcs}\} \\ \forall \mathrm{a} \in \{1..\mathrm{N}_{lines}\} \end{cases} \qquad [5.36]$$

with:

– $J_a^{max,backup}[l]$: the current in the section l of line a in an emergency backup situation;

– I_{th}: the maximum admissible current depending on the type of conductor and the number of conductors per arc;

– V_{min}^{normal} and V_{max}^{normal}: the minimum and maximum voltage in normal operation;

– V_{min}^{backup} and V_{max}^{backup}: the minimum and maximum voltage in emergency backup mode.

5.3.3.5. *Constraints on quality indices*

From the choice of the arcs to supply the distribution stations (solution sub-vectors A_a) and the choice of the location and configuration of the remote-controlled switching devices (solution sub-vectors and C_a), we can divide the electrical system into N_z zones, bounded by switching devices.

As explained in Chapters 1 and 2, it is as important to ensure good power quality for all customers on the power system as it is to ensure that power quality is equivalent from one customer to another. The smallest scale for defining quality indices is the area bounded by two switching devices because all customers in the same area will have the same quality of supply. The calculation of the quality indices has already been introduced in Chapter 2. We define for all zones z, the maximum constraints for the quality indices (equation [5.37]).

$$\begin{cases} 0 \leq SAIDI_z \leq SAIDI_{max} \\ 0 \leq SAIFI_z \leq SAIFI_{max} \\ 0 \leq ENS_z \leq ENS_{max} \end{cases}, \forall \mathrm{z} \in \{1 \dots N_z\} \qquad [5.37]$$

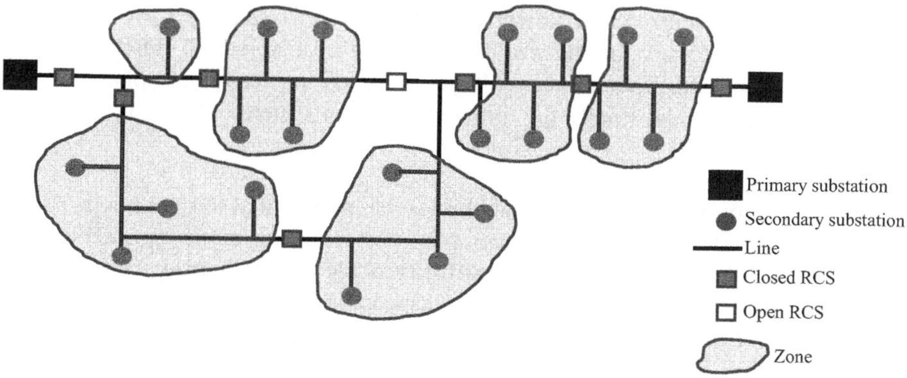

Figure 5.10. *Segmentation of the system into zones through switching devices*

These are the same zones that are used in the objective sub-function f_{PL} which equilibrates the PL product between them. The PL product can also be passed as a constraint to the planning problem, either per zone (equation [5.38]) or per feeder (equation [5.39]). The PL product can also be constrained by limiting the total power per zone or per feeder – which corresponds to the power constraints defined earlier, or by limiting the total length per zone or per feeder.

$$\begin{cases} PL_z \leq PL_{max}^{zone} \\ P_z \leq P_{max}^{zone} \\ L_z \leq L_{max}^{zone} \end{cases} , \forall z \in \{1 \dots N_z\} \qquad [5.38]$$

$$\begin{cases} PL_f \leq PL_{max}^{feeder} \\ P_f \leq P_{max}^{feeder} \\ L_f \leq L_{max}^{feeder} \end{cases} , \forall f \in \{1..N_{s,u}^{feeders}\}, \forall u \in \{1..N_s^{transfo}\}, \forall s \in \{1..N_{PS}\} \qquad [5.39]$$

5.3.4. *Load distribution calculation*

5.3.4.1. *Definition of the problem*

The load distribution calculation, also called load flow or power flow, is a central element for the planning and operation of electrical systems, from transmission to distribution. This calculation consists of calculating the unknown electrical quantities at any node of the system, based on the known electrical quantities (voltages, currents or power extracted/injected). It allows for estimating indicators such as losses for the evaluation of the cost objective function f_{costs}, cable congestion and operational voltage margins to verify compliance with technical constraints, and finally the initial state of the system before any transitory event.

The calculation for load flow, in addition to being essential to the various stages of distribution system planning, is a complex problem with numerous methods of resolution: Newton-Raphson's backward-forward algorithm, hybrid methods, etc. The mathematical problem is well known, especially in light of Carpentier's first formulation in 1962 (Carpentier 1963), and its solution has been the subject of numerous studies: Stott (1974) proposed a review of existing methods still widely used today. These methods will not be the subject of this book, but it is important to define in this section how to construct the load balancing problem mathematically and to make the link with the modeling of the planning problem proposed in this chapter.

In practice, the load flow calculation is used to compute the vectors $\underline{S}$, $\underline{J}$ and $\underline{V}$ of the decision vector $X_{\delta t}$. Initially, we calculate the vectors $\underline{S}$ and $\underline{V}$, that is, the voltage and power injected/supplied for the $N_{vertices}$ of the system. The vector J, that is, the currents flowing in the sections, is deduced from $\underline{S}$ and $\underline{V}$. Each of the vectors $\underline{S}$ and $\underline{V}$ has $2 \times N_{vertices}$ degrees of freedom because they are complex quantities.

$$\begin{cases} \underline{S_k} = P_k + j \times Q_k \\ \underline{V_k} = V_k \times e^{j\theta_k} \end{cases} \quad k \in \{1 \dots N_{vertices}\}$$

In reality, the number of degrees of freedom of the load flow problem is equal to $4 \times N_{vertices}$ only in the special case where we only calculate one complex voltage and one complex power for each node. The number of degrees of freedom can be larger, under the following assumptions.

– *The neutral regime*: if the system has a neutral conductor, we may need to calculate its potential at any point in the system. Even if it exists, this is not necessarily taken into account in the load flow problem: for example, *Kron's Reduction* assumes that the neutral conductor is perfectly grounded to the earth, which itself has zero potential.

– *Modeling of the "ground conductor"*: the electrical system interfaces with the earth at different nodes of the system, via earthing (at the level of transformers or loads for example) and electrotechnical phenomena represented by shunt admittances. It is often made the arbitrary double hypothesis that the earth if it were modeled as a "*dirt conductor*" or *earth wire*, has a zero and constant potential at any point in the system, and is therefore chosen as a voltage reference. This potential rarely appears in the proposed models (it is chosen equal to 0 and as such disappears from equations).

– *The unbalanced or balanced three-phase regime*: as we have seen in Chapter 1, a balanced three-phase line can be represented by an equivalent single-phase line. The voltage expressed at each node is, for example, the voltage for phase 1 from which we deduct the voltages for phases 2 and 3. The equivalent single-phase line is a four-port system: two for phase 1, and two for a second conductor which is a real or fictitious neutral conductor.

Table 5.16 shows the impact of the assumptions chosen for the load distribution calculation.

Conductors	Regime	Neutral	Ground	k model nodes
Three phases without neutral	Balanced	-	$V_g = 0$	$k \in \{1 \dots N_{vertices}\}$
	Unbalanced		$V_g = 0$	$k \in \{1 \dots N_{vertices}\} \times \{1,2,3\}$
			$V_g \neq 0$	$k \in \{1 \dots N_{vertices}\} \times \{1,2,3,g\}$
Three phases + neutral	Balanced	$V_n = 0$	$V_g = 0$	$k \in \{1 \dots N_{vertices}\}$
	Unbalanced	$V_n = 0$	$V_g = 0$	$k \in \{1 \dots N_{vertices}\} \times \{1,2,3\}$
		$V_n \neq 0$	$V_g = 0$	$k \in \{1 \dots N_{vertices}\} \times \{1,2,3,n\}$
			$V_g \neq 0$	$k \in \{1 \dots N_{vertices}\} \times \{1,2,3,n,g\}$

Table 5.16. *Assumptions for the load distribution calculation*

In the rest of this section, we keep the notation of $4 \times N_{vertices}$ degrees of freedom for simplicity but this number of course depends on the assumptions stated above.

5.3.4.2. *Putting it into an equation*

5.3.4.2.1. General case: non-zero ground potential

There are therefore $4 \times N_{vertices}$ quantities to be determined. To solve the problem we therefore need to set N_V variables and N_E equations relating them, with $N_V + N_E = 4 \times N_{vertices}$. Initially, we define as an intermediate quantity the vector $\underline{\mathrm{I}_\mathrm{k}}$ of the injected currents at each node.

$$\underline{\mathrm{I}_\mathrm{k}} = \mathrm{I}_\mathrm{k} \times \mathrm{e}^{\mathrm{j}\gamma_\mathrm{k}} \qquad [5.40]$$

Equation [5.41] shows the relationships between the electrical quantities $\underline{S_k}$, $\underline{V_k}$ and $\underline{I_k}$ at each node k of the system.

$$\begin{cases} \underline{S_k} = \underline{V_k}.\underline{I_k^*} \\ \underline{I_k} = \sum_{h=1}^{n} \underline{y_{kh}}.\left(\underline{V_k} - \underline{V_h}\right) \end{cases} \quad [5.41]$$

$\underline{y_{kk'}}$ is the admittance between nodes k and k', in Siemens, which depends on the structure of the system (whether nodes k and k' are connected and by what line length) and on the characteristics of the equipment used (number of conductors in parallel, material and cross-section, type of insulation, overhead line or underground cable, etc.). The complex admittance is defined by its real part $g_{kk'}$ called conductance, and its imaginary part $b_{kk'}$ called susceptance.

$$\underline{y_{kk'}} = g_{kk'} + j.b_{kk'} \quad [5.42]$$

Physically, the admittance $\underline{y_{kk'}}$ can represent shunt admittance elements between a conductor and the ground, transverse capacitances between two conductors, series admittance elements between two nodes connected by a line, or mutual inductances between two conductors. Table 5.17 shows the different formulations for the impedance and admittance between the c = 1 phase conductor of vertex k and the c' phase conductor of vertex index k'. We denote $\underline{z}$ and $\underline{y}$ for the sake of simplification. The parameters of resistance R, self-inductance L, mutual inductance M, conductance G and capacitances C (between phases and between phases and the ground) depend on the solution vector X, which determines the length and type of each conductor.

Node 1	Node 2	Conductor	Admittance $\underline{y}$
Vertex k Conductor $c' = 1$	$k' = k$	$c' = c$	n/a
		$c' \in \{2,3,n\}$[3]	$j\omega.C_{pp}(X_{\delta t}) + \frac{1}{j\omega.M(X_{\delta t})}$
		$c' = g$	$G(X_{\delta t}) + j\omega.C_{pt}(X_{\delta t})$
	$k' \neq k$	$c' = c$	$\frac{1}{R(X_{\delta t}) + j\omega L(X_{\delta t})}$
		$c' \in \{2, 3, n, g\}$	n/a

Table 5.17. $\underline{y_{kk'}}$ *coefficients*

3 In this case, the mutuals between conductors are neglected. To take them into account, we must express the line currents as a function of the voltages in the nodes in order to obtain the generalized line matrices defined in Kersting (2012).

For example, in Figure 5.11, admittances $\underline{y}_{12}$, $\underline{y}_{13}$ and $\underline{y}_{23}$ represent the resistive and inductive elements connecting nodes 1, 2 and 3, while admittances $\underline{y}_{1g}$, $\underline{y}_{2g}$ and $\underline{y}_{3g}$ represent the shunt capacitances between nodes 1, 2 and 3 and the ground. In Figure 5.11, each node has a single potential (the case of a single-phase equivalent model for a balanced three-phase line, for example), and an associated ground potential. The nodal rule equations according to equation [5.41] yield equation [5.43].

$$\begin{aligned}\underline{I}_1 &= \underline{y}_{12}(\underline{V}_1 - \underline{V}_2) + \underline{y}_{13}(\underline{V}_1 - \underline{V}_3) + \underline{y}_{1g}(\underline{V}_1 - \underline{V}_{1g})\\ \underline{I}_2 &= \underline{y}_{12}(\underline{V}_2 - \underline{V}_1) + \underline{y}_{23}(\underline{V}_2 - \underline{V}_3) + \underline{y}_{2g}(\underline{V}_2 - \underline{V}_{2g})\\ \underline{I}_3 &= \underline{y}_{13}(\underline{V}_3 - \underline{V}_1) + \underline{y}_{23}(\underline{V}_3 - \underline{V}_2) + \underline{y}_{3g}(\underline{V}_3 - \underline{V}_{3g})\end{aligned} \quad [5.43]$$

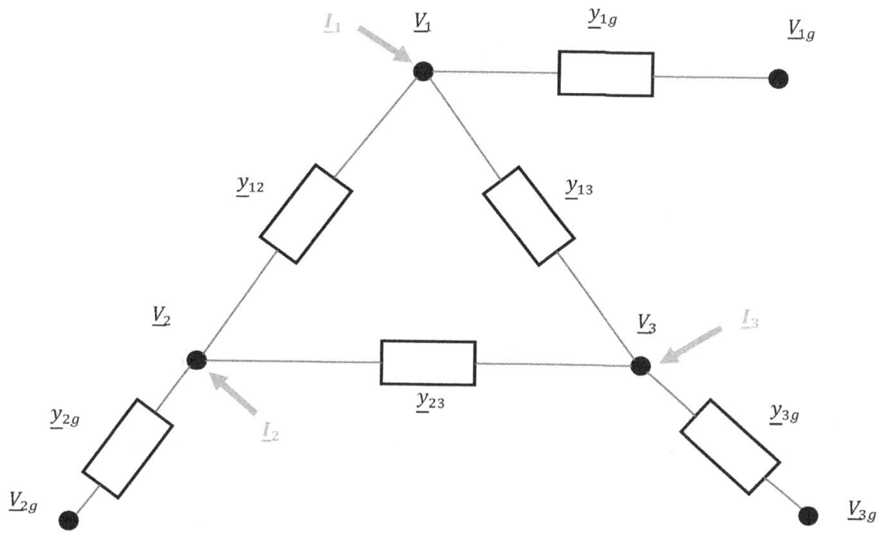

Figure 5.11. *Example of admittances for a three-node system*

By injecting the expression for I_k into the expression for $\underline{S_k}$ in equation [5.41], we obtain the following sequence of equations before arriving at the final expression in equation [5.45] for P_k and Q_k as a function of $g_{kk'}$, $b_{kk'}$ and the voltages in each node (modulus and phase).

$$\underline{S_k} = \underline{V_k}.\left(\sum_{k'=1}^{n} \underline{y_{kk'}}.\left(\underline{V_k} - \underline{V_{k'}}\right)\right)^*$$

$$\underline{S_k} = \sum_{k'=1}^{n} \left[\underline{y_{kk'}}.\underline{V_k}.\underline{V_k}^* - \underline{y_{kk'}}.\underline{V_k}.\underline{V_{k'}}^*\right]$$

$$\underline{S_k} = \sum_{k'=1}^{n} \left[(g_{kk'} + j.b_{kk'}).V_k^2 - (g_{kk'} + j.b_{kk'}).V_k.e^{j\theta_k}.V_{k'}.e^{-j\theta_{k'}}\right] \quad [5.44]$$

$$P_k + j.Q_k = V_k \sum_{k'=1}^{n} \left[g_{kk'}.V_k + j.b_{kk'}.V_k - (g_{kk'} + j.b_{kk'}).V_{k'}.e^{j(\theta_k - \theta_{k'})}\right]$$

$$\begin{cases} P_k = -V_k \sum_{k'=1}^{n} [V_{k'} \times (g_{kk'} \cos(\theta_k - \theta_{k'}) + b_{kk'} \sin(\theta_k - \theta_{k'})) - V_k.g_{kk'}] \\ Q_k = -V_k \sum_{k'=1}^{n} [V_{k'} \times (g_{kk'} \sin(\theta_k - \theta_{k'}) - b_{kk'} \cos(\theta_k - \theta_{k'})) + V_k.b_{kk'}] \end{cases} \quad [5.45]$$

5.3.4.2.2. Zero ground potential

The previous formulation considers that the ground potential is not constant at any point in the system; however, in the literature, the assumption of zero ground potential at any point in the system is commonly adopted, as illustrated in Figure 5.12.

The equation at the nodes will instead be written, renaming the admittance $\underline{y_{1g}}$ to $\underline{y_{11}}$ (idem for nodes 2 and 3):

$$\begin{aligned} \underline{I_1} &= \underline{y_{12}}(\underline{V_1} - \underline{V_2}) + \underline{y_{13}}(\underline{V_1} - \underline{V_3}) + \underline{y_{11}}\underline{V_1} \\ \underline{I_2} &= \underline{y_{12}}(\underline{V_2} - \underline{V_1}) + \underline{y_{23}}(\underline{V_2} - \underline{V_3}) + \underline{y_{22}}\underline{V_2} \\ \underline{I_3} &= \underline{y_{13}}(\underline{V_3} - \underline{V_1}) + \underline{y_{23}}(\underline{V_3} - \underline{V_2}) + \underline{y_{33}}\underline{V_3} \end{aligned} \quad [5.46]$$

which is written in the matrix form:

$$\begin{pmatrix} \underline{I_1} \\ \underline{I_2} \\ \underline{I_3} \end{pmatrix} = \begin{pmatrix} \underline{y_{11}} + \underline{y_{12}} + \underline{y_{13}} & -\underline{y_{12}} & -\underline{y_{13}} \\ -\underline{y_{12}} & \underline{y_{12}} + \underline{y_{22}} + \underline{y_{23}} & -\underline{y_{23}} \\ -\underline{y_{13}} & -\underline{y_{23}} & \underline{y_{13}} + \underline{y_{23}} + \underline{y_{33}} \end{pmatrix} \begin{pmatrix} \underline{V_1} \\ \underline{V_2} \\ \underline{V_3} \end{pmatrix} \quad [5.47]$$

$$\begin{pmatrix} \underline{I_1} \\ \underline{I_2} \\ \underline{I_3} \end{pmatrix} = \underline{Y}.\begin{pmatrix} \underline{V_1} \\ \underline{V_2} \\ \underline{V_3} \end{pmatrix} \quad [5.48]$$

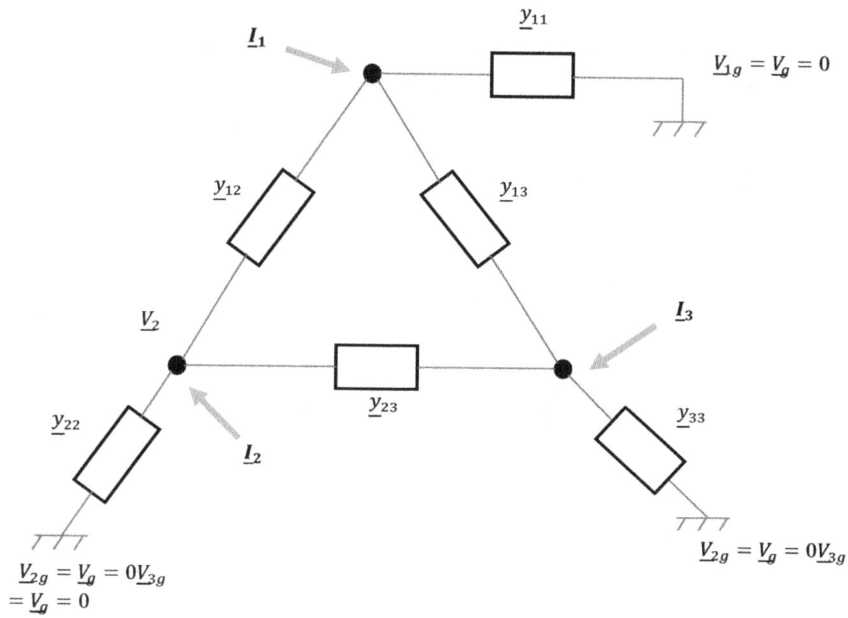

Figure 5.12. *Example of admittances for a three-node system and the assumption of constant and zero Vg*

$\underline{Y}$ is the *admittance matrix* of the system, which expresses the injected currents as a function of the potentials, and is constructed generically according to equation [5.49]. If vertex k' and h are not connected to vertex k, then $\underline{\mathrm{y_{kk'}}} = \underline{\mathrm{y_{kh}}} = 0$.

$$\begin{cases} \underline{Y_{kk}} = \sum_{k'=1}^{n} \underline{\mathrm{y_{kk'}}} = \mathrm{G}_{kk'} + j.\mathrm{B}_{kk'} \\ \underline{Y_{kh}} = -\underline{\mathrm{y_{kh}}} = \mathrm{G}_{kh} + j.\mathrm{B}_{kh} \end{cases} \quad [5.49]$$

Equation [5.41] shows the relationships between the electrical quantities $\underline{S_k}$, $\underline{V_k}$ and $\underline{I_k}$ at each node k of the system can then be rewritten using the admittance matrix $\underline{Y}$, via equation [5.50].

$$\begin{cases} \underline{S_k} = \underline{V_k}.\underline{I_k^*} \\ \underline{I_k} = \underline{Y_{kk}}.\underline{V_k} = \sum_{k'=1}^{n} \underline{Y_{kk'}}.\underline{V_{k'}} \end{cases} \quad [5.50]$$

We can then express the powers $\underline{S_k}$ for each node k entirely in terms of the voltages at any node in the system, via equation [5.51].

$$\underline{S_k} = \underline{V_k}.\left(\sum_{k'=1}^{n} \underline{Y_{kk'}}.\underline{V_{k'}}\right)^* \quad [5.51]$$

$$P_k + j.Q_k = \underline{V_k}.\left(\sum_{k'=1}^{n} \underline{Y_{kk'}}.\underline{V_{k'}}\right)^* \quad [5.52]$$

$$P_k + j.Q_k = V_k \times e^{j\theta_k}.\sum_{k'=1}^{n} \underline{Y_{kk'}}^*.\left(V_{k'} \times e^{-j\theta_{k'}}\right)^* \quad [5.53]$$

$$P_k + j.Q_k = V_k \times e^{j\theta_k}.\sum_{k'=1}^{n} (G_{kk'} - j.B_{kk'}).\left(V_{k'} \times e^{-j\theta_{k'}}\right) \quad [5.54]$$

$$P_k + j.Q_k = V_k \sum_{k'=1}^{n} (G_{kk'} - j.B_{kk'}).\left(V_{k'} \times e^{j(\theta_k - \theta_{k'})}\right) \quad [5.55]$$

$$P_k + j.Q_k = V_k \sum_{k'=1}^{n} V_{k'} \times (G_{kk'} - j.B_{kk'}) \times (\cos(\theta_k - \theta_{k'}) + j.\sin(\theta_k - \theta_{k'})) \quad [5.56]$$

$$P_k + j.Q_k = V_k \sum_{k'=1}^{n} V_{k'} \times (G_{kk'} \times \cos(\theta_k - \theta_{k'}) + j.G_{kk'} \times \sin(\theta_k - \theta_{k'}) - j.B_{kk'} \times \cos(\theta_k - \theta_{k'}) + B_{kk'} \times \sin(\theta_k - \theta_{k'})) \quad [5.57]$$

We thus obtain the $2 \times N_{vertices}$ equations for the load flow calculation of equation [5.58] relating the powers to the voltages at any point in the system. This formulation is the most commonly used in the literature.

$$\begin{cases} P_k = V_k \sum_{k\prime=1}^{n} [V_{k'} \times (G_{kk'} \cos(\theta_k - \theta_{k'}) + B_{kk'} \sin(\theta_k - \theta_{k'}))] \\ Q_k = V_k \sum_{k\prime=1}^{n} [V_{k'} \times (G_{kk'} \sin(\theta_k - \theta_{k'}) - B_{kk'} \cos(\theta_k - \theta_{k'}))] \end{cases} \quad [5.58]$$

5.3.4.3. *Choice of known variables*

There is still $2 \times N_{vertices}$ unknown variables in the system to solve. In order to solve the system, we fix for each node one or more quantities (amongst V_k, θ_k, P_k, Q_k, I_k and φ_k, these six variables are dependent: if we fix two of them, we can determine the other four) and/or add new equations.

First, we define the *slack bus* or *swing bus* node, also called the $V\theta$ node, for which the modulus and phase of the voltage are known. This node corresponds to the vertex representing the transmission system upstream of the primary substations (for a study of the MV system) and the distribution system located upstream of the secondary substations (for a study of the LV system). This node serves as a phase reference for the rest of the system.

$$\underline{V_{slack}} = V_{slack} \times e^{j\theta_{slack}} \quad [5.59]$$

The reference θ_{slack} is chosen arbitrarily, usually at a value of zero. The reference voltage value V_{slack} is chosen according to the type of the source and its characteristics. For example, if the voltage source is the secondary of the HV/MV transformer of a primary substation, the V_{slack} value can be set to a reference value equal to a certain percentage of the nominal voltage. This setpoint will be provided by the on-load tap changer of the HV/MV transformer (see Table 5.5).

For the remaining nodes, to reduce the degree of freedom of the system to zero, we can:

– define *constant power* system loads with two types of nodes:

- *PQ load* nodes for which the injection and extraction power values are known,

- *PV generation* nodes for which the active power injected and the voltage modulus are known;

– define the active and reactive powers as a function of the voltage variable, in which case we speak of flexible loads whose injected or extracted power is modulated according to the voltage constraints:

- $P = f_1(V)$ and Q constant,

- P constant and $Q = f_2(V)$,

- $P = f_1(V)$ and $Q = f_2(V)$;

– define the *constant current* system loads by setting $\underline{I_k}$;

– define the *constant impedance* loads by adding an equation to the system, for example, a three-phase star-connected load:

$$\underline{V_{k,a}} - \underline{V_{k,n}} = \underline{Z_{k,a}}.\underline{I_{k,a}};$$

– add another equation linking several variables, such as the nodal rule for a node in the system with a neutral conductor that is modeled:

$$\underline{I_{k,a}} + \underline{I_{k,b}} + \underline{I_{k,c}} + \underline{I_{k,n}} = 0.$$

It is possible to define the known variables and the equations of the system to be solved for the load flow calculation by mixing these different approaches or by favoring one of them.

For example, a constant impedance approach allows for linearizing the equations and directly using linear optimization methods.

After calculating the vectors $\underline{S}$ and $\underline{V}$, vector $\underline{J}$ can be calculated for each segment l, oriented from node k to node k':

$$\underline{J_l} = y_{kk'}.[(V_k \cos\theta_k - V_{k'} \cos\theta_{k'}) + j \times (V_k \sin\theta_k - V_{k'} \sin\theta_{k'})]$$

$$if\ A[L] = 1, with \begin{cases} l \in \{1 \ldots N_{arcs}\} \\ E_l = (k, k') \\ k, k' \in \{1 \ldots N_{vertices}\}^2 \end{cases} \quad [5.60]$$

5.3.4.4. *Optimal power flow*

When fewer quantities are known then there are degrees of freedom in the system, that is, $N_V + N_E < 4 \times N_{vertices}$, the problem is underdetermined. The unknown variables are then constrained by minimum and maximum values, and so, we add an objective function: this then becomes a nonlinear optimization problem.

This problem is called *economic dispatching* because the objective function to be minimized is generally the production cost or the losses, known as optimal power flow (OPF).

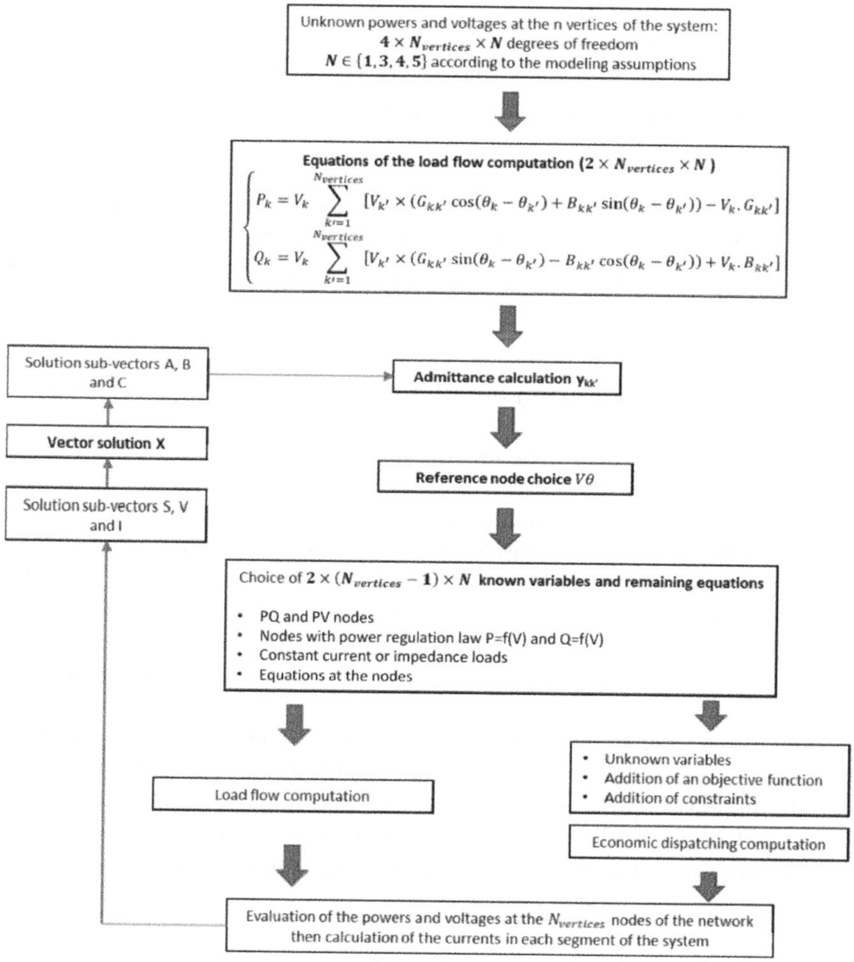

Figure 5.13. *Step-by-step summary of the load distribution calculation*

This more general form of the load balancing problem allows for various planning levers to be incorporated into the optimization, such as:

– optimizing the voltage plan by adjusting V_{slack} at the reference node, or Q_k at the capacitor bank nodes;

– optimization of producers' allocation by varying the injections of power P_k to maximize voltage operating margins or minimize technical losses, with or without reactive power control (via Q = f(V) or a constant $tan(\varphi)$);

– optimization of load curtailment by varying the extracted powers P_k;

– reconfiguring the power system by adding $y_{kk'}$ admittances for the Boolean variables $\underline{e_{kk'}}$ ($\underline{e_{kk'}} = 1$ if nodes k and k' are connected by admittance $\underline{y_{kk'}}$), to minimize technical losses, for example, or to optimize the voltage plan.

Figure 5.13 summarizes the various steps in the calculation of classical load flow and economic dispatch. When possible, simplifying assumptions (such as a small phase shift between voltages, $\theta_k - \theta_{k'} \approx 0$) allows for the linearization of equation [5.58] of the load flow calculation to change the nonlinear optimization problem into a simpler-to-solve linear problem (see section 5.4).

5.4. Algorithms for optimizing the planning of distribution systems

5.4.1. *Analysis of the optimization problem*

In the previous section, the optimization problem was formulated mathematically for the planning of electrical distribution systems whose first objective is to minimize the total discounted cost of the system, and secondly, to guarantee a balance in the quality of supply between users, while respecting a set of constraints related to the thermal limits of the equipment, to the legal obligations of the distributor, to the quality objectives, or to the topology of the area being served.

In order to understand the optimization algorithms and methods for solving this problem, we evaluate the characteristics that contribute to its complexity. The first source of complexity is the presence of *constraints* (topological, electrical, power balancing and power quality). Problems with constraints are more complicated to solve than problems without constraints, and the nature of the constraints also influences the possible solution methods.

The second source of complexity is the *combinatorial* (or *discrete*) nature of the problem. As we saw in section 5.2, many of the decision variables are either Boolean, such as the choice of arcs for energy routing, or the decision to open them or not under normal operation, or discrete, such as the type of line for each arc selected to be part of the structure. The discrete nature of the problem does not allow for the use of classical methods for continuous optimization problems (Newton's method, least squares methods, global optimization, etc.), which are generally simpler to solve, depending on the differentiability of the objective function, the convexity of the admissible set of solutions or the presence of constraints.

The discrete decision variables form a discrete and finite set of solutions. Among such a set, the optimal solution can be found by trying all solutions and comparing the value of their different objective functions; we refer here to an *exhaustive method*. However, as is the case for most *combinatorial optimization* (or *discrete optimization*) problems, the number of solutions is very large and as such their enumeration is not feasible, even with substantial computing power. We refer here to a *combinatorial explosion*. This explosion makes the exhaustive approach to finding the best solution among the set of possible solutions in a reasonable time very complex, even impossible. This becomes even more difficult given that the objective function and the evaluation of the constraints are computationally expensive. Moreover, since the set of feasible solutions is defined implicitly, it is also sometimes very difficult to find even one feasible solution.

For an example of a combinatorial explosion, let us consider the planning problem of an MV secured feeder structure for which the general characteristics are already fixed, that is, the number of feeders, $N_{feeders}$, and the sources to which they are connected. The objective is to determine which distribution substations each feeder must supply, which thereby allows for optimization of the layout and sizing of each line. The following formula gives the number of possible solutions to this problem.

$$N_{solutions} = \left(\frac{N_{PS}(N_{PS}+1)}{2}\right)^{N_{feeders}} \times {N_{feeders}}^{N_{SS}} \qquad [5.61]$$

with:

– $\frac{N_{PS}(N_{PS}+1)}{2}$: the number of possible configurations for feeders connecting two primary substations among the N_{PS} primary substations, while considering the possible looping of a branch on the same primary substation;

– ${N_{feeders}}^{N_{SS}}$: the number of possible solutions for the distribution of the secondary substations between the different feeders. Considering for example:

- two primary substations, three feeders and 50 secondary substations: ~1.9 x 10^{25} solutions,

- three primary substations, five feeders and 100 secondary substations: ~6.1 x 10^{73} solutions,

- four primary substations, 10 feeders and 200 secondary substations: ~10^{210} solutions.

The planning problem is also subject to several *nonlinearities*, either in the resolution of the load distribution problem or in the evaluation of some sub-function objectives such as the calculation of Joule effect losses. The problem is also mixed (discrete variables, as we have seen previously, but also continuous with regard to current and voltage levels). It is therefore a *mixed integer nonlinear programming* (MINLP) problem, which is the most general form that we can define in mathematical programming; all the other families of problems are derived from it.

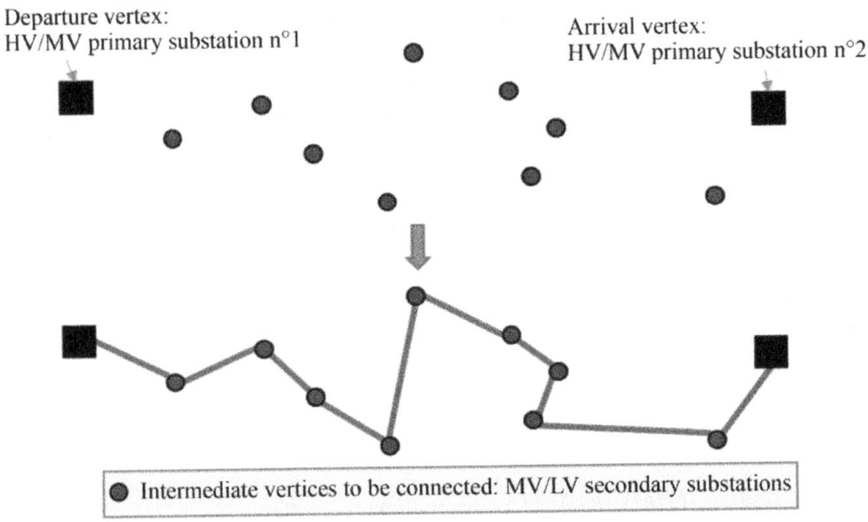

Figure 5.14. *Illustration of the traveling salesman problem for planning*

Another way to appreciate the complexity of the planning problem can be to look at the complexity of the optimization sub-problems that form our general problem and which have already been studied in the literature. We can take as an example the *traveling salesman problem*, which consists of determining the shortest path between different vertices of a graph by imposing the departure and arrival vertices. In planning, this corresponds to determining the route of a secured feeder to connect different secondary substations, using the primary substations supplying the secured feeder as the start and end points. Figure 5.14 shows an example of nine MV/LV substations to be connected. If this example is graphically feasible by a human being, or in a greedy way[4], we can, on the other hand, apprehend its complexity for a large number of vertices to be connected and for the weight of the arcs of the graph that does not correspond to distances *as the crow flies*, but instead to the distances that have to follow a realistic route (streets, natural obstacles, etc.), possibly

4 That is to say, by making a succession of optimal choices until convergence, without having to go back. The resulting solution is usually much better than the initial one, but the guarantee of having reached the global optimum is rarely assured.

weighted by arc coefficients. This problem is considered *NP-hard* (Laporte and Martello 1990), that is, we do not know of an algorithm that can solve it in polynomial time. The fact that a sub-problem of the planning problem is NP-hard illustrates its complexity.

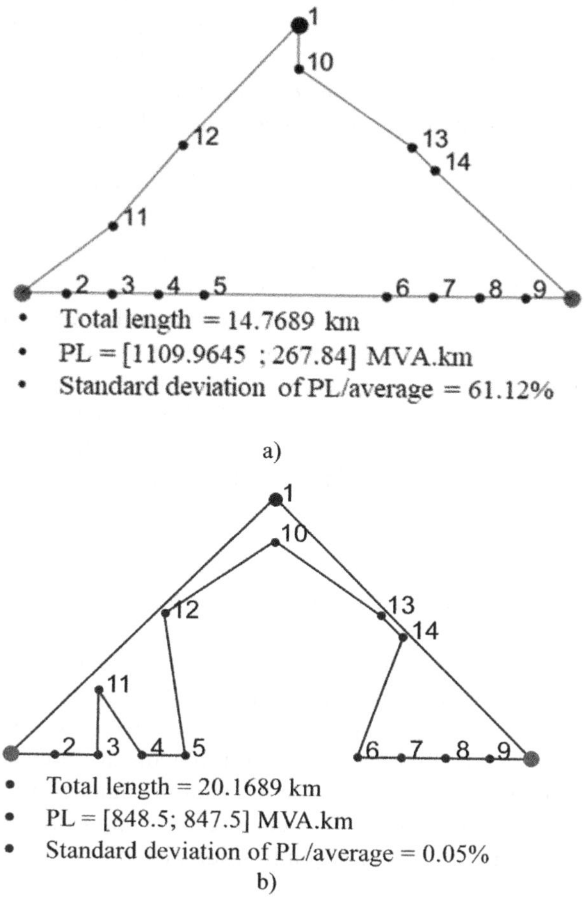

Figure 5.15. *Illustration of the contradictory character of a) the minimization of L and b) the balancing of PL products (source: Alvarez-Herault (2009))*

Finally, the problem is *multi-objective* as shown in equations [5.17]–[5.20]. However, some objectives may be contradictory. For example, the objective function f_{costs} is very much related to decreasing the total system length, as we will see in a case study in this chapter. By minimizing the structural lengths in order to reduce investment and operational costs, the resulting solution will not necessarily go in the same direction as balancing PL products (f_{PL}). The example in Figure 5.15 illustrates this point well. In both cases, it is a secured feeder-type structure with two main feeders. The objective is to find the paths of the two main secured feeders that minimize the total system length (case no. 1) and balance of the PL products (case no. 2). We can see that balancing the PLs in this case causes an increase in length by 37% because the load distribution is not homogeneous. A possible solution is to determine the structure that minimizes the total length of the system or minimizes f_{costs}, and then perform a second optimization phase to render a second solution that minimizes f_{PL} while remaining as close as possible to the result previously obtained.

We can also find a contradiction between the minimization of CAPEX and OPEX, for example between the minimization of the investment in remote-controlled switching devices $\sum_{a=1}^{N_{lines}} \sum_{k=1}^{N_{vertices}} D_a(k) \times C_{RCS}$ (equation [5.10]) and minimizing the costs of energy not supplied f_{ENS} and outages f_{PO} in case of incidents. Indeed, placing fewer RCSs on the system decreases the initial investment but increases the energy not supplied because the possibilities of reconfiguration in the backup regime are less. We can overcome this contradiction by imposing constraints on the quality indices (directly on the ENS or indirectly by playing with the SAIFI and SAIDI indices that influence the ENS and the power cut in the event of incidents), or by using the weighting coefficients w_h of the different objective sub-functions.

5.4.2. *Breakdown of sub-problems to be optimized*

Solving the planning problem of a distribution power system is usually not achieved by using a single algorithm from the list proposed in section 5.4.3, but rather through *decomposition methods* (Gardeux 2012). In this approach, the problem is divided into a combination of sub-problems that are easier to model and less prone to a combinatorial explosion in terms of the number of solutions, for each of which an already existing solution method can be applied, as far as is possible. We can differentiate three types of decomposition methods.

– *Dynamic programming*[5] if the problem can be divided into several steps that can be optimized in series. The optimization of each step, independently of the others, thus leads to the construction of the global optimum. The order of solution of the steps is generally from the smallest to the biggest problem, storing the intermediate results.

5 The term "programming" refers to the planning and scheduling of steps and not to its computer equivalent.

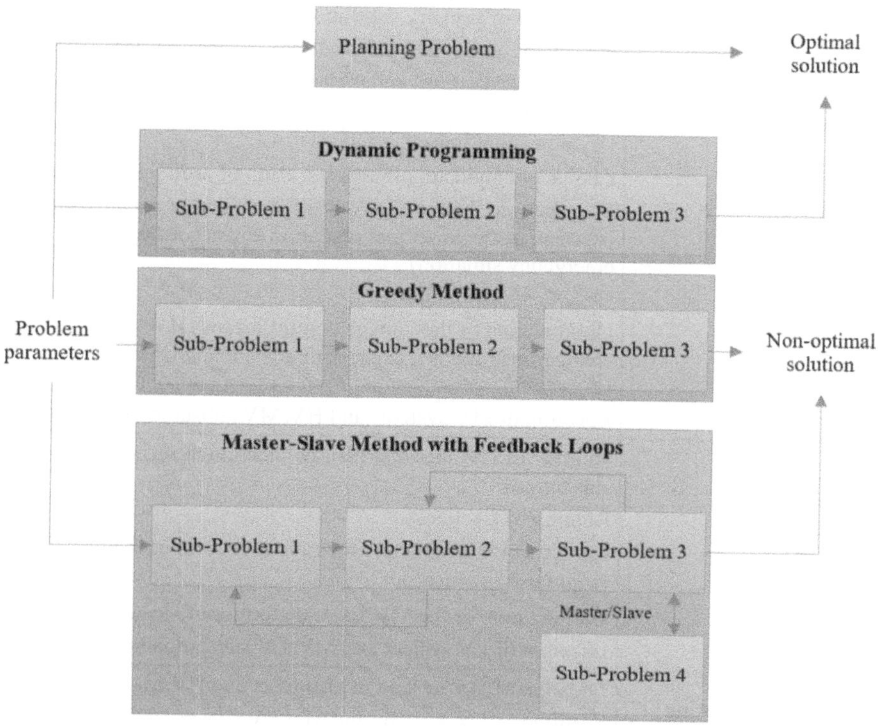

Figure 5.16. *Possible breakdowns of the planning problem*

– The *greedy method* (or *greedy algorithm*), if the problem is divided into several non-independent steps, and whose serial optimization will not lead to the optimal solution for the whole problem. The advantage here is that it makes the problem easier to model and allows for the use of methods and algorithms adapted to each step. The sequence of steps can be determined empirically, by the experience of the distributor, or by going from the smallest to the biggest problem. At each step, the optimal solution is determined. The steps are linked in a single sequence, so if a "sub-optimal decision" was made at any step, it is kept within the intermediate solutions until the ultimate solution is constructed.

– The *master-slave method* with feedback loops, provided it is possible to add feedback loops in the optimization sequence of the planning sub-problems. The solution of a sub-problem can also be conditioned by the solution of another sub-problem; for example, a load flow calculation must be re-run each time a structure has been built to check compliance with current and voltage constraints. We refer here to a *master-slave* relationship.

MV planning	Placement of primary substations Placement of secondary substations Number of MV lines and their connection to primary substations Layout of MV lines Choice of conductors for each section of each line Placement of the remote-controlled switching devices (number, position, status in normal operation and according to the emergency situation) Load flow calculation Optimization of the voltage control means (HV/MVon-load tap changer instructions, capacitor banks, P and Q control of generators) Location of MV systems and HV/MV substation investments
LV planning	Number of LV lines and their connection to secondary substations Routing of LV lines Choice of LV conductors for each section of each line Load flow calculation Optimization of the MV/LV transformer off-load tap changer Location of LV system and MV/LV substation investments
MV and LV planning	Location of MV system investments and HV/MV Location of LV systems and MV/LV substation investments

Table 5.18. *Examples of distribution system planning sub-problems*

The general distribution system planning problem can be broken down into three sub-problems: MV-only planning, LV-only planning and joint MV and LV planning. These three sub-problems can themselves be split into several sub-problems, a non-exhaustive list of which is given in Table 5.18.

5.4.3. *Summary of optimization methods used in planning*

The optimization methods used for distribution system planning problems are numerous, constantly revisited and often hybridized with each other to create new ones. Lists have been compiled numerous times in the literature, such as that by Willis (2004). The methods listed below can be applied to the general system planning problem or to different sub-problems to be solved independently.

Exhaustive search or brute-force search has already been mentioned: it examines all solutions to find the best one. It is robust and reliable but with a very long resolution time. Current computing resources allow this method to be applied to smaller problems, for example some of the sub-problems mentioned above. Exhaustive search is applicable to any type of problem provided the time needed to explore the space of possibilities is of a suitable duration.

Linear programming (or *optimization*) refers to methods for solving a problem whose objective functions and constraints are defined by linear functions. In this case, it is mathematically proven that the best solution belongs to the subset of solutions defined by the vertices of the admissible set (Jacquet-Lagreze 1998). For example, for a three-dimensional problem, we visualize an example of a solution, represented in Figure 5.17 by a polyhedron. The solution gives the optimal value to the studied function, which is located on one of the vertices of the polyhedron of admissible solutions.

This solution can be found quickly, for example in polynomial time using an interior point algorithm. If the variables can only take integer values, we speak of *integer linear programming* (or *optimization*). This second case is considered NP-hard and can be solved by splitting and evaluating it, or through intersecting plane methods (Gardeux 2012).

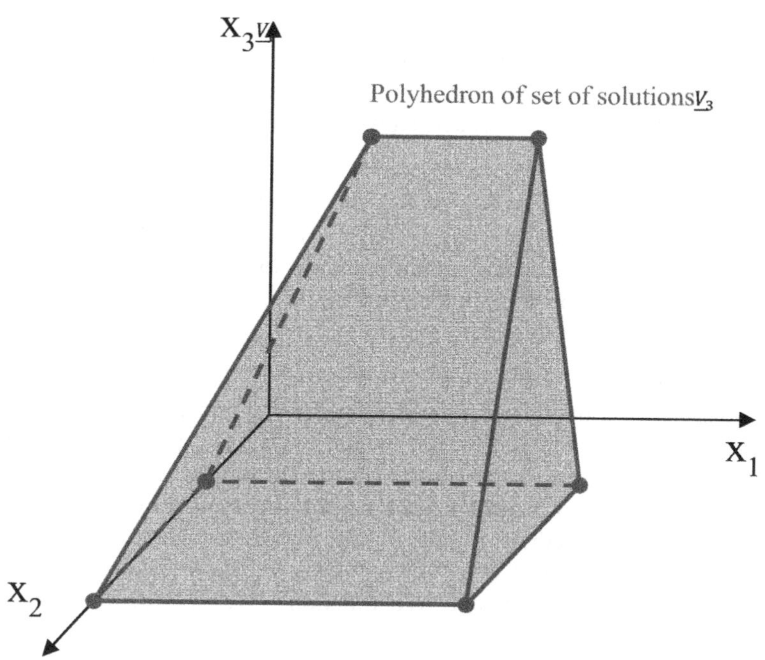

Figure 5.17. *Linear optimization for a three-dimensional problem*

In the case of nonlinear constraints, we speak of *nonlinear optimization*. If the functions defining the problem are differentiable as many times as is necessary, there are several algorithms to determine the optimal solution: Newton's method, quasi-Newton's method, conjugate gradient method, etc. The problem will be easier to solve if the objective function is convex or concave and if the set of constraints is convex. For one-dimensional problems (one decision variable), it is possible to apply the dichotomy method or the secant method (Gardeux 2012); these are simple methods to apply, are robust, and converge relatively quickly.

When linear optimizations, integer linear optimizations, or nonlinear optimizations are no longer applicable with the corresponding methods mentioned above, there are other classes of optimization methods to be considered.

Heuristic methods (Gardeux 2012) consist of a succession of deterministic decisions to approach the optimal solution. The rules for decision-making are inspired by empirical and/or theoretical findings and are based on the experience of the distributor and its community. We cite the following examples:

– *Branch and Bound*: this approach starts with the systematic estimation of all solutions by exploring the different branches of the solution tree. As the solutions are examined, the characteristics of an optimal solution are identified. These characteristics, coupled with heuristic rules, derived from the distributor's experience, make it possible to discriminate between certain branches in the exploration of solutions. For example, not using a cable with a cross-section smaller than 240 mm² at the head of the feeder. However, this technique remains difficult to apply to large problems.

– *Branch exchange* is also one of the first widely-used heuristics for power system planning. It consists of starting with a feasible solution, adding a line so as to form a looped mesh, then removing an old line so as to improve the chosen objective function, all the while respecting the constraints. The operation is repeated until the solution can no longer be improved.

– *Metaheuristics* (Siarry and Dréo 2003) present a higher level of abstraction than heuristics because they offer a general theoretical framework that makes them adaptable to problems of a different nature. We can classify metaheuristics according to their performance (value of the objective function) but also their adaptability (number of parameters, ease of setting them according to the case study), their variance (whether or not the same values are found by executing the same algorithm several times) or their computational time (in terms of time and number of simulations for the objective function). We also distinguish between distributed algorithms if several solutions are studied in parallel, or neighborhood algorithms if only one solution is being studied, which we make evolve. A specificity of metaheuristics is their great adaptability: we can hybridize them, optimize in parallel several contradictory objectives, search for several local optimums, etc.

– *Evolutionary algorithms*, developed in the 1970s, are inspired by natural evolutionary processes and are among the first metaheuristics. *Genetic algorithms*, for example, use the concepts of inheritance, mutation and crossover, for generations of individuals corresponding to solutions, until an optimal solution is obtained. They are very adaptive algorithms, easy to program and reasonably fast, but they do not provide any indication of the quality of the best solution found with respect to the true optimum, be it local or global.

– The *tabu search* is a procedure that explores the solution space locally by analyzing the immediate neighborhood of the current solution. The use of memory structures and adapted selection rules prevent the algorithm from getting stuck on a local optimum. This algorithm is also one of the first metaheuristics: Kirkpatrick et al. (1983).

– *Biomimetic metaheuristics*: here we cite as an example the ant colony algorithm, reproducing the exploration of a solution environment via the dispersion of pheromones that indicate paths to the local optima. This algorithm is similar to the *particle swarm optimization algorithms* created in 1995 by James Kennedy, based on the collaboration of several individuals (birds, bats, bees, etc.), of which we find many examples in the literature.

– *Simulated annealing* reproduces the process of the same name used in metallurgy; as a molten metal cools, the constraints accumulating in the core of its material are minimized in order to reach minimum energy in the steady state. This method is suitable for discrete and large solution spaces. The performance of these algorithms depends on the complexity of the problem and the empirical rules transferred from the user's experience.

For each distribution system planning sub-problem (see Table 5.18), there are *dedicated heuristic methods* (not applicable to any other sub-problem). We can cite the *sequential branch opening* method developed by Merlin and Back (1975) which addresses the optimal RCS configuration for a radial system sub-problem in normal operation. In this method, we consider the fully meshed system to be the one that satisfies the condition of minimum joule losses, which will form the initial solution to the optimization problem. At each iteration of the algorithm, we open the branch through which the lowest current flows until we obtain a radial system. We therefore open the branches that are the least far from the optimal system with regard to joule losses.

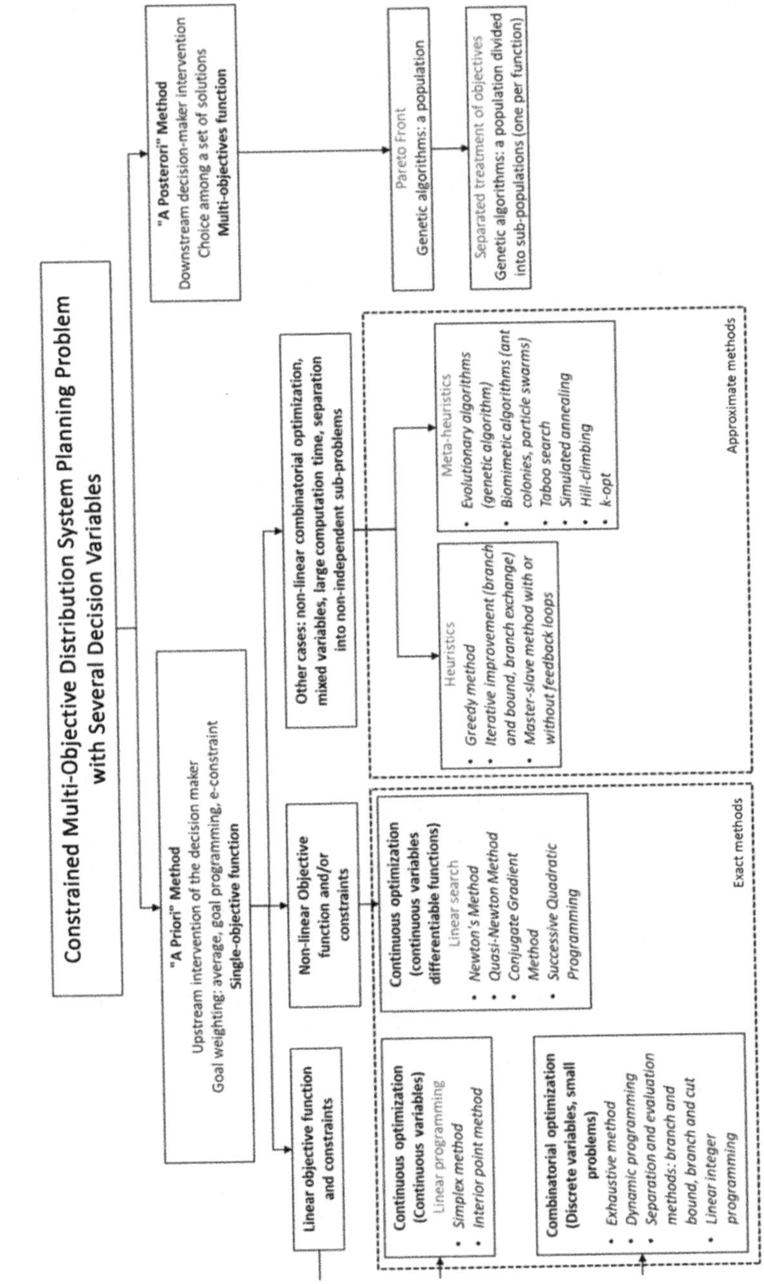

Figure 5.18. *Synthesis of optimization methods for the planning problem*

Figure 5.18 summarizes the optimization methods for planning electrical distribution systems.

5.4.4. *Integration of uncertainties in planning*

Whatever the time scale of the planning, there are uncertainties on the different parameters of the system that are more or less controlled. These uncertainties decrease as the spatial scale increases: there are more uncertainties about the consumption of an LV customer than about the consumption of an MV/LV distribution substation (the phenomenon of diversity). However, these increase with the time scale: the uncertainties on the predictability of the day ahead of renewable energy production are much less important than those on the production in thirty years, since, in addition to meteorological hazards, there are uncertainties on the type and location of renewable production. As mentioned in section 5.2.2, the uncertain parameters in long-term planning are:

– uncertainties on the location and power curves of uses, in particular, for DERs: location of MV and LV generators, electric vehicles (at home, fleets, recharging stations, etc.), the evolution of traditional consumption usages (electric heating vs. thermal), deployment of MV and LV storage systems, individual and collective self-consumption;

– uncertainties on the development of flexibilities and their economic model: consumption and/or production curtailment service.

These uncertainties are usually taken into account in the formulation of the planning problem using three different methods: the probabilistic analytical method, the probabilistic Monte Carlo method and the possibilistic method. The real options method has started to appear in recent years. Table 5.19 shows the advantages and disadvantages of these methods.

	Advantages	**Disadvantages**	**Applied example**
Worst-case method (fit and forget, copper plate)	Risk close to 0 Simpler in terms of calculation	Oversizing	Sizing of the conductors
Analytical probabilistic method	Very fast	Independent normal laws Linearization of the problem Strong dependence on the data used The positive effect of uncertainties is not valued	Load balancing calculation Hosting capacity Cost–benefit analysis of flexibility

Monte Carlo method	Simple to implement Can be used in co-simulation	High computation time Strong dependence on the data used The positive effect of uncertainties is not valued	Load balancing calculation Hosting capacity Cost–benefit analysis of flexibility
Possibilistic method	Very fast Several types of laws can be defined Simple arithmetic laws	Linearization required Strong dependence on the data used The positive effect of uncertainties is not valued	Load flow calculation
Real option method	Robust Allows values for the different options considered Makes it possible to value the arrival of new information Accurate resolution	Complex (needs to model all equations) Requires simplifying assumptions	Only for investment decision problems

Table 5.19. *Advantages and disadvantages of methods used for problems with uncertainties*

5.4.4.1. *The probabilistic analytical method*

This method is based on the strong assumption that the input variables are independent of each other and represented by normal laws. The uncertain variables of the problem are represented by a probability density function, and the output variables are computed by linearizing the different equations of the planning problem and using normal law arithmetic operations (Schilling et al. 1990). This method is often applied to perform load flow calculations with uncertainties.

5.4.4.2. *The Monte Carlo method*

This method is widely used when the problem studied cannot be linearized and/or when the input variables cannot be represented by independent, normal laws. Very simple to implement, it consists of performing a set of random simulations of the uncertain input variables for a range of variations, defined beforehand, and then calculating the objective function in order to find the output variables. This process is repeated several times until convergence is reached (Ellard 2003). This iterative method is very time-consuming but used widely because of its simplicity, the possibility of using "black box" models, and its potential coupling with other software.

5.4.4.3. *The possibilistic method*

Zadeh (1965) was the first to introduce this method via the theory of fuzzy sets. In fuzzy logic, a proposition belongs to a fuzzy set [0,1] and is no longer to the binary set {0,1}. Several degrees of satisfaction can thus be defined to introduce a more subjective approach that is more compatible with human behavior and does not decide its actions through a systematic binary approach. Let us take the case of an electric line: instead of associating the state {unconstrained (the current is lower than the limit current); or constrained (the current is higher than the limit current)}, we can choose to associate a load state corresponding to the percentage of the load in relation to the limit current. An uncertain variable is thus modeled by a fuzzy number on a set of confidence intervals according to a distribution that is usually normal, triangular or trapezoidal (Zadeh 1987). As with the analytical method, the output variables are then obtained via arithmetic operations, which are simpler.

5.4.4.4. *Emerging method: real options theory*

The evolution of the literature on investment strategies has moved away from a static framework, based on net present value (NPV) calculations of costs and benefits without the possibility of changing the characteristics of a project in the future to a dynamic framework in the decision-making process. The real options theory emerged in the 1980s (Myers 1977; Brennan and Schwartz 1985; Trigeorgis 1993; Dixit and Pindyck 1994; McDonald and Siegel 2021) as an evaluation technique, particularly appropriate for investments with a high degree of uncertainty. In long-term power system planning, it is difficult to acquire all the elements necessary to formalize the problem in a deterministic context due to uncertainty. Power system operators must be able to adapt their investment strategies in the coming years as and when new information becomes available. This flexibility has value, and it must be taken into account in the evaluation of investments. In this way, the value of a project is the sum of its discounted cash flows plus the value of the option to invest at the most favorable time, that is, when an uncertain variable – such as the value of a technology, the cost of a resource, or the market price – reaches a critical value that can be determined. The wait-and-see option, the growth option, the multi-sequence investment option, the abandonment option and the exchange option are widely discussed in the literature and applied in the energy field. To date, a limited number of studies have used the option approach in power system planning (Mancarella and Moriarty 2016; Schachter and Mancarella 2016; Schachter et al. 2016; Alvarez-Herault et al. 2021). Nevertheless, these studies highlight so many directions to explore for future research in this area.

5.5. Conclusion

The planning problem is a complex one, mathematically, but also because not all variables and constraints are modelable. Therefore, we cannot completely automate the planning of electrical distribution systems. However, it is possible to develop decision support tools according to the objectives and constraints defined by the planner and the data they have at their disposal. Therefore, in this chapter, we have tried to give a mathematical formulation of the planning problem that is as general as possible and that the reader can use to solve their own case studies. In the first step, a set of preliminary definitions allowed us to model the electrical system in the form of a graph, thus proposing an expression for the decision vector. Then, several objective functions (cost and reliability) and constraints (topological, structural and technical) were defined, as well as the equations for the load flow calculation, linking the electrotechnical parameters of the system together. This first step has allowed us to analyze the planning problem in order to propose a set of optimization methods applicable to each case study, and also to the presence of uncertainties. Some of these methods will be illustrated in the following chapter for the case study presented in Chapter 4.

5.6. References

Afnor (2011). Installations de branchement à basse tension, NF C14-100.

Alvarez-Hérault, M.C. (2009). Architectures des réseaux de distribution du futur en présence de production décentralisée. PhD Thesis, Institut National Polytechnique de Grenoble, Grenoble.

Alvarez-Hérault, M., Ionescu, O., Dib, J., Raison, B. (2021). Long-term planning of electrical distribution grids: How load uncertainty and flexibility affect the investment timing? Working document.

Brennan, M. and Schwartz, E. (1985). Evaluating natural resource investment. *Journal of Business*, 58(2), 135–157.

Carpentier, J. (1963). Contribution à l'étude du dispatching économique. *Bulletin de la Société Française des Electriciens*, 3(8), 431–447.

Coiffier, A. (2013). Schémas directeurs de développement des réseaux électricité de distribution MT (D4211). *Techniques de l'ingénieur Génie électrique*, 1–28.

Dixit, A. and Pindyck, R. (1994). *Investment under Uncertainty*. Princeton University Press, Princeton.

Doulet, A. and Horson, J.P. (2008). Réseaux de distribution – enfouissement. *Techniques de l'Ingénieur*, D4225, 11.

Ellard, D. and Ellard, P. (2003). SQ Course Book [Online]. Available at: http://ellard.org/dan/www/Courses/sq98_root.pdf

Gardeux, V. (2012). Conception d'heuristiques d'optimisation pour les problèmes de grande dimension : Application à l'analyse de données de puces à ADN. PhD Thesis, Université de Paris-Est Créteil, Créteil.

Jacquet-Lagreze, E. (1998). *Programmation linéaire: Modélisation et mise en œuvre informatique.* Economica, Paris.

Kersting, W. (2012). *Distribution System Modeling and Analysis.* CRC Press, Boca Raton.

Kirkpatrick, S., Gelatt, C.D., Vecchi, M.P. (1983). Optimization by simulated annealing. *Science,* 220(4598), 671–680.

Laporte, G. and Martello, S. (1990). The selective travelling salesman problem. *Discrete Appl. Math.,* 26(2/3), 193–207.

Mancarella, P. and Moriarty, J. (2016). Real options modelling and tool architecture for flexible network investment. Report, ENWL, Warrington.

McDonald, R. and Siegel, D. (2021). The value of waiting to invest. *Quarterly Journal of Economics,* 10, 707–727.

Merlin, A. and Back, H. (1975). Search for a minimal-loss operating spanning tree configuration in an urban power distribution system. In *Proc. 5th Power Syst. Comp.* Cambridge.

Myers, S. (1977). Determinants of corporate borrowing. *Journal of Financial Economics*, 5, 147–175.

Schachter, J. and Mancarella, P. (2016). A critical review of real options thinking for valuing investment flexibility in smart grids and low carbon energy systems. *Renew. Sustain. Energy Rev.*, 56, 261–271.

Schachter, J., Mancarella, P., Moriarty, J., Shaw, R. (2016). Flexible investment under uncertainty in smart distribution systems with demand side response: Assessment framework and practical implementation. *Energy Policy*, 97, 439–449.

Schilling, M.T., Leite da Silva, A.M., Billinton, R., El-Kady, M.A. (1990). Bibliography on power system probabilistic analysis. *IEEE Trans. Power Syst.,* 5, 1–11.

SéQuélec (2016). Guides 6 à 10 : Réalisation de postes HTA/BT de distribution publique. Report, Enedis, Conseul, Capeb.

Siarry, P. and Dréo, J. (2003). *Métaheuristiques pour l'optimisation difficile.* Eyrolles, Paris.

Siemens Energy Sector (2016). *Power Engineering Guide.* Siemens, Erlangen.

Stott, B. (1974). Review of load-flow calculation methods. *Proceedings of the IEEE,* 62(7), 916–929.

Trigeorgis, L. (1993). Real options and interactions with financial flexibility. *Financ. Manag.* 22(3), 202–224.

Willis, H.L. (2004). *Power Distribution Planning Reference Book*. Marcel Dekker, New York.

Zadeh, L.A. (1965). Fuzzy sets. *Inf. Control*, 8, 338–353.

Zadeh, L.A. (1987). *Fuzzy Sets as a Basis for a Theory of Possibility. Fuzzy Sets and Applications: Selected Papers*. John Wiley & Sons, New York.

6

Mathematical Tools for Planning: Application to Case Studies

6.1. Introduction

Given the complexity of the problems to be studied when planning, due to both temporal and spatial dimensions, the methods chosen are generally of the greedy or master-slave type with feedback loops (see Chapter 5). The various case studies covered in Chapter 4 are classified here according to their approach to the global optimization problem (greedy and master-slave), the chosen optimization methods and the consideration of uncertainties.

Table 6.1 summarizes for each case study, the optimization method used, the spatial and temporal scale and how uncertainties are considered, where appropriate.

These case studies fall into four categories:

– optimization of the traditional architecture without calling into question historical choices;

– design and optimization of new system architectures more adapted to the integration of new uses such as renewable generation and electric vehicles;

– integration of flexibility levers in investment decisions under uncertainty;

– development of planning tools for rural electrification (microgrids).

For a color version of all the figures in this chapter, see www.iste.co.uk/alvarezherault/distribution.zip

Case study	Optimization method	Spatial scale	Time scale	Integration of uncertainties
No. 1: Optimization of the existing traditional architecture	Decomposition method master-slave type with a feedback loop and use of metaheuristics	MV lines, secondary substations and primary substations	Long-term target: one sizing case (maximum consumption)	Monte Carlo method a posteriori (uncertainty on generation)
No. 2: New topologies for power distribution systems				
No. 2a: Hybrid loops (meshing on the same transformer)	Greedy decomposition method	MV lines, secondary substations and primary substations	Long-term target: one sizing case (maximum consumption)	Monte Carlo method a posteriori (uncertainty on generation)
No. 2b: Looped secured feeder (mesh between two different transformers)	Brute-force search	MV lines, secondary substations and primary substations	Long-term target: two sizing cases (maximum consumption and maximum production)	Optimization method integrated into a Monte Carlo algorithm (generation uncertainties)
No. 3: Solutions based on non-wire alternatives				
No. 3a: Integration of load curtailment of consumption in the planning process to postpone system investments	Linear programming with continuous variables	MV lines, secondary substations and primary substations	Study period of 10 years with a time interval of 30 min for each Monte Carlo simulation (about 100), that is, about 17,520,000 time intervals	Optimization method integrated into a Monte Carlo algorithm (uncertainties on production and consumption)
No. 3b: Integration of generation curtailment to postpone a primary substation investment	No optimization, direct resolution	Primary substation	Study period of 5 years with a time interval of 10 min	Probabilistic method

Case study	Optimization method	Spatial scale	Time scale	Integration of uncertainties
No. 4: NWA to integrate DERs in rural LV distribution systems	Monte Carlo method	LV customers and MV/LV substation	Study period of 15 years with a time interval of 30 min for each Monte Carlo simulation (about 50) or about 13,140,000 time intervals	Optimization method integrated into a Monte Carlo algorithm (uncertainties on production and consumption)
No. 5: Use of storage to defer investments in LV systems	Nonlinear programming (quadratic objective function and linear constraints)	LV customers and MV/LV substation	Study period of 15 years with a time interval of 30 min for each Monte Carlo simulation (about 50) or about 13,140,000 time intervals	Optimization method integrated into a Monte Carlo algorithm (uncertainties on production and consumption)
Case study no. 6: Rural electrification - Cambodia				
Phase balancing problem	Integer linear programming	LV customers and MV/LV substation	Long-term target: one sizing case (maximum consumption)	None
Placement and sizing of storage and production units	Nonlinear programming (quadratic multi-objective function and nonlinear constraints → metaheuristics	LV customers and MV/LV substation	Long-term target: one sizing case (maximum consumption)	None

Table 6.1. *Summary of case studies*

6.2. Master-slave decomposition method with a feedback loop and use of metaheuristics: case study no. 1

Case study no. 1 in Chapter 4 proposes a method for constructing the secured feeder architecture to meet the main planning objectives (customer supply, consumption evolution and supply continuity) and to optimize the existing traditional approach, that is, without DER. We focus here on the achievement of sub-objective no. 1, which is to supply all customers within the power system. For this case study, the spatial scale is all MV lines in the MV distribution system and the temporal scale is the long-term target. At this stage, the target chosen to be minimized is the total length of the system, called L_{tot}, and the decision variables are only the subvectors A_a, that is, the location of the sections of each line. As we have seen before, the final solution A is purely combinatorial, and there is a combinatorial explosion of the number of solutions (see section 5.4.1 of Chapter 5). For this combinatorial problem, we choose to model the line layout differently, by introducing another solution vector X_{rac}.

$$X_{rac} = \begin{pmatrix} x_1 \\ x_2 \\ \vdots \\ x_d \\ \vdots \\ x_{N_{SS}} \end{pmatrix} \text{ with } x_d = a \text{ if the substation d is connected to the feeder } a \qquad [6.1]$$

$\forall i \in \{1 \ldots N_{SS}\}, \forall a \in \{1 \ldots N_{lines}\}$

We also define beforehand the configuration of the secured feeders between the different primary substations: for example, the secured feeder $a = 1$ is fed by the primary substations $s = 2$ and $s = 3$. To determine the vector A from the vector X_{rac}, the following steps are performed.

– Step 1: the graph G containing all the vertices (primary substations, secondary substations and intermediate vertices) and the possible arcs are constructed (see section 5.2.1.1 of Chapter 5);

– Step 2: the shortest distance between each primary substation vertex or secondary substation within the graph G (and thus taking into account topological constraints such as the layout of streets or natural obstacles) is calculated with the Dijkstra algorithm (Dijkstra 1959);

– Step 3: we construct a second graph G' consisting only of the primary or secondary substations which is complete, as we can see in Figure 6.1;

– Step 4: for each secured feeder a, the order in which the loads are connected is obtained by solving the non-looped traveling salesman problem (Lawler 1985), taking as starting and ending vertices the primary substations feeding the secured feeder, and using as distances between the vertices the weight of the arcs of the graph G'. The algorithm used is the Christofides approximation algorithm (Christofides 1976);

– Step 5: once the route of the secured feeders is determined within graph G', we can determine their route in the initial graph G and thus reconstruct vector A by identifying the arc used for each section of each line.

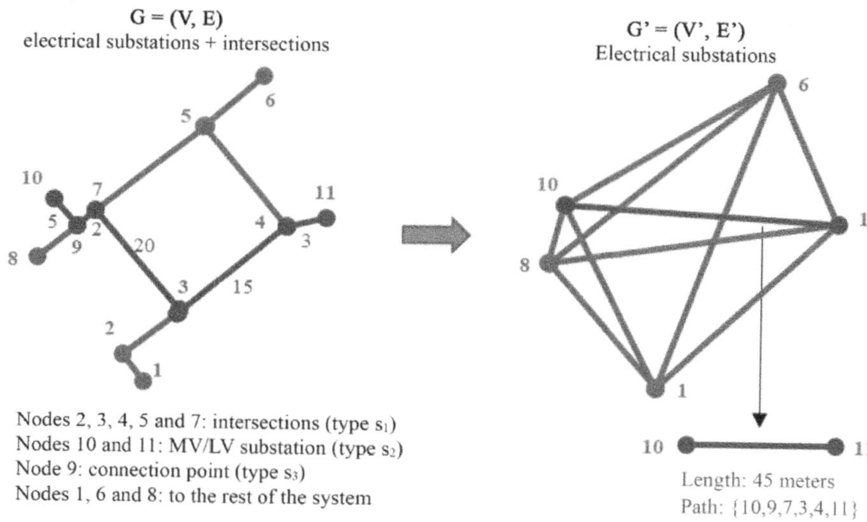

Figure 6.1. *Transformation of the input graph G into a complete graph G'*

The formulation via vector X_{rac} is adapted to the use of different heuristics. We choose to study four of them: the genetic algorithm, simulated annealing, tabu search and the ant colony algorithm. We compare these four heuristics on the IEEE test system with 69 nodes and three primary substations (used in Glamocanin (1993) and then Cai (2012)). We report in Table 6.1 the evaluation for the performance of these heuristics (lowest L_{tot} value obtained), the variance of the result, the number of parameters and their ease of tuning and performance (computation time and calls to the objective function). On the basis of the results observed in Table 6.2, we see that simulated annealing offers the best performance and a lower variance in terms of the results. It is also very simple to implement and requires few parameter adjustments when compared to the other methods.

Algorithm used	Genetic algorithm	Simulated annealing	Taboo search	Ant colony
Type	Distributed	Neighborhood	Neighborhood	Distributed
Solution range	197.89–213.09 km	166.53–188.23 km	202.22–272.07 km	201.27–221.54 km
Average	207.65 km	172.72 km	231.40 km	210.35 km
Standard deviation	5.94 km 2.86%	5.24 km 3.03%	20.94 km 9.05%	5.73 km 2.72%
Setting parameters	Seven ($N_{individuals}$, α, β, γ, p_S, selection and mutation method)	Three (τ_0, α, N_{trials})	Five (hash function, α, t_1, t_2, Δt_2)	Three (k, ρ, formation of pheromones)
Ease of adjustment	- -	+ +	-	-
Calculation time	20 h and 25 min	29 min	8 h and 5 min	10 h and 21 min
Calls to the function	2,000,000	121,708	100,001	1,000,000

Table 6.1. *Comparison of different heuristics for the construction of the architecture*

Simulated annealing assimilates our system – the power system – to a metal to which we alternate cooling and heating processes. In metallurgy, this process allows for a decrease in the thermodynamic free energy on which some important properties of materials depend. The process is described in Figure 6.3. In our case, the free energy corresponds to the L_{tot} length of the system, which we try to minimize.

T_{anneal} is a fictitious temperature that reflects the "excited" state of the current solution of length L_{tot}. The higher the T_{anneal}, the more likely an elementary change on the solution vector X_{rac} (equation [6.2]) that would lead to degradation (increase) of the target L_{tot} to be realized, according to the probability $p_{degradation}$ (equation [6.3]). This allows the solution space to be explored efficiently at the beginning of the simulation. The temperature gradually decreases with a coefficient α (typically set to 0.9 according to Dréo (2005)) to reduce the number of accepted changes, until converging at a local minimum. The initial temperature is set so that the initial acceptance rate is equal to 50% (Dréo 2005).

$$X_{rac}(d) = a \text{ with} \begin{cases} d \in \{1 \dots N_{SS}\} \\ a \in \{1 \dots N_{lines}\} \setminus X_{racc}(d) \end{cases} \quad [6.2]$$

$$\text{p}_{degradation} = exp\left(\frac{-\Delta L}{T_{anneal}}\right) \quad [6.3]$$

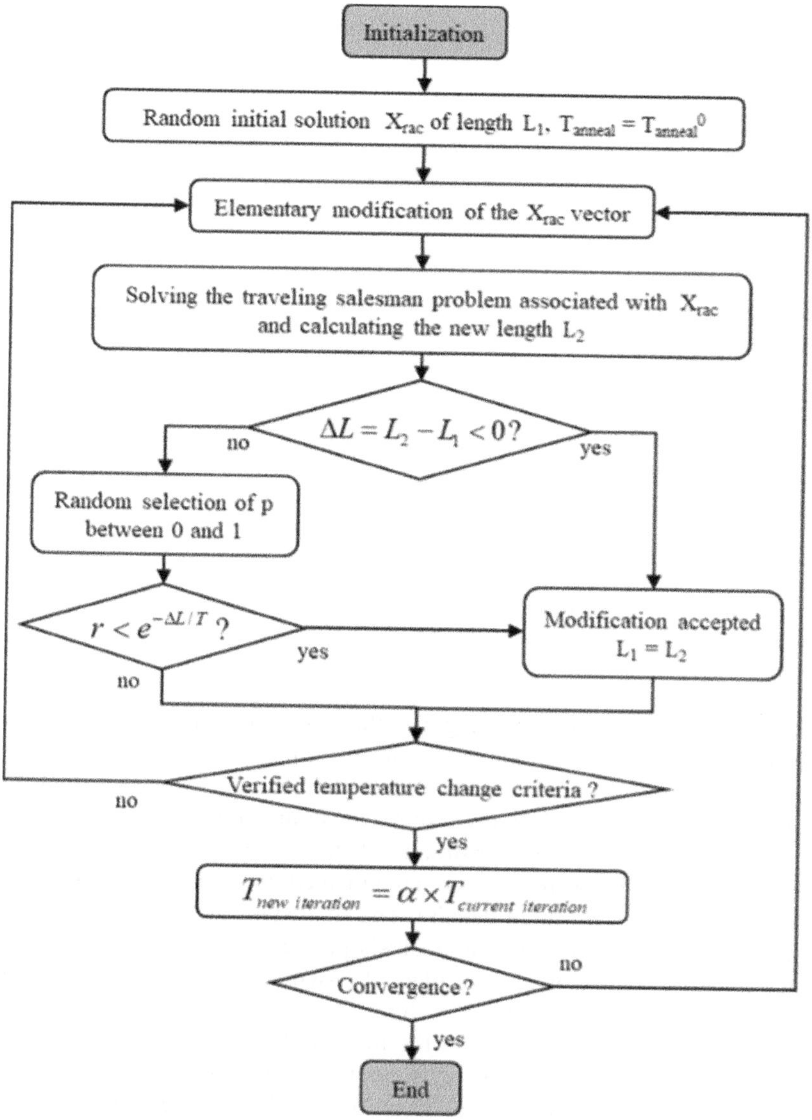

Figure 6.2. *Simulated annealing flowchart*

Figure 6.3 shows the evolution of the temperature steps, the length L_{tot} of the current solution and the length L_{tot} of the best solution for the example of the IEEE system with 69 nodes and three primary substations.

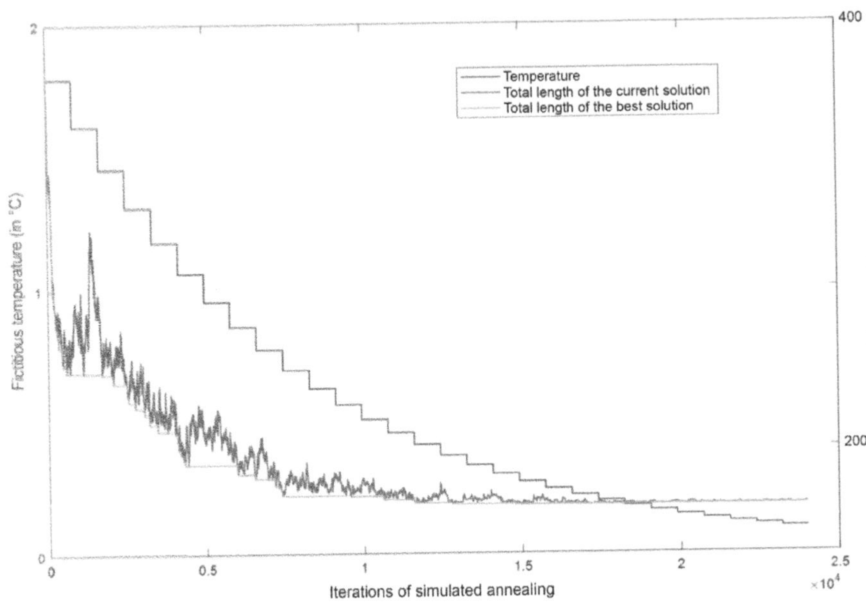

Figure 6.3. *Convergence of the simulated annealing on the 69-node IEEE example*

The minimization of the total length is the first step in the construction of the architecture. The obtained solution is optimized for the length criterion to which the investment (cables and trenches) and operational (technical losses and incidents) costs are directly correlated. However, some constraints may not be respected, such as the power and thermal limit constraints of the system elements. A greedy load exchange algorithm between lines is implemented to correct the optimized L_{tot} solution (see Chapter 4 for more details). The final steps are to size the conductors using a brute-force techno-economic analysis per section (OPEX and CAPEX are evaluated for each conductor), and to place the switching devices along the feeders with a greedy heuristic adapted to this sub-problem. Figure 6.4 shows the arrangement of sub-problems forming the main architecture construction problem. The method is greedy because there is no feedback loop, and leads to a local optimum. Some sub-problems are called by others via a master-slave relationship, such as the optimized line routing at each evaluation of L_{tot} for simulated annealing, or the load flow computation to check compliance with electrical constraints when sizing the conductors. The reader can find more information on this case study and the method used in Chapter 5, in Gouin (2015) and Gouin et al. (2015b, 2017).

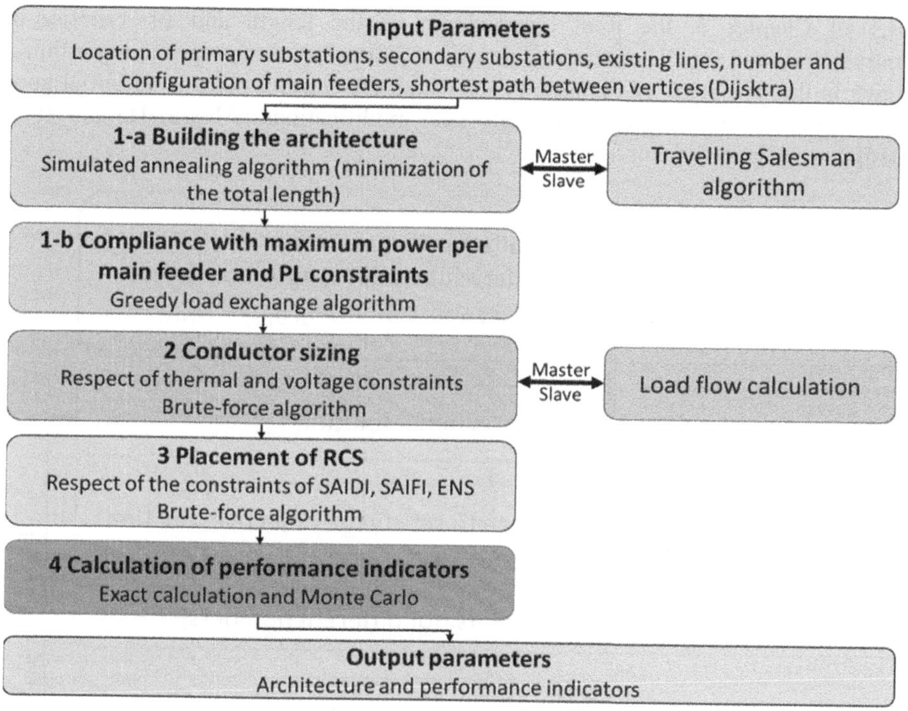

Figure 6.4. *Decomposition of the architecture construction into sub-problems*

6.3. Greedy decomposition method

6.3.1. *Heuristics: case study no. 2a*

Case study no. 2a of Chapter 4 proposes a hybrid MV loop architecture based on a secured feeder connecting MV/LV substation loops. The objective is to find the right compromise between the cost of the solution and the reliability criteria. We verify a posteriori that this architecture is favorable to the introduction of generation thanks to a hosting capacity (HC) calculation. For this case study, the spatial scale is the set of MV lines of the MV distribution system, and the temporal scale is the long-term target.

The objective function, in this case, is therefore composed of f_{CAPEX}, f_{OPEX} and f_{PL} as defined in section 5.3 of Chapter 5, without weighting and subject to topological, power, voltage and current level constraints. As explained in section 5.4.1 of Chapter 5, the joint optimization of the length and PL criterion is contradictory and the PL criterion is nonlinear. Therefore, the proposed algorithm is shown in the five steps of Figure 6.5, for which the starting point is an initial good solution that is then improved through local modifications (Alvarez-Hérault et al. 2010).

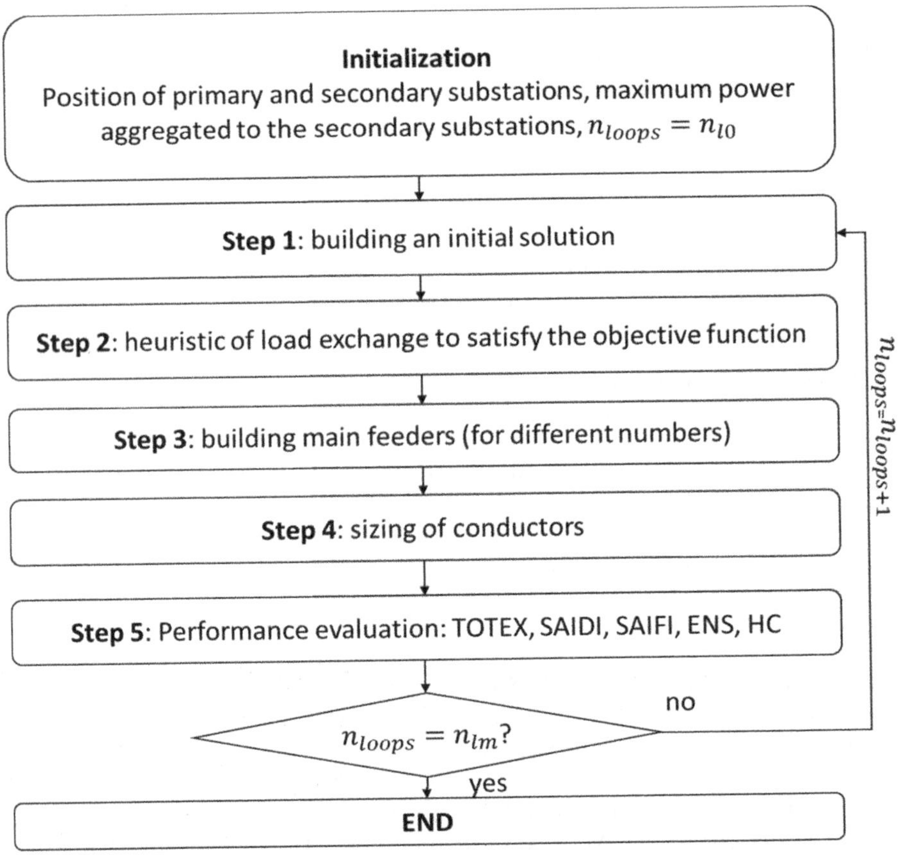

Figure 6.5. *Algorithm for the construction of hybrid loops*

6.3.1.1. *Step no. 1: building an initial solution*

The objective here is to build a good initial solution to obtain a realistic starting point for the optimization heuristic. This initial solution consists of two steps:

– Step no. 1a – Partitioning of the studied perimeter into n_{loop} zones: several methods can be considered, such as minimizing the distance through the application of clustering algorithms, or obtaining loops for a given number of homogeneous loads (checkerboard partitioning), in order to have an initial solution that is easily modifiable This latter option has been selected for this case study, the steps of which are detailed in Table 6.2;

Average number of secondary substations per loop:	
$N_{avg\,SS} = rou\left(\frac{N_{SS}}{N_{lps}}\right)$ with "rou" for rounding	[6.4]
Number of lines in the checkerboard (N_{CR}):	
$N_{CR} = arr\left(\sqrt{N_{lps}}\right)$	[6.5]
Number of columns in the checkerboard (N_{CCavg}) for the first N_{CR}-1 lines:	
$N_{CCavg} = arr\left(\frac{N_{lps}}{N_{CR}}\right)$	[6.6]
Number of secondary substations per checkerboard line for the first N_{CR}-1 lines:	
$N_{SS\,lines\,avg} = N_{avg\,SS} \times N_{CCavg}$	[6.7]
Number of secondary substations in the last row of the checkerboard, $N_{SS\,last\,line}$:	
$N_{SS\,last\,line} = N_{SS} - (N_{CR} - 1) \times N_{SS\,lines\,avg}$	[6.8]
Number of columns in the last row of the checkerboard, $N_{CC\,last}$:	
$N_{CClast} = N_{lps} - (N_{CR} - 1) \times N_{CCavg}$	[6.9]

Table 6.2. *Steps for checkerboard partitioning*

– Step no. 1b – Loop construction: once the secondary substations have been grouped into N_{lps} groups, they are connected in loops in order to minimize the distance. This problem, very classical in mathematics, is the vehicle routing problem (VRP). In this instance, the Christofides looping algorithm (Christofides 1976) has been implemented here.

Example: Step nos. 1a and 1b are illustrated in the example shown in Figure 6.6, consisting of two primary substations (red squares) and 31 secondary substations (black circles) that we wish to group into five loops.

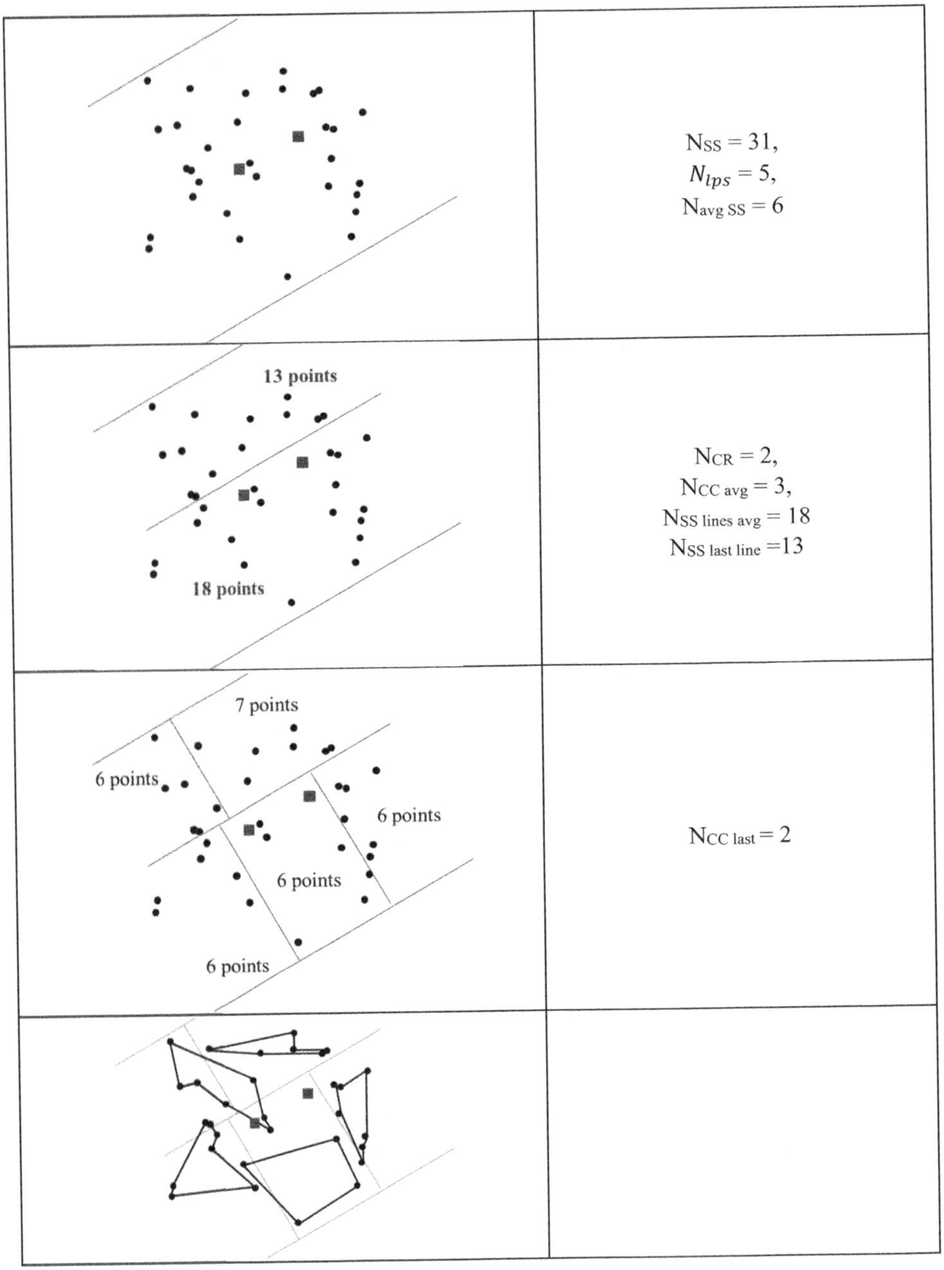

Figure 6.6. *Example of partitioning on a system with 31 secondary substations (red squares = primary substations and black circles = secondary substations)*

6.3.1.2. *Step no. 2: load exchange heuristics to satisfy the objective function*

The implemented heuristic consists of performing several iterations during which a loop gives a load to another loop until f_{PL} reaches a value fixed by the DSO. The loop selection process is done in the following way and illustrated through the previous example shown in Figure 6.7:

– selection of the $B_{sel\ 1}$ loop furthest from the average PL;

– selection of the $B_{sel\ 2}$ loop by random selection according to a law weighted by the minimum distance of each $B_{sel\ 1}$ loop;

– selection of the point for the $B_{sel\ 2}$ loop by random selection according to a distribution weighted by the minimum distance between each point of $B_{sel\ 2}$ and the points of $B_{sel\ 1}$.

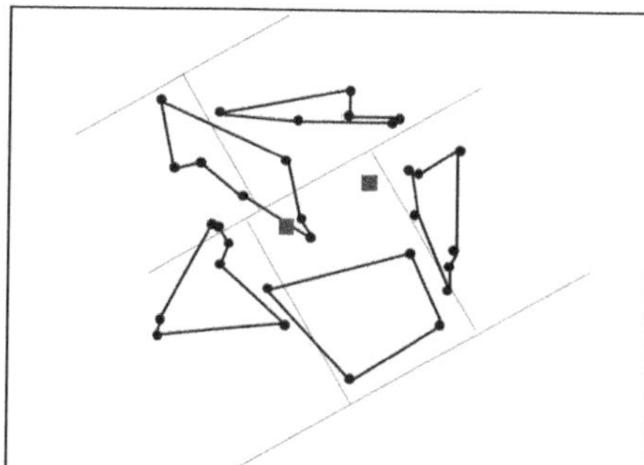

$$PL = [10.4; 13.2; 7; 5.5; 13.8]\ MVA.km$$
$$PL_{avg} = 10\ MVA.km$$
$$f_{PL} = 73.83\%$$

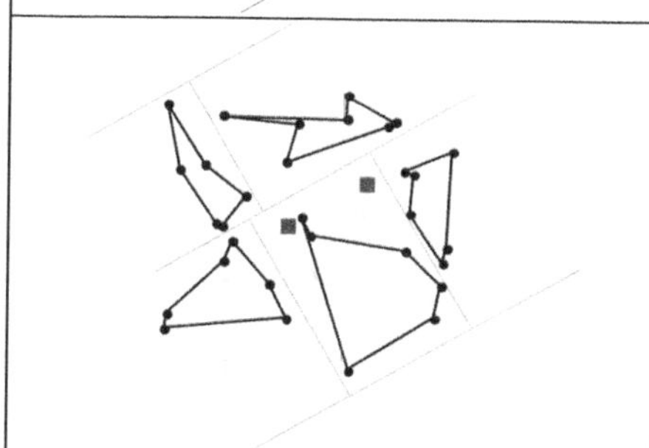

$$PL = [10.7; 10.7; 10.9; 10.8; 10.5]\ MVA.km$$
$$PL_{avg} = 10.77\ MVA.km$$
$$f_{PL} = 2.73\%$$

Figure 6.7. *Position of loops before and after PL optimization*

6.3.1.3. *Step no. 3: construction of the main feeders*

For each loop, the point closest to the primary substations is selected to be the loop head substation, that is, the one that will be connected to the main feeder. For this example, these points are shown in green, as depicted in Figure 6.8.

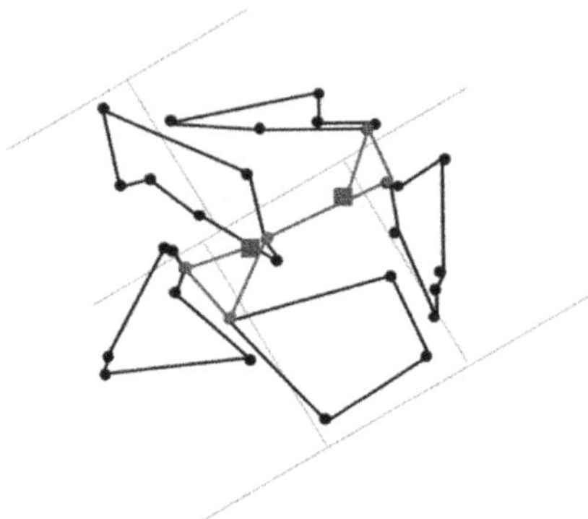

Figure 6.8. *Main feeder connecting the loops to the primary substations*

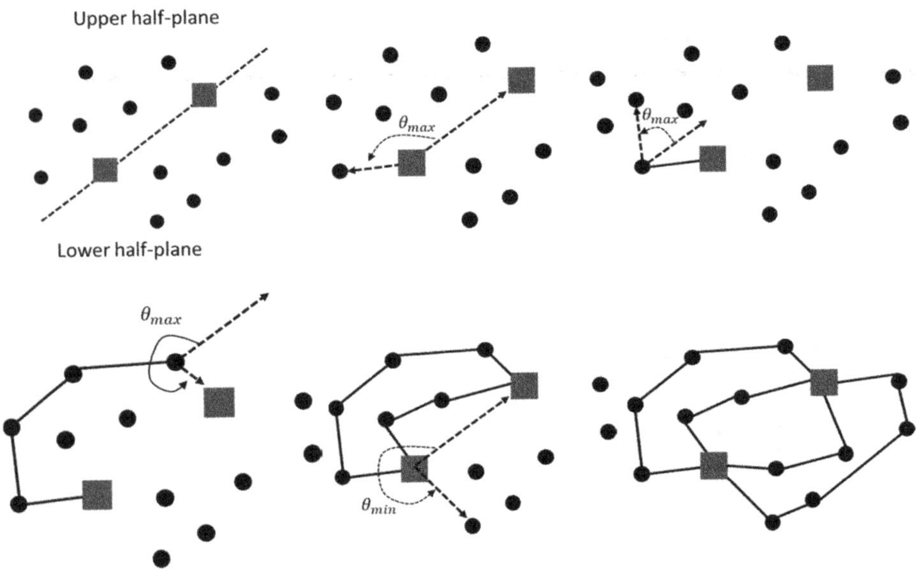

Figure 6.9. *Construction of the initial solution on angular criteria*

The algorithm used to connect the head substations of the loop to the primary substations follows Steps 1 and 2 of the general algorithms, with a few differences:

– The initial solution is determined using an angular criterion: the secured feeders are formed successively by selecting the secondary substations that form a maximum (respectively minimum) angle with regards to the line, connecting the two primary substations for the upper (respectively lower) half-plane, as illustrated in Figure 6.9.

– The load exchange heuristic is done with the secured feeders closest to the selected secured feeder, so there is only one wheel biased on the loads to be exchanged.

– The reconstruction of the main feeders after load exchange consists of connecting a primary substation to another by passing through the loads selected in the previous step. This problem is known in mathematics as the traveling salesman problem (TSP). The looped Christofides algorithm (Christofides 1976) is used here.

6.3.1.4. *Step nos. 4 and 5: sizing of the conductors and HC calculation of the built system*

Once the structure is obtained, the conductors are sized to meet the N-1 criterion at the feeder and loop levels. The economic section (see Chapter 2) is chosen in order to minimize the TOTEX, and the fulfillment of constraints is verified thanks to a load flow. The performance of the architecture thus obtained is then evaluated by calculating the TOTEX, SAIDI, SAIFI, ENS and HC using the formulas and methods described in this chapter.

6.3.2. *Brute-force search: case study no. 2b*

We consider a distribution system architecture with N_{RCS} remote-controlled switching devices. Regardless of the topologies used in the world, MV architectures all have a few remote-controlled switching devices allowing them to reconfigure the system in case of a fault to resupply customers located outside the faulted area. Some LV systems in dense urban areas may also have a few RCSs. The number and location of these RCSs are the result of a compromise between cost and reliability made by each DSO. In case study no. 2 of Chapter 4, we highlighted the benefit in meshing distribution systems to increase the HC of DERs. In particular, in case study no. 2b, it is proposed to mesh the distribution system by closing one or more RCSs. The question that arises regards the optimal number of RCSs to close, as well as their location in the system, in order to optimize a given objective function (in this case study, the renewable generation HC). For this case study, the spatial scale is the MV distribution system and the temporal scale varies according to the chosen objective function: for example, in the short term for minimizing technical losses, to the long term for maximizing the integration of distributed generation.

The brute-force search can be applied if the number of RCSs in the studied system is reasonable. Indeed, the total number of possible cases is $2^{N_{RCS}}$. For a number of RCSs equal to 10, the number of possible cases is 1,024, however, this increases to 1,048,576 for 20 RCSs. In case study no. 2b of Chapter 4, there are seven RCSs, that is 128 possible combinations of open or closed RCSs. The brute-force search algorithm is thus as follows:

– $i_{combination} = 1$;

– until all RCS combinations have been studied;

– for the RCS combination no. $i_{combination}$:

- closure of the corresponding RCSs,

- evaluation of the objective function;

– $i_{combination} = i_{combination} + 1$.

For case study no. 2b, the objective function is the HC of distributed generation, the calculation method of which will be explained later on in this chapter. Figure 6.11 represents this HC as a function of the number of closed RCSs and their location in the system. Therefore, for one closed RCS there are seven possible combinations (the seven points represented on the graph), for two closed RCSs there are 21 possible combinations, etc. In Alvarez-Hérault et al. (2014), greedy methods are presented for the case where the number of RCSs would be too large and a brute-force search is impossible.

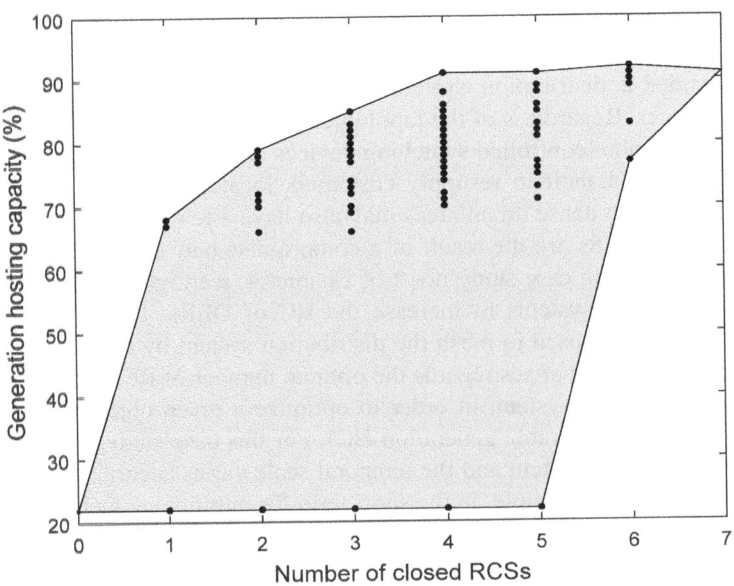

Figure 6.10. *Generation HC according to the number of closed remote-controlled switches and their location in the system*

Table 6.3 gives for each number of RCSs that the DSO wishes to close the number of possible combinations and the minimum and maximum HC of generation obtained according to the chosen RCSs. The fully looped system is the one that maximizes capacity. If the DSO does not wish to close the entire system or wishes to do so progressively, Table 6.3 outlines the optimal sequence for closing the RCSs.

	RCS 1	RCS 2	RCS 3	RCS 4	RCS 5	RCS 6	RCS 7
Number of possible combinations	7	21	35	35	21	7	1
HC$_{min}$	22%					77%	91%
HC$_{max}$	68%	79%	85%	91%	91%	92%	91%[1]

Table 6.3. *Results of the brute-force RCS selection algorithm*

6.4. Linear programming

6.4.1. *Consumption curtailment (demand response): case study no. 3a*

Case study no. 3 in Chapter 4 shows the implementation of a load curtailment strategy on a secured feeder MV line to avoid reinforcement due to current constraints on conductors. The study is performed over a 10-year period with 30-min time intervals and allows for rebound and deferral effects, as shown in Figure 6.11. The architecture remains unchanged, that is, vectors A, B, C and D of the solution vector are fixed. The decision variables are those of vector $S_{\delta t}$ which determine the power extracted at each MV/LV substation and at each time δt. The voltage $V_{\delta t}$ and current $I_{\delta t}$ values are determined via a load flow calculation, taking as inputs the system topology (vectors A, C and D), the conductor characteristics (vector B) and the power extracted (vector $S_{\delta t}$). For the following, we denote each period of the study t_p, with $p \in \{1 \dots N_{\delta t}\}$ where $N_{\delta t}$ is the total number of study periods.

$$X_{\delta t} = \begin{bmatrix} A \\ B \\ C \\ D \\ S_{\delta t} \\ I_{\delta t} \\ V_{\delta t} \end{bmatrix}$$

1 Using an additional RCS, regardless of its position, does not improve the HC. The fact that the maximum capacity with seven RCSs is slightly lower than with six RCSs is due to the random and approximate nature of the Monte Carlo method.

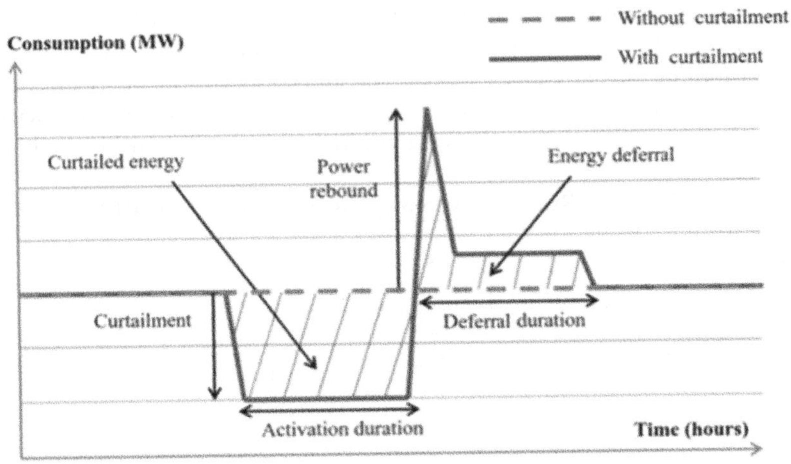

Figure 6.11. *Schematic diagram of curtailment*

The load curtailment scheme in Figure 6.11 is discretized into 30-min time intervals via the W_{cur} vector. For this case study, the reference values used are a power rebound of 140% for 30 min and an energy deferral of 100% for 2 h. The W_{cur} is thus constant and constitutes an input parameter for our optimization problem.

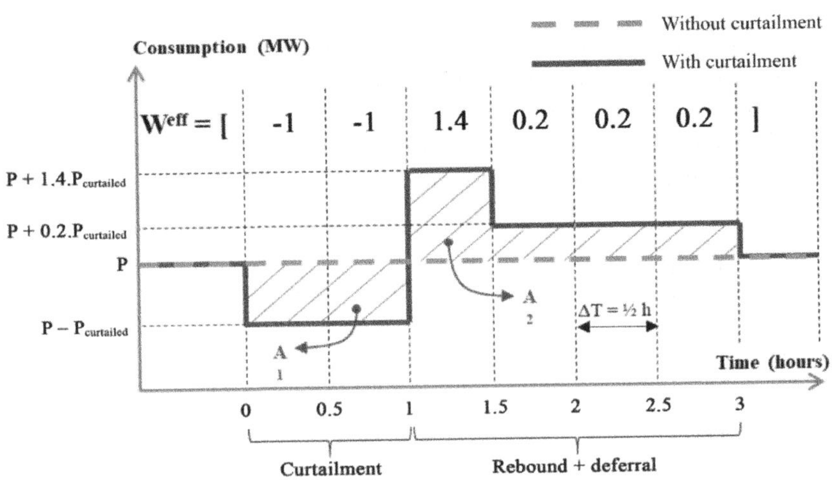

Figure 6.12. *Curtailment vector*

We define the decision vector X_{cur} for each time δt and each MV/LV secondary substation. The curtailment is expressed in terms of the active powers extracted. To determine the vector $S_{t_p} = P_{t_p} + j.Q_{t_p}$, we assume a constant power factor for each substation. The vector is defined by equation [6.10], where $P_d(t_p)$ is the active power extracted at the time t_p by substation d, and $P_d^{cur}(t_p)$ is the active power curtailed at the time t_p by substation d.

$$x_d^{cur}(t_p) = \frac{P_d^{cur}(t_p)}{P_d(t_p)} \times 100(\%) \begin{cases} \forall d \in \{1 \ldots N_{SS}\} \\ \forall p \in \{1 \ldots N_{\delta t}\} \end{cases}$$

$$X_{cur} = \begin{pmatrix} x_1^{cur}(t_1) \\ x_2^{cur}(t_1) \\ \vdots \\ x_{N_{SS}}^{cur}(t_1) \\ \vdots \\ x_1^{cur}(t_{N_{\delta t}}) \\ x_1^{cur}(t_{N_{\delta t}}) \\ \vdots \\ x_{N_{SS}}^{cur}(t_{N_{\delta t}}) \end{pmatrix} \qquad [6.10]$$

The vector X_{cur} is used to calculate the new values of the powers extracted after curtailment, with equation [6.11].

$$P_d(t_p) = P_d(t_p) + \sum_{l=0}^{T_{cur}-1} P_d(t_{p-l}).x_d^{cur}(t_{p-l}).W_{cur}(l+1) \qquad [6.11]$$

The curtailment time T_{cur} is expressed as the number of time intervals and corresponds to the size of the vector W_{cur}, that is, for our case study, six time intervals. This formula is illustrated in Figure 6.13 using the values of the vector W_{cur}. We seek to calculate the power consumed at the time interval t_7. Suppose there are six successive curtailments at t_2, t_3, t_4, t_5, t_6 and t_7, then $x_d^{cur}(t_{p-l})$.

$$\begin{aligned} P_d(t_7) = P_d(t_7) &+ P_d(t_7).x_d^{cur}(t_7).W_{cur}(1) \\ &+ P_d(t_6).x_d^{cur}(t_6).W_{cur}(2) \\ &+ P_d(t_5).x_d^{cur}(t_5).W_{cur}(3) \\ &+ P_d(t_4).x_d^{cur}(t_4).W_{cur}(4) \\ &+ P_d(t_3).x_d^{cur}(t_3).W_{cur}(5) \\ &+ P_d(t_2).x_d^{cur}(t_2).W_{cur}(6) \end{aligned} \qquad [6.12]$$

$$P_d(t_7) = P_d(t_7) - P_d(t_7).x_d^{cur}(t_7) - P_d(t_6).x_d^{cur}(t_6) + 1.4 \times P_d(t_5).x_d^{cur}(t_5) + 0.2 \times \left(P_d(t_4).x_d^{cur}(t_4) + P_d(t_3).x_d^{cur}(t_3) + P_d(t_2).x_d^{cur}(t_2)\right) \quad [6.13]$$

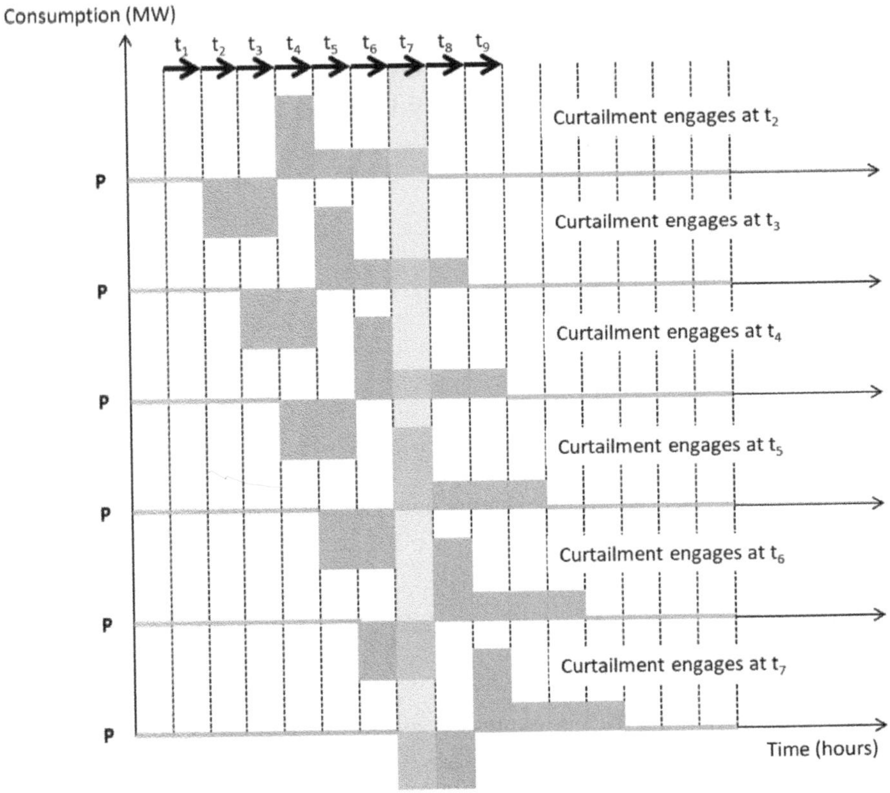

Figure 6.13. *Example timeline for the calculation of consumption after curtailment*

The load curtailment problem is defined as a linear optimization problem, defined by the equation below. The weight vector C_p of the objective function is constructed from the consumption of MV/LV substations at different times during the study period: the objective function $C_p.X_{cur}$ therefore corresponds to the total energy curtailed during the study period. We therefore posit $C_p = [\mathrm{P}_1(\mathrm{t}_1) \ldots \mathrm{P}_{\mathrm{N_{SS}}}(\mathrm{t}_1) \cdots \mathrm{P}_1(\mathrm{t}_{N_{\delta t}}) \ldots \mathrm{P}_{\mathrm{N_{DP}}}(\mathrm{t}_{N_{\delta t}})]$.

$$\min_{X_{eff}} C_p.X_{cur} \text{ with } \begin{cases} A'.X_{cur} \leq B' \\ 0 \leq X_{cur} \leq \tau_{max} \end{cases} \qquad [6.14]$$

$$C_p.X_{cur} = \sum_{d=1}^{\mathrm{N_{SS}}} \sum_{p=1}^{N_{\delta t}} \mathrm{x_d^{cur}(t_p).P_d(t_p)} = \sum_{d=1}^{\mathrm{N_{DP}}} \sum_{p=1}^{N_{\delta t}} \mathrm{P_d^{cur}(t_p)} \qquad [6.15]$$

τ_{max} is the maximum authorized curtailment rate, and depends on the nature and characteristics of the loads supplied by the MV/LV substations. The matrix A' and the vector B' reflect the power constraints in the various system elements: MV/MV transformers, MV feeders, MV sections, MV/LV transformers or LV sections.

In this case study, the constraints considered are those of MV conductors. Unlike equation [5.35] in Chapter 5 where the thermal limit of a conductor is expressed in amperes, here it is expressed in watts, with the variable $P_j^{margin}(\mathrm{t_p})$, which is the maximum allowable power in section j at the time $\mathrm{t_p}$. The calculation of $P_j^{margin}(\mathrm{t_p})$ is a linearization of the current constraints in order to model the problem as a linear optimization problem.

However, the formula in equation [6.16] leads to an overestimation of the available margins because the voltage drops along the line are not taken into account. A feedback loop, not described in this section, is implemented in order to correct the value of $P_j^{margin}(\mathrm{t_p})$ via a correction coefficient if, after optimization, the verification via a load flow shows that the constraints are not respected.

$$P_j^{margin}(\mathrm{t_p}) = I_{th}(l) \times U \times \sqrt{3}\cos\varphi \qquad [6.16]$$

with φ: the angle of the current with respect to the voltage assumed to be constant in MV.

To construct the matrix A' and the vector B', in addition to $P_j^{margin}(\mathrm{t_p})$, we define Ω_j^{SS} for each section j, which is the set of MV/LV secondary substations it supplies. At any time $\mathrm{t_p}$, the sum of the powers flowing through each section j cannot exceed its power limit.

$$\sum_{d\epsilon\Omega_j^{SS}} \mathrm{P_d(t_p)} \leq P_j^{margin}(\mathrm{t_p}) \qquad [6.17]$$

Taking curtailment into account, the equation thus becomes:

$$\sum_{d\epsilon\Omega_j^{SS}}\left[\mathrm{P_d(t_p)} + \sum_{l=0}^{T_{cur}-1} P_d(t_{p-l}).x_d^{cur}(t_{p-l}).W_{cur}(l+1)\right] \le P_j^{margin}(\mathrm{t_p})$$

$$\sum_{d\epsilon\Omega_j^{SS}}\sum_{l=0}^{T_{cur}-1} P_d(t_{p-l}).x_d^{cur}(t_{p-l}).W_{cur}(l+1) \quad [6.18]$$

$$\le P_j^{margin}(\mathrm{t_p}) - \sum_{d\epsilon\Omega_j^{SS}} \mathrm{P_d(t_p)}$$

$$A'_j(\mathrm{t_p}).X_{cur} \le B'_j(\mathrm{t_p})$$

Equation [6.18] shows by identification how to construct the matrices $A'_j(\mathrm{t_p})$ and the vectors $B'_j(\mathrm{t_p})$, and hence A' and B', as defined by equation [6.19].

$$\begin{pmatrix} A'_1(\mathrm{t_1}) \\ \vdots \\ A'_{N_{sections}}(\mathrm{t_1}) \\ \vdots \\ A'_1(\mathrm{t}_{N_{\delta t}}) \\ \vdots \\ A'_{N_{sections}}(\mathrm{t}_{N_{\delta t}}) \end{pmatrix}.X_{eff} \le \begin{pmatrix} B'_1(\mathrm{t_1}) \\ \vdots \\ B'_{N_{sections}}(\mathrm{t_1}) \\ \vdots \\ B'_1(\mathrm{t}_{N_{\delta t}}) \\ \vdots \\ B'_{N_{sections}}(\mathrm{t}_{N_{\delta t}}) \end{pmatrix} \quad [6.19]$$

Modeling of curtailment with consideration of the rebound and deferral effect is detailed in Gouin (2015) and Gouin et al. (2015a).

6.4.2. Phase balancing problem – integer linear programming: case study no. 6

In case study no. 6 for the creation of an off-grid LV microgrid, the first step is to determine the topology of an LV system. All the loads must be connected in single-phase to a three-phase transformer, which must be balanced in terms of power. For this case study, the spatial scale is the LV system, the customers, while the temporal scale is the long-term target. The objective function to be minimized is given in equation [6.20] and is subject to the constraints outlined by equations [6.21] and [6.22].

$$\min_{X} \sum_{c=1}^{3} \varepsilon^{+}(c) + \varepsilon^{-}(c) \quad [6.20]$$

$$\forall c \in [1;3], \sum_{i=1}^{N_{cust}} P_{load}(i) \times X(i,c) + \varepsilon^{+}(c) - \varepsilon^{-}(c) \leq \frac{\sum_{i=1}^{N_{cust}} P_{load}(i)}{3} \quad [6.21]$$

$$\forall i \in [1; N_{N_{cust}}], \sum_{c=1}^{3} X(i,c) = 1 \quad [6.22]$$

$$\forall c \in [1;3], 0 \leq \varepsilon^{+}(c) \quad [6.23]$$

$$\forall c \in [1;3], 0 \leq \varepsilon^{-}(c) \quad [6.24]$$

with:

– X: the solution vector such that $X(i,c) = 1$, if the load u is connected to the phase c and 0 if not;

– $\varepsilon^{+}(c)$ and $\varepsilon^{-}(c)$: the deviation variables to be minimized to respect equation [6.21];

– $P_{load}(i)$: the power of load i.

6.5. Nonlinear programming

6.5.1. *Storage to remove system constraints: case study no. 5*

In case study no. 5, the benefit in using storage units to solve system constraints as an alternative to reinforcement was described. The objective is to use as little storage as possible (to minimize OPEX costs) to solve constraints on the electrical power system. For this case study, the spatial scale is the LV system and the customers, and the temporal scale is the medium-term target (15 years). We assume here that the location of the storage units is known and we want to estimate the total amount of energy injected or absorbed by the storage so as to relieve pressure on the grid. By making assumptions about the cost of paying for storage, we can deduce whether it is more economically attractive than reinforcement. In mathematical terms, we seek to determine the power profile of the storage units in order to minimize the total energy injected and absorbed by the storage to solve system constraints. Equation [6.25] represents the objective function to be minimized subject to the constraints outlined by equations [6.26]–[6.35] (Hadjsaid 2018; Hadjsaid et al. 2020).

$$\min_{P_{battery}} \sum_{t=1}^{T_s} \sum_{k=1}^{N_{peaks}} \left[P_{charge}(t,k) + P_{discharge}(t,k)\right] + \gamma \sum_{t=1}^{T_s} \sum_{j=1}^{N_{arcs}} \left(P_{margin}(t,j)\right)^2 \quad [6.25]$$

$$SOC(t+1,k) = SOC(t,k) + \left(P_{load}(t,k) \times \frac{\eta}{SOC_{max}(k)} - P_{discharge}(t,k) \times \frac{1}{\eta \times SOC_{max}(k)}\right) \times \Delta t \quad [6.26]$$

$$SOC_{min}(k) \leq SOC(t,k) \leq SOC_{max}(k) \quad [6.27]$$

$$SOC(0,k) = SOC(T_s,k) \quad [6.28]$$

$$P_{charge}(t,k) > 0 \quad [6.29]$$

$$P_{discharge}(t,k) > 0 \quad [6.30]$$

$$P_{battery}(t,k) = P_{charge}(t,k) - P_{discharge}(t,k) \quad [6.31]$$

$$\left|P_{battery}(t,k)\right| \leq P_{max}^{battery}(k) \quad [6.32]$$

$$P_{margin}(t,j) = P_{admissible}(t,j) - P_{br}(t,j) \quad [6.33]$$

$$|P_{br}(t,j)| \leq P_{admissible}(t,j) \quad [6.34]$$

$$\begin{bmatrix} P_{br}(t,1) \\ \cdots \\ P_{br}(t,m) \end{bmatrix} = A'.\begin{bmatrix} P_{res}(t,1) + P_{battery}(t,1) \\ \cdots \\ P_{res}(t,n) + P_{battery}(t,n) \end{bmatrix} \quad [6.35]$$

with:

– T_s: the periodicity of storage (24 h for a day's storage);

– N_{peaks}: the number of nodes in the system;

– N_{arcs}: the number of electrical sections of the system;

– Δt: the time interval being considered (for example, the hour);

– $P_{charge}(t,k)$ and $P_{discharge}(t,k)$: the charge and discharge power of the battery at time interval t and node k;

– $P_{battery}(t,k)$: the battery power at node k and time interval t (positive if the battery is charging and negative if the battery is discharging);

– $P_{max}^{battery}$: the maximum charging and discharging power of the battery;

– $P_{br}(t,j)$: the power flowing on section j at time interval t;

– $P_{admissible}(t,j)$: the maximum allowable power of section j at time interval t without creating current and voltage constraints[2];

– $P_{margin}(t,j)$: the remaining power margin on section j at time interval t;

– $P_{res}(t,k)$: the resulting power at node k at time interval t (consumer, producer or the difference between the two);

– $SOC(t,k)$: the state of charge of the battery at node k at time interval t;

– $SOC(0,k)$: the initial state of the battery charge at node k;

– $SOC(T_s,k)$: the final state of the battery charge at node k;

– γ: the weighting coefficient to ensure the batteries are not used more than necessary;

– η: the battery efficiency;

– $SOC_{min}(k)$ and $SOC_{max}(k)$: the minimum and maximum state of the battery charge;

– A' is a matrix that makes it possible to define the power transmission in the lines (excluding joule losses). Because the system is radial, this power is equal to the sum of the powers of the nodes located downstream of the line. The matrix A' is composed of 0 or 1 such that $A'(l,m) = 1$ if the node m is downstream of line l, and 0 otherwise.

COMMENT.– The first term of the objective function reflects the desire to minimize battery usage while the second term encourages the battery to recharge when power flows in the lines are low. It is thus a multi-objective function in which the setting of γ should not be too large, at the risk of overusing storage. Other mathematical formulations are possible depending on the objective functions and constraints related to the use of the storage.

2 These power limits can be calculated by a preliminary statistical study and represent the amount of power that can flow through the line without creating thermal and voltage constraints at the end of the line. They depend on the topology of the system and the customers present (consumers or producers).

6.5.2. *Placement and sizing of storage and production units: case study no. 6*

On a given system, the problem of the placement and sizing of generation and/or storage units is a problem with discrete (unit location) and continuous (unit power) variables, with an objective function that can be linear or quadratic depending on whether we are looking to minimize a cost or losses, for example. In case study no. 5 presented in Chapter 4, the objective is to maximize the local consumption of energy as well as the losses. For this case study, the spatial scale is the LV system and customers, and the temporal scale is the medium-term target. Equation [6.36] gives the objective function used with two terms:

– The first term, P_{slack}, represents the power remaining to be supplied to ensure the generation–consumption balance once the generation and storage units have been placed. In our case study, the location of a future MV/LV substation has been defined as the slack node where a storage unit and a diesel generator will be placed to ensure the voltage reference.

– The second term, P_{losses}, represents the losses of the electrical system which vary according to the number, location and power of the production and storage units.

Minimizing $P_{balance}$ means ensuring that the entire consumption of the grid is provided by the PV storage systems, however, depending on their number and location, the value of the joule losses will not be the same. The weighting coefficients a and b, therefore, emphasize one or the other term. This objective function is subject to nonlinear constraints (equations [6.37] and [6.38]) since the system voltages and currents depend on the power through a nonlinear relationship, due to the voltage/current phase shifts of the loads and the inductive parameters of the lines.

$$\min_{X}\left(a \times \sum_{t=1}^{T_S}\sum_{c=1}^{3} P_{slack}^2(X) + b \times \sum_{t=1}^{T_S}\sum_{c=1}^{3} P_{losses}(X)\right) \qquad [6.36]$$

$$\forall c \in [1;3], k \in [1;n], 0.9 \leq V(k,c) \leq 1.1 \qquad [6.37]$$

$$\forall c \in [1;3], j \in [1;m], I(j,c) \leq I_{max}(j,c) \qquad [6.38]$$

with:

– X: the vector of size n+2 such that for $k \in [1;n]$, $X(k) = 1$ if there is a PV storage system at node k and 0 otherwise. $X(m+1)$ is the maximum power of a PV unit and $X(m+2)$ is the maximum charge and discharge power of a storage unit. We recall that in this case study, we regard the location of PVs and storage units as identical and all PVs (and storage units, respectively) have the same installed power;

– P_{losses}: total system losses;

– $P_{balance}$: total power injected or absorbed at the slack node;

– $V(k,j)$: the voltage at node k and phase j;

– $I(j,c)$: the current in line j of phase c;

– $I_{max}(j,c)$: the maximum admissible current in line j of phase c.

6.6. Integration of uncertainties

6.6.1. *Monte Carlo method applied to the calculation of the DER HC and the technical and economic interest of flexibilities*

Figure 6.14 shows the general algorithm for determining the HC of DERs (e.g. generation and electric vehicles) with or without flexibility mechanisms. This algorithm is based on the dichotomy principle, that is, the HC is initialized to 100% of the maximum conventional consumption of the scope of study and denoted τ_{int}.

A Monte Carlo-based algorithm then calculates the probability, $p(\tau_{\text{int}})$, that the user-defined technical constraints of the system are respected.

If this probability is strictly less than p_{target} then the HC is reduced according to a dichotomy principle, otherwise, it is increased. p_{target} can be set to 100% or can take into account a margin of error denoted m_{error}, that is, $p_{target} = 1 - m_{\text{error}}$. A stopping criterion, ε, set by the user, is derived from a compromise between the desired accuracy and the required computation time.

Table 6.4 gives an example of how the algorithm works with its initialization settings. For a production penetration rate (τ_{int}) of 100%, the probability of constraint compliance is less than p_{target}, which implies that τ_{int} of iteration 1 is equal to 50%. The constraint compliance probability of this new rate is still less than p_{target}, so τ_{int} of the second iteration will be equal to 25%, and τ_{sup} of the second iteration will equal 50%. This time, the probability of constraints is greater than

p_{target}, so τ_{int} of the third iteration will equal 37.5% and τ_{inf} of the third iteration will equal 25%. As the iterations proceed, the interval $[\tau_{inf}; \tau_{sup}]$ shrinks until it becomes smaller than the stopping criterion, ε, set by the user, which in this example is equal to 4%, reached at the sixth iteration.

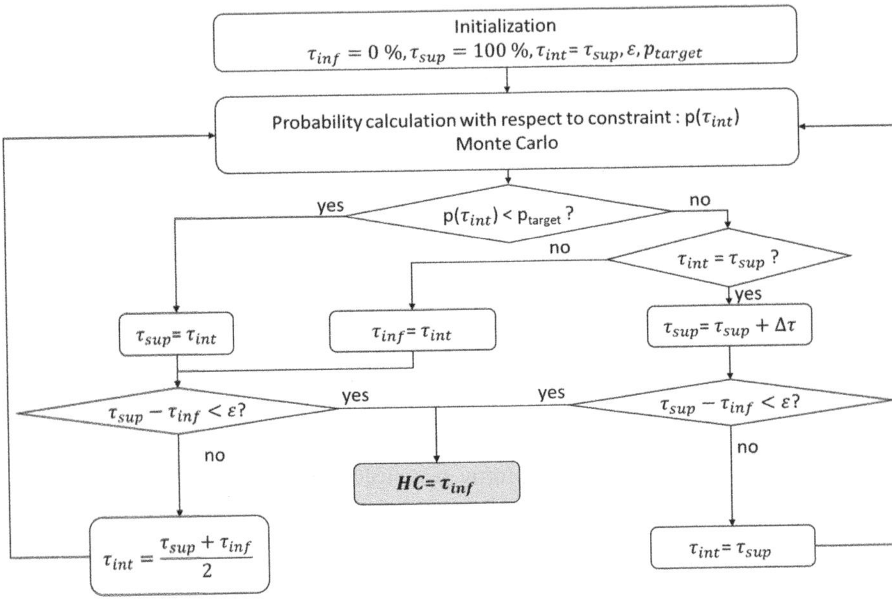

Figure 6.14. *General algorithm for estimating the maximum DER HC of a system (dichotomy + Monte Carlo)*

The methodology for calculating the probability of compliance with the constraints is given in the flow chart shown in Figure 6.15 with n_{MC} being the incremental index of Monte Carlo iterations and Res, the result vector, to store the results.

There are several ways to define the convergence criterion of this algorithm depending on the desired accuracy and the acceptable computation time. We can, for example, set a number of iterations as determined by convergence objectives, or fix a confidence interval objective and a margin of error.

	Initialization	Iteration 1	Iteration 2	Iteration 3	Iteration 4	Iteration 5	Iteration 6
$[\tau_{inf}; \tau_{sup}]$ (%)	[0;100]	[0;100]	[0;50]	[25;50]	[25;37.5]	[31.25;37.5]	[34.375;37.5]
$\tau_{sup} - \tau_{inf}$ (%)	100	100	50	25	12.5	6.25	3.125
τ_{int} (%)	100	**50**	**25**	37.5	**31.25**	34.375	STOP

$p(\tau_{int}) < p_{target}$ $p(\tau_{int}) < p_{target}$ $p(\tau_{int}) \leq p_{target}$ $p(\tau_{int}) < p_{target}$ $p(\tau_{int}) \leq p_{target}$ $p(\tau_{int}) \leq p_{target}$

Table 6.4. *Example of how the DER HCdetermination algorithm works with ε = 4%*

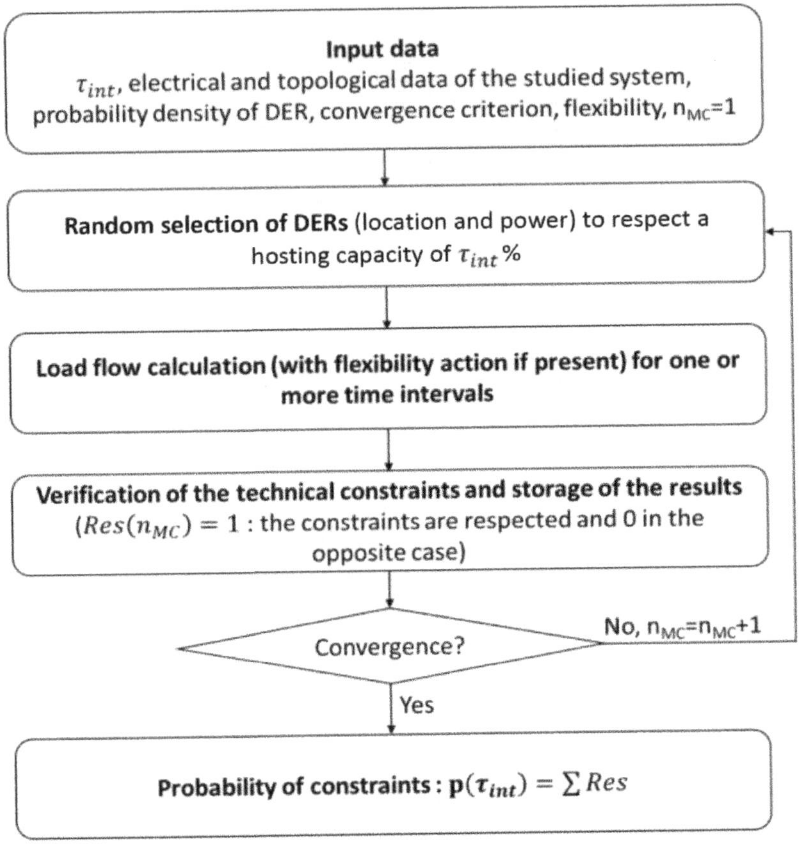

Figure 6.15. *Algorithm for calculating the probability of compliance with constraints (Monte Carlo)*

6.6.1.1. *Method 1: fixed number of Monte Carlo iterations*

According to the *central limit theorem* (Laplace 1809), if the number of Monte Carlo iterations performed is sufficiently large, then the distribution obtained as an output for each variable approaches a Gaussian distribution. We can then determine (Ellard 2003):

– the margin of error, e, defined by equation [6.39];

– the confidence level: the probability that the result is within the margin of error e:

$$e = \frac{z}{2 \times \sqrt{N_{trials}}} \quad [6.39]$$

with:

– e: the margin of error;

– $N_{trials\ s}$: the number of attempts of the Monte Carlo simulation;

– z: confidence interval variable that has already been calculated in the literature (Alvarez-Hérault et al. 2010).

Table 6.5 gives some examples of the confidence level and associated margin of error as a function of the number of Monte Carlo iterations. For example, for a number of simulations that is equal to 50, there is a 95% chance that the result obtained by the Monte Carlo method is ±13.86% of the actual result. If the user wants to reduce this error to 4.38%, then the number of iterations must be increased to 500.

Level of confidence	z	***e***		
		$N_{attempts} =$ **50**	$N_{attempts} =$ **100**	$N_{attempts} =$ **500**
95%	1.96	13.86%	9.80%	4.38%
99%	2.58	18.24%	12.9%	7.77%
99.9%	3.29	23.26%	16.45%	7.36%

Table 6.5. *Confidence level and error margins as a function of the number of Monte Carlo simulations*

6.6.1.2. *Method 2: variable number of iterations*

The user defines $\beta\%$ confidence intervals with a margin of error of e, such that there is a $\beta\%$ chance that the actual probability $p(\tau_{int})$ falls within an interval $[p_{inf}(\tau_{int}); p_{sup}(\tau_{int})]$ that is included within the interval $[1-e;1]$. For example, if the user chooses $\beta = 95\%$ and $e = 1\%$ then we look for $p_{inf}(\tau_{int})$ and p_{sup} (τ_{int}), such that there is a 99% chance that $p(\tau_{int})$ is within an interval $[p_{inf}(\tau_{int}); p_{sup}(\tau_{int})]$ $\subset[0.99;1]$ to be a success, and within $[0;0.99]$ to be a failure.

Let $\Omega_{\tau_{int}}$, defined by equation [6.40], be the set of functions $f_{\tau_{int}}$ that allocate to each vertex v_k of the grid a DER power and a DER type such that the total sum of the installed DER powers is equal to $\tau_{int}\%$ of the total conventional grid consumption, P_C.

$$\Omega_{\tau_{int}} = \begin{cases} f_{\tau_{int}}: \Omega_G \rightarrow (\{0\} \cup [P_{min}^{DER}; P_{max}^{DER}]) \cup \Omega_{type} \\ v_k \rightarrow f_{\tau_{int}}(v_k) = \frac{P_k^{DER}}{\sum_{k=1}^{n} P_k^{DER}} = \frac{P_k^{DER}}{\tau_{int} \times P_C}, k \epsilon [1; n] \end{cases} \quad [6.40]$$

with:

- Ω_{type}: the set of possible DERs;
- production: wind, solar, hydraulic, etc.;
- electric vehicles: low power terminals, recharging stations;
- storage: batteries, flywheels, micro PHES, etc.;
- P_{min}^{DER}: the minimum installed power range;
- P_{max}^{DER}: the maximum installed power range;
- P_k^{DER}: the DER power installed at node k.

Let $X_{\tau_{int}}$ be the random variable defined by equation [6.41] that has a given function $f_{\tau_{int}} \in \Omega_{\tau_{int}}$ to which we associate the value 0 or 1. Then, $X_{\tau_{int}}$ is a Bernouilli random variable of parameter $p(\tau_{int})$. Let N_{trials} be the number of Monte Carlo iterations performed for a penetration rate of $\tau_{int}\%$ then $X_{\tau_{int}}(1), \dots, X_{\tau_{int}}(N_{trials})$ are Bernouilli random variables of parameter $p(\tau_{int})$, independent and identically distributed. The random variable $Y_{\tau_{int}}$ defined by equation [6.42] follows a binomial distribution with parameter N_{trials} and probability $p(\tau_{int})$. The probability that there are r successes over N_{trials} is defined by equation [6.43].

$$X_{\tau_{int}}: \begin{cases} \Omega_{\tau_{int}} \rightarrow \{0\} \cup \{1\} \\ f_{\tau_{int}} \rightarrow \begin{cases} 1 \text{ if constraints are respected} \\ 0 \text{ otherwise} \end{cases} \end{cases} \quad [6.41]$$

$$Y_{\tau_{int}} = \sum_{u=1}^{N_{trials}} X_{\tau_{int}}(u) \quad [6.42]$$

$$P(Y_{\tau_{int}} = r) = C_{N_{trials}}^{r} \times p(\tau_{int})^r \times (1 - p(\tau_{int}))^{N_{trials}-r} \quad [6.43]$$

Clopper and Pearson calculated the confidence intervals $[p_{inf}(\tau_{int}); p_{sup}(\tau_{int})]$ at $\beta\%$ (Agresti 2013) where the probability that $p(\tau_{int})$ is greater than $p_{inf}(\tau_{int})$ is given by equation [6.44], and the probability that $p(\tau_{int})$ is smaller than $p_{sup}(\tau_{int})$ is given by equation [6.45]. They showed that $((p_{inf}(\tau_{int}); p_{sup}(\tau_{int}))$ is the unique solution for the system in equation [6.46].

$$P\left(\mathrm{p}(\tau_{int}) > p_{inf}(\tau_{int})\right) = \sum_{u=r}^{N_{trials}} C_{N_{trials}}^{u} \times p_{inf}(\tau_{int})^{u} \times \left(1 - p_{inf}(\tau_{int})\right)^{N_{trials}-u} = \frac{1-\beta}{2} \qquad [6.44]$$

$$P\left(\mathrm{p}(\tau_{int}) < p_{sup}(\tau_{int})\right) = \sum_{u=o}^{\mathrm{r}} C_{N_{trials}}^{u} \times p_{sup}(\tau_{int})^{u} \times \left(1 - p_{sup}(\tau_{int})\right)^{N_{trials}-u} = \frac{1-\beta}{2} \qquad [6.45]$$

$$\begin{cases} \sum_{u=r}^{N_{trials}} C_{N_{trials}}^{u} \times p_{inf}(\tau_{int})^{u} \times \left(1 - p_{inf}(\tau_{int})\right)^{N_{trials}-u} = \frac{1-\beta}{2} \\ \sum_{u=o}^{\mathrm{r}} C_{N_{trials}}^{u} \times p_{sup}(\tau_{int})^{u} \times \left(1 - p_{sup}(\tau_{int})\right)^{N_{trials}-u} = \frac{1-\beta}{2} \end{cases} \qquad [6.46]$$

It is therefore possible to determine the minimum number of trials to make, as well as a table to determine the number of additional iterations to make according to the number of successes and failures. Let us return to the previous example, where we assume we want a 99% confidence interval with a 1% margin. The top of Figure 6.16 gives the 99% confidence intervals obtained as a function of the number of iterations executed, for the case where we always have one success. We can see that if the number of simulations is less than 459, then the breadth of the confidence interval is greater than the 1% error margin. If this margin is set at 5%, then the minimum number of simulations needed increases to 152. The top figure shows the confidence intervals as a function of the number of failures for 459 simulations. We can see that if there are 0 failures across the 459 simulations, then the margin of error is satisfied and we can conclude to a general success. If there are more than 11 failures, then the confidence interval is in the range [0; 0.99] allowing us to conclude to a general failure. On the other hand, if the number of failures is between 1 and 10, we cannot conclude and thus need to increase the number of simulations. This method allows us to optimize the computation time and quantify the errors obtained from the results (Alvarez-Hérault et al. 2013).

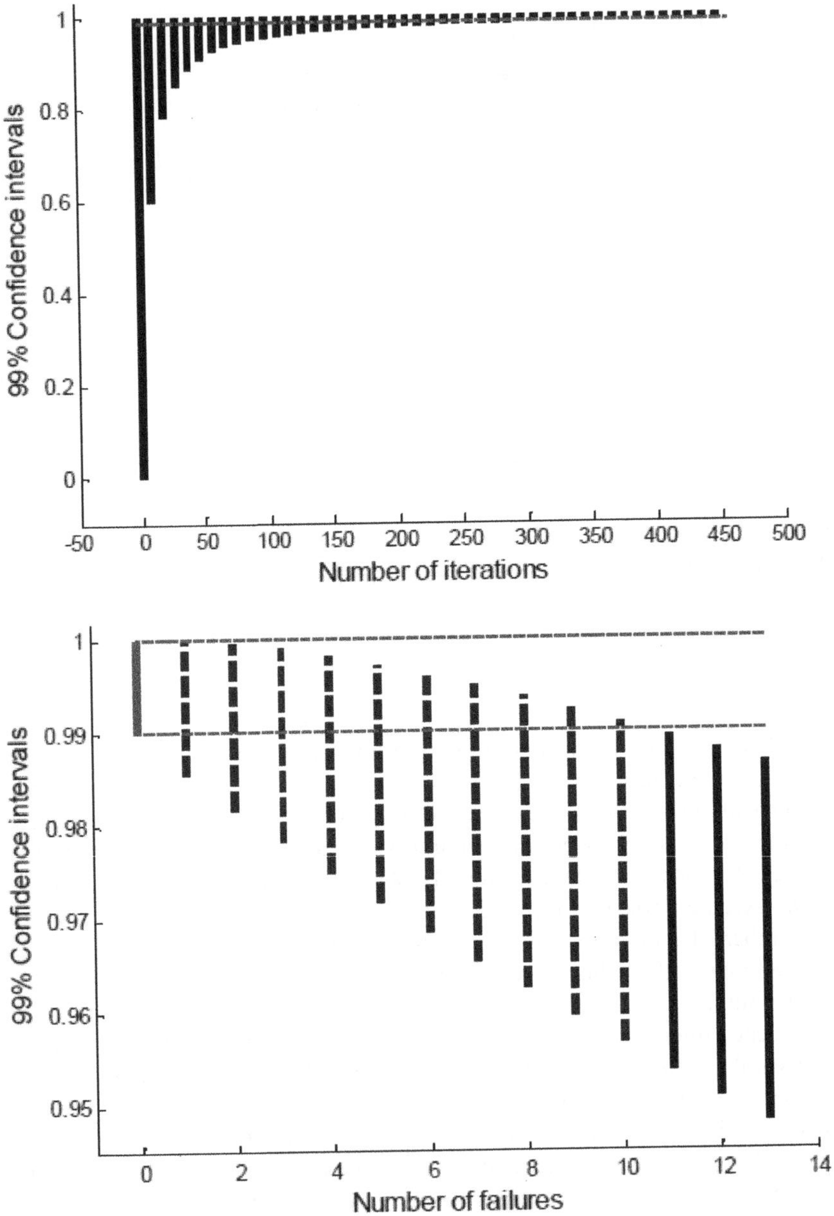

Figure 6.16. *Number of iterations to perform according to the number of failures*

In Chapter 4, case studies no. 1 and 2 evaluated the performance of structural modifications to the system in terms of generation HC. The "constraint probability calculation" subpart of the algorithm was also used to determine the flexibility needs, as well as their technical and economic performances for three case studies:

– case study no. 3a: flexibility through load curtailment, MV perimeter (Gouin et al. 2018);

– case study no. 4: comparison of flexibilities (load control, pro-generation curtailment and reactive control of producers), LV perimeter (Bernier et al. 2019);

– case study no. 5: flexibility through storage, LV perimeter (Hadjsaid et al. 2020).

6.6.2. *Probabilistic method applied to the technical and economic interests of flexibilities: case study no. 3b*

In case study no. 3b, a probabilistic approach is used to integrate generation curtailment as an alternative to primary substation reinforcement for generator connection requests. Figure 6.18 shows the algorithm developed and applied to two generators, but which is generalizable to several types of generators (Garry et al. 2018). The installed powers P1 and P2 of generators 1 and 2 are initialized to P10 and P20, the initial values at the beginning of the study period. The occurrence of a technical constraint depends on the scope of the study.

– The case of an MV/LV primary substation (case study no. 3b) or MV/LV: the occurrence of constraints are defined by the boundary of equation [6.47].

$$P_2 = P_{max}^{PS} - P_{min} + P_1, P_1 \in [P_{10}; P_{1max}] \; and \; P_2 \in [P_{20}; P_{2max}] \quad [6.47]$$

with:

- P_{max}^{PS}: the maximum admissible power of the primary substation (case study no. 3b) or secondary substation;

- P_{min}: the minimum consumption of the loads supplied by the substation;

- P_{1max} and P_{2max}: the maximum variation range of the installed powers of generators 1 and 2.

– The case of an MV and/or LV distribution system: the occurrence of constraints is defined by the result of a linearized load flow calculation which allows us to see if the voltage across all the vertices of the system and the current in all the conductors of the system respect the admissible ranges.

If a constraint occurs, the annual energy to be curtailed is calculated using the LIFO method (see Chapter 3). The average boundary between curtailment and reinforcement can thus be defined by equation [6.48].

$$F_m = \left\{(P_1, P_2) \text{ such that } E_c(P_1, P_2) = \frac{C_{annuity}}{p}\right\}$$

$$C_{annuity} = \frac{1 - \dfrac{1}{1+i}}{1 - \dfrac{1}{(1+i)^{N_{annuity}}}} \times (C_{inv}^2 - C_{inv}^1) \qquad [6.48]$$

with:

- F_m: the average curtailment/reinforcement boundary;
- E_c: the total energy to be curtailed (kWh);
- $C_{annuity}$: the investment annuity (k€);
- C_{inv}^1: the investment cost of solution no. 1 (left of the boundary) (k€);
- C_{inv}^{21}: the investment cost of solution no. 2 (right of the boundary) (k€);
- p: the cost of remuneration for load curtailment (k€/kWh).

Solution no. 1 is either curtailment alone (in which case $C_{inv}^1 = 0$) or curtailment and reinforcement (such as adding a 10 MW transformer, in which case $C_{inv}^1 \neq 0$). Solution no. 2 requires a larger reinforcement (with or without curtailment) than solution no. 1 (such as the addition of a 20 MW transformer). Indeed, initially for low values of P1 and P2, there is no constraint and thus no need for reinforcement and/or curtailment.

When P_1 and P_2 increase, the constraints appear gradually. Curtailment is then more interesting than reinforcement since it is not used very much. A threshold then appears for which reinforcement becomes more promising than curtailment, which is then no longer activated. A second threshold appears for which the reinforcement performed is no longer sufficient and must be completed by a curtailment, and then followed by a more significant reinforcement, and so on.

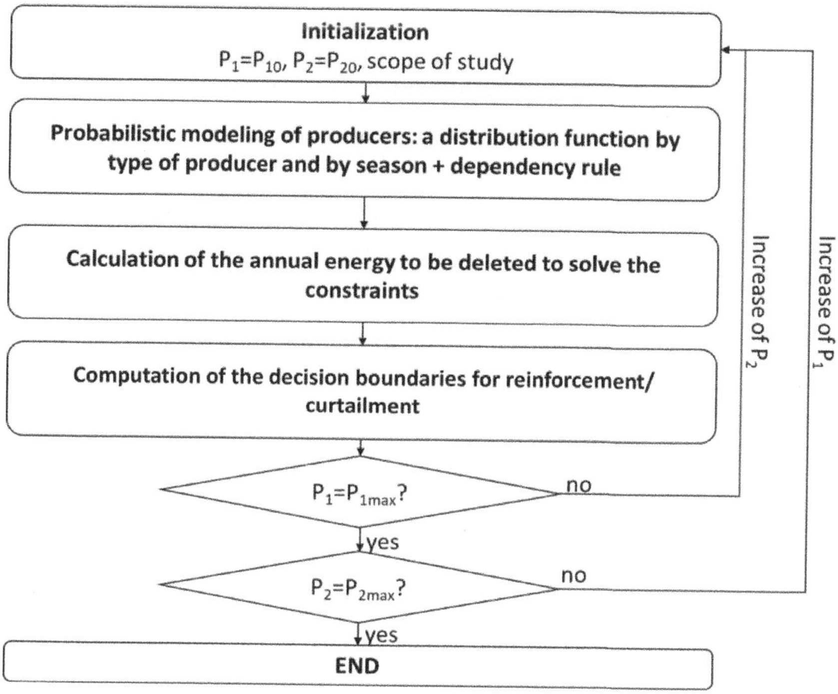

Figure 6.17. *Probabilistic algorithm for making investment decisions*

6.6.2.1. *Probabilistic modeling of producers: identification of seasons*

A statistical analysis of the data sets allowed the identification of "seasons" over which we estimate that the normalized distribution function of a given type of production is equal to the average of the distribution functions constructed with the available data.

We assume here that the load curves are known by the DSO. These seasons are shown for each generation type in Table 4.7 in Chapter 4. To identify the seasons for a given generator, we use the following metrics:

– power normalized by the maximum annual power (per producer or group of producers);

– average (and per-quantile) load factor: the ratio of the actual average (or per-quantile) energy produced over a given period of time to the theoretical energy produced, if the producer were able to deliver its maximum power continuously. The time period can be, for example, an hour, a month or a year. Figure 6.18 shows an example of the average monthly load factor in the case of hydroelectric generation from 1,202 2-year generation curves at 10-min time intervals. In the case of a hydroelectric system, we can see low production for the months of July to October, and higher activity outside of these months but with greater variability;

– filtering using sliding averages over a given period when the variability is too great, as illustrated in Figure 6.19 for wind power;

– percentage of active producers: in France, in the case of cogeneration, a tariff incentive mechanism encourages cogeneration producers to produce in winter, thus creating two distinct seasons.

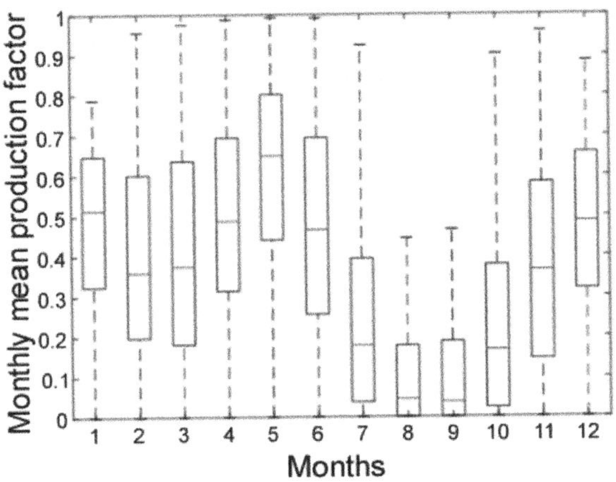

Figure 6.18. *Average monthly load factor (1,202 hydraulic production curves)*

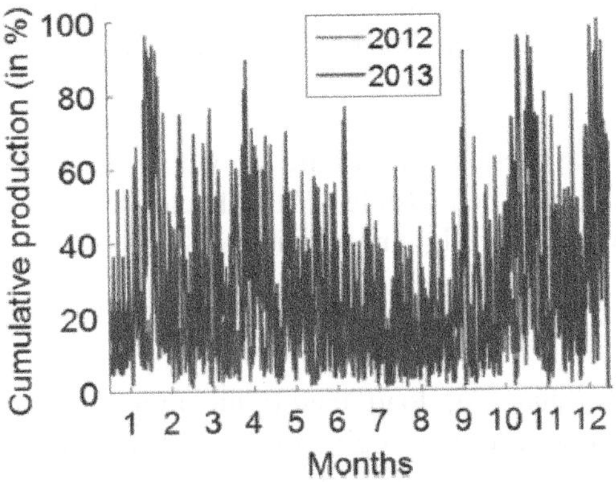

a)

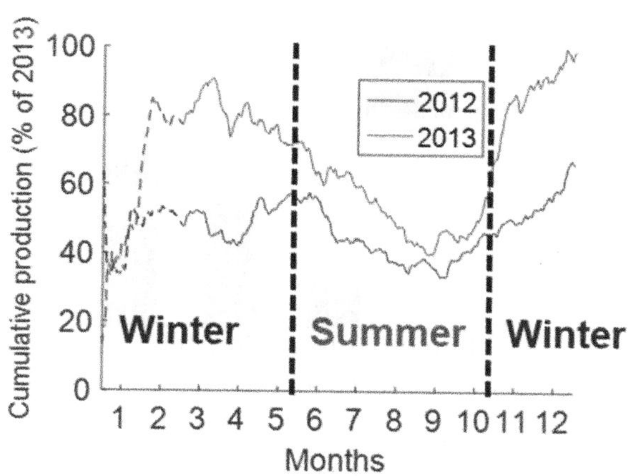

b)

Figure 6.19. *Normalized cumulative power (806 wind generation curves: a) unfiltered and b) filtered)*

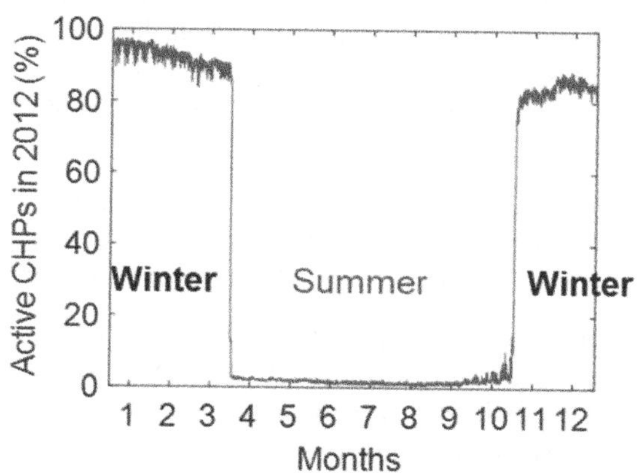

Figure 6.20. *Percentage of active and inactive cogeneration*

6.6.2.2. *Probabilistic modeling of producers: dependence and calculation rules*

For a given type of production, there may be dependencies on a daily or seasonal basis. There may also be dependencies between two identical or different types of production.

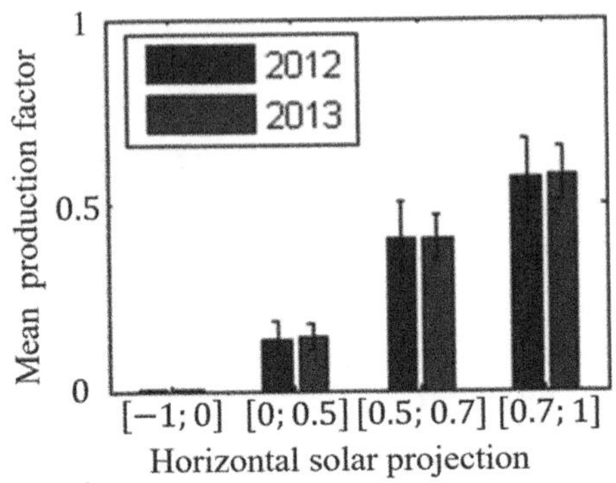

Figure 6.21. *Average load factor per solar season (676 PV production curves)*

Indicators to define the type of dependency can be as follows:

– Daily dependence: average power per hour of a year. In the case of photovoltaics, four seasons at the day scale can be determined according to the position of the sun: at night production is zero; when the sun is low, production is low; during the day, production is high; and at zenith, production is maximum. Figure 6.21 shows the average load factor in 2012 and 2013 of PV generation for the four seasons listed above.

– Seasonal dependence: rate of non-production (respectively maximum production) over a season: number of hours when production is equal to 0 (or respectively when it is at maximum) over the total number of hours in the season.

– Dependence between two types of production (identical or not): the dependence laws between two types of production and between production and a load, given in Table 4.8 of Chapter 4, were determined using two statistical criteria. Let F_X and F_Y be two distribution functions. We wish to determine the distribution function of X+Y noted F_Z:

- criterion no. 1: Spearman's rho, ρ_n, is used to find non-necessarily linear dependencies between two random variables, because it looks at the correlation coefficient between the ranks of these variables sorted in ascending order, and not, between their values. Let $(x_j)_{1\leq j\leq n}$ and $(y_j)_{1\leq j\leq n}$ be the ranks of two random variables P_1 and P_2, ρ_n is given by equation [6.49].

$$\rho_n = 1 - \frac{6\sum_{j=1}^{n}(x_j - y_j)^2}{n \times (n^2 - 1)} \qquad [6.49]$$

- if ρ_n is close to 0, then Y and X are uncorrelated and the distribution function F_Z can be represented by the independent model (model 2 in Chapter 4), that is, the one in equation [6.50];

- if ρ_n is close to 1, then X and Y are very strongly correlated and F_Z can be represented by the model without diversity (model 1 in Chapter 4), that is, the one in equation [6.51];

- for other values of ρ_n, we use a Gaussian copula model, that is, the one given by equation [6.52].

$$F_Z(t) = \int_{-\infty}^{t} f_Z(u)du\,, t \in [0;1]\ and\ f_Z(s) = f_X * f_Y \qquad [6.50]$$

$$F_Z = F_X + F_Y \qquad [6.51]$$

$$\forall (u,v)\epsilon[0;1]^2, C_p(u,v) = \int_{-\infty}^{\varphi^{-1}(u)} \int_{-\infty}^{\varphi^{-1}(v)} \frac{1}{2\pi\sqrt{1-\rho^2}} e^{\frac{s^2-2\rho st+t^2}{2\sqrt{1-\rho^2}}} dsdt \qquad [6.52]$$

with:

– ρ: the linear correlation coefficient between u and v;

– φ: the distribution function of the normal distribution;

– results for producers of different types and producers and load: Spearman's rho is very close to 0 (maximum mean value of 0.14) indicating an independent model;

– results for producers of the same types: Table 6.6 gives the ranges for variation in the mean values and standard deviations of ρ_n across producers and seasons for each type of production;

	PV	**Wind**	**Cogeneration (winter)**	**Hydroelectric**
Average value	[0.85; 0.93]	[0.81; 0.86]	0.05	[0.31; 0.37]
Standard deviation	[0.09; 0.19]	[0.17; 0.22]	0.16	[0.32; 0.37]
Model selected	ρ_n close to 1 Model without diversity	ρ_n close to 0 Independent model	ρ_n close to 0 Independent model	Gaussian copula model

Table 6.6. *Spearman's rho for production of the same type*

– the selected models are then verified using the Kolmogorov–Smirnov test, which quantifies whether a distribution function, F_X, is different from a given distribution F.

$$\lim_{n\to\infty} P\left[K > \frac{c}{\sqrt{n}}\right] = \alpha(c) \qquad [6.53]$$

with:

– K: $\|F - F_X\|_\infty = \max_{[0;1]}|F - F_X|$;

– c: $\sqrt{\frac{-1}{2}\ln\frac{\alpha}{2}}$;

– $\alpha(c) = 2\sum_{r=1}^{+\infty}(-1)^{r-1}e^{-2c^2r^2}$;

– n: the number of independent trials to construct F_X.

For example, when $c = 1.36$ then $\alpha(c) = 5\%$, that is, the probability that the maximum deviation between F_X and F is greater than $\frac{1{,}36}{\sqrt{n}} = 9.6\%$ (for $n = 200$) is 5%, so if K < 9.6% then F_X can be approximated by F with a 5% margin of error. This test is validated for all the models proposed above except for the cogeneration/hydroelectric power and cogeneration/PV.

The reader can find further details on this statistical analysis in Garry (2016) and Garry et al. (2019).

6.7. Conclusion

The main conclusion of this chapter is that the planning problem, which is already associated with the significant complexities of historical planning studies, is becoming more complex, with new uses and flexibilities, as well as with the many growing uncertainties. It is necessary to develop adapted methods to make the best decisions at the right time. There are many deterministic and stochastic optimization methods used in domains other than power systems, which we can draw from to answer concrete planning problems. For the moment, the input data for all of these planning algorithms are mainly based on historical models, such as load profiles and those taken from joule loss estimation methods. The development of smart meters and other smart measurement devices will provide access to large amounts of data, the analysis of which will improve knowledge of the system in real time, but which will also validate or modify the models and assumptions used in planning. Chapter 7 provides perspectives on new trends and challenges that are currently emerging, particularly at the level of distribution systems.

6.8. References

Agresti, A. (2013). *Categorical Data Analysis*. John Wiley & Sons, Hoboken.

Alvarez-Hérault, M.-C., Picault, D., Caire, R., Raison, B., Hadjsaid, N., Bienia, W. (2010). A novel hybrid system architecture to increase DG insertion in electrical distribution systems. *IEEE Trans. Power Syst.*, 26(2), 905–914.

Alvarez-Hérault, M.-C., Battegay, A., Gandioli, C., Descloux, J., Hadjsaid, N., Tixador, P. (2013). Maximal DG indicator to quantify the efficiency of Smart Grid solutions regarding renewable penetration. In *IEEE PowerTech*, Grenoble.

Alvarez-Hérault, M.-C., N'Doye, N., Gandioli, C., Hadjsaid, N., Tixador, P. (2014). Mecurtailed distribution system vs reinforcement to increase the distributed generation connection. *Sustain. Energy Grids Syst.*, 1, 20–27.

Bernier, M., Alvarez-Hérault, M.-C., Cadoux, F., Hadjsaid, N., Lagouardat, A. (2019). Comprehensive framework for PV integration with an OLTC in a rural distribution grid within the SMAP project. *CIRED 2019 – The 25th International Conference and Exhibition on Electricity Distribution*, Madrid.

Cai, Y. (2012). Urban medium voltage distribution system planning based on ASAGA and TSP path optimization methods. *CICED*, 1–4.

Christofides, N. (1976). Worst-case analysis of a new heuristic for the travelling salesman problem. Report 388, Graduate school of Industrial Administration, Carnegie Mellon University, Pittsburgh.

Dijkkstra, E. (1959). A note on two problems in connection with graphs. *Numerische Mathematik*, 1, 269–271.

Dréo, J. (2005). *Metaheuristic for Hard Optimization*. Springer, Berlin.

Ellard, D. and Ellard, P. (2003). SQ course book [Online]. Available at: http://ellard.org/dan/www/Courses/sq98_root.pdf

Garry, A. (2016). Intégration des incertitudes liées à la production et à son effacement sur les méthodes de planification des réseaux. PhD Thesis, Université Grenoble Alpes, Grenoble.

Garry, A., Alvarez-Hérault, M.-C., Cadoux, F., Hadjsaid, N. (2018). Risk aversion model of distribution system planning rules considering distributed generation curtailment. *Int. J. Electr. Power Energy Syst.*, 99, 385–393.

Garry, A., Alvarez-Hérault, M.-C., Cadoux, F., Hadjsaid, N. (2019). Probabilistic modelling of renewable generation to account for uncertainties in interconnection studies. *Int. Trans. Electr. Energy Syst.*, 29(10), 1–19.

Glamocanin, V. (1993). Open loop distribution system design. *IEEE Trans. Power Deliv.*, 8(4), 1900–1906.

Gouin, V. (2015). Évaluation de l'impact du Smart Grid sur les pratiques de planification en cas d'insertion de production décentralisée et de charges flexibles. PhD Thesis, Université Grenoble Alpes, Grenoble.

Gouin, V., Alvarez-Hérault, M.-C., Raison, B. (2015a). Benefits of load curtailment for distribution grids investments planning. In *IEEE Power Energy Soc. Gen. Meet.*, Denver.

Gouin, V., Alvarez-Hérault, M.-C., Raison, B. (2015b). Optimal planning of urban distribution network considering its topology. *Proceedings of CIRED 2015 – The 23nd International Conference on Electricity Distribution*, Lyon.

Gouin, V., Alvarez-Hérault, M.-C., Raison, B. (2017). Innovative planning method for the construction of electrical distribution system master plans. *Sustain. Energy Grids Syst.*, 10, 84–91.

Gouin, V., Alvarez-Hérault, M.-C., Raison, B. (2018). Stochastic integration of demand response and reconfiguration in distribution system expansion planning. *IET Generation, Transmission & Distribution*, 12(20), 4536–4545.

Hadjsaid, A. (2018). Intégration du stockage dans les méthodes de planification des réseaux électriques basse tension. PhD Thesis, Univeristé Grenoble Alpes, Grenoble.

Hadjsaid, A., Alvarez-Hérault, M.-C., Debusschere, V., Caire, R. (2020). Integrating storage in planning of LV distribution systems with PV production. *IET Gen. Trans. Dist.*, 14(1), 79–86.

Laplace, P. (1809). *Mémoire sur les approximations des formules qui sont fonctions de très grands nombres et sur leur application aux probabilités*, Gallica.

Lawler, E. (1985). *The Traveling Salesman Problem: A Guided Tour of Combinatorial Optimization*. John Wiley & Sons, Hoboken.

7

New Trends and Challenges

7.1. Introduction

Power systems are undergoing technological and economic changes to support the energy transition and the integration of DERs. In addition to the trends already discussed in Chapter 3, new trends are emerging that will impact power system planning in terms of methodologies, objectives and performance indicators. We can group these trends into the three following topics:

– New architectures and new products: after a century of dominance by AC-based architectures, new architectures are hybridizing with DC, and in some cases are even natively becoming DC only. This hybridization allows for new optimization functions by enabling the connection of different voltage levels or portions of systems. These new architectures are accompanied by new operating values as well as new products and components. Developments such as virtualization and recyclability will impact technical and economic models and therefore planning methodologies.

– New integrated planning tools (design and management): with the deployment of smart meters and other measurement devices, the available data are growing and impacting both the planning and operation of systems. Control platforms allowing for interoperability and an increasingly advanced use of data are being developed to design, predict, optimize, monitor and maintain the system. Software environments are becoming data centric and creating digital twins.

– New economic actors and business models: many software tool developers are emerging in order to offer more or less integrated planning tools for the various stakeholders (DSOs, suppliers, government agencies, etc.) whose objectives vary

---

This chapter was co-authored with Heitor FARIAS DE BARROS. Section 7.2.1.4 was reviewed by Thierry CORMENIER.

(development and management of the system, flexibility services, development of territories, etc.).

All of the technologies and concepts presented in this chapter presuppose a freedom of thought, economic trajectories and investments that may not be compatible with existing systems and the efforts to implement them. Without seeking to be exhaustive, the objective of this chapter is to give a vision of the different possible trajectories for the evolution of electrical distribution systems in the medium to long term.

7.2. New architectures and new products

7.2.1. *A new set of values*

7.2.1.1. *Evolution of system codes*

In the coming years, the penetration of renewable energies interfaced with electronic power components will continue to develop rapidly. These production devices are connected to the power system historically built with centralized high-power production using synchronous machines to ensure the stability of the global system. In this logic, system codes have been defined giving rules on frequency and voltage ranges, in normal and degraded modes, depending on the size of the system (continental or island system), in order to ensure the security of the global system and to avoid a blackout. However, in a 100% renewable energy scenario, other strategies need to be defined, such as:

– synchronous condensers: use of synchronous machines from old no-load power plants (solutions currently envisaged by the French TSO, RTE (2020b));

– solution for the control of renewable energy production based on power electronics: these production devices currently behave like current sources (*grid following*), that is, they inject power at the frequency imposed by the system. The objective is to develop their control so that these production devices behave like synchronized voltage sources (*grid forming*). This strategy, envisaged by several European TSOs, including RTE, is analyzed in the MIGRATE project (MIGRATE 2020), which compares three control strategies: the droop control, the virtual oscillator and the virtual synchronous machine (TenneT TSO GmbH 2021);

– decentralization of inertia: the MIGRATE project also analyzes the local dimension of inertia by presenting the power system as made up of interconnected local centers of inertia. Synchronized measurements could detect areas with higher imbalances in order to initiate local frequency stabilization actions. This concept could be extended to the level of distribution systems, in collaboration with the

DSOs, whereby the temporary creation of off-grid microgrids would be a complementary solution toward maintaining system stability;

– evolution of historical system codes (see Chapter 1, section 1.5.5.2): greater flexibility on the acceptable voltage and frequency variation ranges, in normal and degraded regimes, could bring more flexibility to the overall system and thereby reduce investments. Indeed, the notion of frequency is not as important in a system with a high penetration rate of power electronics. In the work of Banjar (2019), a control strategy with a constant voltage-to-frequency ratio associated with load curtailment has been experimentally validated through a case study of a 100% renewable energy microgrid with a power electronics interface.

7.2.1.2. *Evolution of the sizing power*

As explained in the previous chapters, traditional distribution system planning methods have been implemented in order to provide consumers with the lowest possible costs and to guarantee good quality and continuity of supply. In a simple and structured past context (well-separated roles of consumers and producers, few other actors involved, absence of the notion of flexibility, limited future uncertainties), the planning solutions were determined by investment (CAPEX-based solutions), by taking a more or less significant margin, according to the advice of the DSO, and with a well-known and controlled risk. However, in the current context of energy transition, environmental protection, better control of consumption and significant development of new technologies, a range of distribution system development opportunities are emerging. It is now possible to develop optimized planning methods that make the most of the services offered by new emerging players and new technologies (TOTEX-based solutions) rather than investments. These developments are changing the relationship to risk, its quantification and the sharing of responsibilities between the different actors (planners and service providers).

In particular, the assumptions concerning the maximum system sizing powers used in planning must integrate the flexibility options (see Chapter 3). Currently, these alternatives to reinforcement are considered on a case-by-case basis by the operator, for a short- to medium-term perspective, but never a long-term development perspective. The systems of developed countries are sized so as to transmit the maximum power that has a given probability of occurring. For example, in France, the DSO sizes its system by considering the maximum power, which has a probability of being greater than the maximum power served over the last 30 years, that is, 0.3% greater (see Chapter 2). This maximum power does not usually occur over a long cumulative period in the year, which ultimately leads to a large investment to ensure a low occurrence of constraint. Figure 7.1 (top) shows an example of a single-phase distribution substation, at hourly time intervals over 1 year, measured in the Greenlys project (Enedis 2021a) and expressed as a percentage of the annual maximum power. In Figure 7.1 (bottom), the powers have been grouped into 10% increments of the substation's maximum power to show their occurrence expressed as a percentage of hours in the year. We can see that the substation power is less than 70% of the maximum power for 85% of the year. Therefore, the integration of alternative solutions upstream in the planning phase could allow for a review of the maximum power assumptions used for the development of the distribution systems. It is then important to associate a notion of risk management, and in particular, to define the sharing of responsibilities between the planner and the flexibility service providers. Moreover, these assumptions could also impact the historical choices of authorized voltage variation ranges or even voltage levels. For example, the planner can anticipate a future increase in the voltage level of the existing system by pre-investing in equipment designed to operate at a higher voltage than the current one, so as to accommodate the new reinforcements installed down the line.

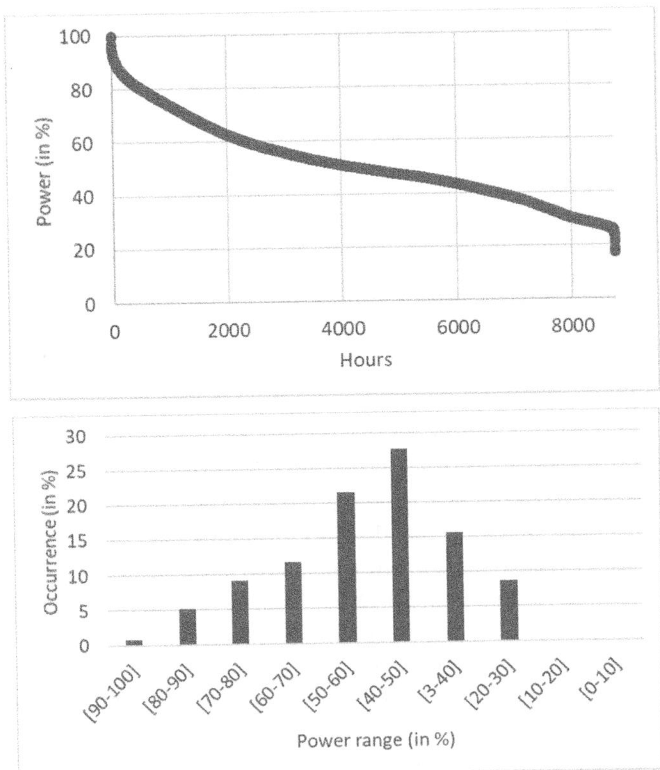

Figure 7.1. *Example of a real load monotone of a MV/LV secondary substation. For a color version of this figure and all subsequent figures in this chapter, see www.iste.co.uk/alvarezherault/distribution.zip*

7.2.1.3. *Decentralizing resiliency*

7.2.1.3.1. Historical resiliency

Another important development concerns the reliability currently quantified by SAIDI, SAIFI and energy not supplied (see Chapter 2). These indicators are currently not evolving since they depend on the structure of the global system and its regulations, which do not change much. For example, for the time being, in most countries the DSO is not allowed to operate a portion of the power system as an island (off-grid microgrids), in case of faults or maintenance operations, thanks to the use of local resources such as DERs. Second, these indicators are yearly averages per consumer without any particular distinction. These indicators must integrate a more accurate idea of prioritization to guarantee a high level of reliability, especially for what is really vital, in order to adapt to increasingly frequent exceptional events. Currently, the only existing prioritization is linked to critical consumers (hospitals, banks and data centers), which generally have dedicated backup solutions (private generators, dedicated feeders, for example). Finally, planning flexible distribution systems can benefit the transmission system and even the overall stability of the system. Therefore, the historical reliability of distribution systems is expanding toward a notion of resiliency perceived as an additional service, as mentioned in particular in Klingel (2020), who introduced the concept of resiliency as a service (RaaS), which is starting to be implemented in some countries. The Florida Public Service Commission changed the regulations in June 2019 to favor projects that seek to improve the resiliency of the power system (including the distribution side). Distributors could include in the electricity bill of its system's users, an additional fee to improve the resiliency of the system, for example, a *storm hardening fee* to bury lines in areas impacted by regular hurricanes, the development of smart meters and the development of infrastructure for the monitoring and control of the systems (system automation). An example of a flexibility service for system resiliency is the preventive load curtailment, carried out by PG&E in October 2019, of about 800,000 customers so as to avoid the spread of fires during dry season. A second example is Elenia, the second largest Finnish DSO, which very clearly prioritizes supply security aspects in their development plan (Myllymäki 2020). The solutions are based on burying lines (target of 75% burial by 2028), but also the automation and use of flexibilities such as load curtailment and storage.

This notion of resiliency also exists on two scales: local and global. Indeed, several countries are orienting their power system planning toward structures that can be partitioned into clusters of autonomous or partially autonomous microgrids that are less dependent on the interconnected power system. The notion of local resiliency is then linked to the optimal sizing and control of microgrids. This not only reduces the investments of the overall system, but also the risk of instability, thereby increasing the overall resiliency. For example, the Finnish utility Elenia is testing the use of batteries, installed to provide balancing services (primary reserve) in normal operation, for the automatic *islanding* of a portion of the 20 kV system in the event of a fault. A second example can be found in Australia (see Chapter 4), which plans to switch just over 15,000 people to individual or collective autonomous power supply with batteries. In California, PG&E is planning to deploy pilot microgrids of about 20 kW each, to supply isolated sites connected to the power system by very long lines with a high risk of outages due to fires.

7.2.1.3.2. Digital before power

The need for electricity has evolved with the rise of digital technology in both developed and developing countries. Historically, these needs were primarily related to food preservation (refrigerators, freezers), but today they are more related to information and communication with the outside world. These new priorities impact the way the system is planned, making data centers appear as critical infrastructures. These loads, although sensitive to energy interruptions, benefit from storage systems that allow them to be resilient to interruptions. Finally, in order to optimize resiliency management, it would be appropriate to approach this notion with a global/local vision and to analyze the cost borne by the distribution infrastructure in relation to the cost and capacity of the loads to be supported during power outages.

7.2.1.4. *Consideration of material efficiency to minimize the environmental footprint*

As the components and equipment that make up the distribution system evolve, the methodologies and materials used to make them evolve to include notions pertaining to material efficiency, such as life extension and/or recyclability to minimize environmental impact. For example, the 109 IEC Guide (International Electrotechnical Commission, 2003-6) provides the framework and methodology for evolving standards for electrical products and equipment so as to address environmental concerns. These aspects cover the whole lifecycle and propose the drafting of a standard to take into account resource protection, energy efficiency, pollution prevention and waste minimization by promoting material efficiency as well as recyclability.

New regulations such as eco-design (EUR-Lex 2021) provide a more directive framework for these aspects of conservation, environmental impact and resource recyclability.

One possible approach is to extend the lifecycle of the equipment, as proposed by the CIRED working group "Lifetime Extension" (CIRED 2021), which proposes to define the determining criteria for equipment replacement and to define and compare strategies for extending and replacing equipment.

In addition to digital equipment monitoring solutions, the deployment of sensors on the system, data analysis tools (analytics) and predictive algorithms can help monitor equipment by collecting data. Several complementary initiatives are emerging:

– collection and analysis by generic software such as West Monroe (Utility Analytics Institute 2021b);

– manufacturers who offer services such as the implementation of the digital solution by domain and equipment experts, such as Schneider Electric's Asset Grid Advisor monitoring solution (Schneider Electric 2021b).

The economic impact on both the lifecycle and the estimated residual value remains to be defined in system planning models. The challenges are to have:

– common data models: databases are beginning to appear on recyclability and carbon impact. PEP (2021) proposes a standardized environmental declaration model called product environmental profile (PEP), which includes information related to the impact categories defined by the EN 50693 Standard and other types of environmental performance, such as the percentage of recyclability according to the method for calculating the recycling and recovery rate of Waste Electrical and Electronic Equipment (WEEE), proposed by ADEME (2019);

– calculation algorithms and software: for example, the "eime" software of LCIE (2021) can be used to calculate the unique environmental impact of a solution;

– a methodology for the recovery of materials, including environmental performance indicators and extending the life span of equipment.

7.2.2. *New objects: virtualization of assets, the virtual lines of the Ringo Project*

As mentioned in Chapter 3, RTE, the French TSO, is currently investigating the Ringo Project (RTE 2020a) for the virtual reinforcement of a transmission line. The objective of the project is to avoid the costly reinforcement of a transmission line needed to evacuate all the power produced by a wind generation site located in Haute-Vienne. The solution imagined and illustrated in Figure 7.2 is to install batteries upstream of the overloaded line to store the surplus electricity that cannot be evacuated. In order for this action to be invisible to the energy market (the TSO cannot produce electricity), other batteries located in Côte-d'Or and Hautes-Alpes must compensate for this energy storage by supplying electricity to the system at the same time. When wind production is low, the energy stored in the batteries in Haute-Vienne is given back to the system, while the other batteries are recharged with the same amount of energy (zero sum game). The deployment of batteries is currently underway with an experiment planned for June 2022 (Spaes 2020). This notion for the virtualization of system components can be extended to the distribution system through the presence of controllable DERs. The DSO could obtain an exemption for the installation of batteries, or lease services to a third-party operator, in order to defer investments.

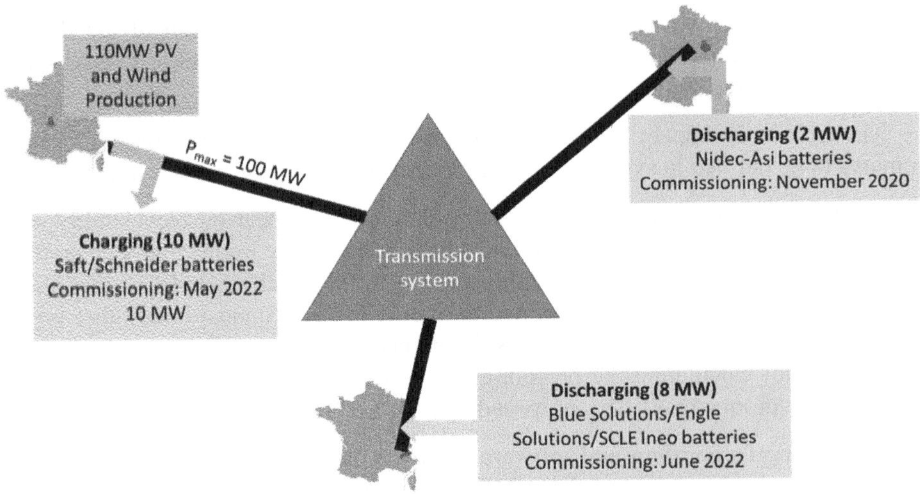

Figure 7.2. *Ringo Project (source: Spaes (2020))*

7.2.3. *Renewed interest for direct current*

In Chapter 1, we detailed the historical stages that saw the major development of AC in electrical systems, pointing out that one of the main obstacles to the development of DC was the late development of power electronics as compared with the transformer. However, in the last few years, we have seen a revival in the use of DC power systems, especially at high voltage, for the connection of offshore wind farms and long-distance connections. This interest is spreading to the medium- and low-voltage level for various reasons (Siemens Energy 2021):

– development of DC components: photovoltaic panels and fixed or mobile batteries (electric vehicles);

– technological progress: DC/DC converters now have better performance when compared to AC/DC converters, and in particular, those accommodating high-power densities (De Doncker 2014);

– improvement of power system resiliency: thanks to DC interconnections, the power systems are decoupled, which makes it possible to stop the propagation of frequency and voltage instabilities and faults;

– connection of isolated sites without production deposits because of the long-distance transmission of electricity in DC;

– reduction of voltage levels for the same power being transported when compared to AC;

– increase in the utilization rate of the lines because of the DC system meshing: this makes it possible to postpone the potential reinforcement of the system due to increases in peak power;

– reduction of the environmental impact: reduced visual pollution.

7.2.3.1. *AC/DC hybridization*

7.2.3.1.1. Interfacing technologies for AC/DC and DC/DC systems

Connection through convertors

As explained in the previous chapters, certain portions of the distribution system are connected to each other via normally open switching devices in order to allow service recovery in the event of a fault. The choice of historical radial operation was primarily based on economic reasons: the reduction of OPEX brought by the looping of systems (better distribution of the power flows) does not compensate for the extra cost due to a change of the protection scheme. However, the use of a technology based on VSC (voltage source converters) in a back-to-back AC/DC/AC topology allows these power flows to transit when decoupling the systems in the event of a fault. In the literature, we find several names for this concept: Soft Open Point (SOP), SIPLINK, DC-link, MV direct current DC link (MVdc-link), Soft Multi-state Open Point (SMOP), Loop Balance Controller (LBC), Soft Normally Open Point (SNOP) or Back-To-Back (BTB) system. This technology provides flexibility when connecting different feeders in a distribution system through the control of these inverters. An SOP can have two or more terminals, and these terminals can be connected via a DC line. Table 7.1 lists the advantages and disadvantages of SOPs based on the work of Cao et al. (2016), Long et al. (2016), and Peng et al. (2017). Therefore, SOPs can be used to reduce joule losses in power systems, better regulate the voltage plan, facilitate generation integration and better distribute power flows throughout the systems. They can also be used to connect areas with different voltage and frequency levels that are connected via less expensive transformers (high-frequency operation to reduce their size and losses) for decoupling the two areas.

Advantages	– Continuous and fast active and reactive power control at each feeder – Provision of voltage regulation services through independent reactive power regulation on each feeder – Connect systems of different voltage levels – Support for fault isolation and restoration
Disadvantages	– Control strategies are not mature – Impacts on power systems not fully explored – Interactions with other control devices (e.g. on-load tap changers and reactive control of generation sources) not fully explored – Expensive technologies and unproven economic benefits

Table 7.1. *Advantages and disadvantages of a Soft Open Point*

The recent state of the art systems, described in Syfullah Fuad et al. (2020), identifies three pilot projects on SOPs in distribution systems:

– In the UK:

- System Equilibrium project (2015–2019) (Western Power Distribution 2014): the DSO, Western Power Distribution, is testing, among other things, a SOP to link two 33 kV AC distribution systems through its flexible power link (FPL) case study. Indeed, the MV systems cannot be looped together under normal circumstances because, being connected to the 132 kV high voltage, there could be significant loop currents during normal operation and in the event of faults. The significance of looping the distribution systems here is to better control the power flows and the voltage plan, so as to increase the DER capacity and reduce investment in the distribution and transmission system (e.g. FACTS). A 20 MW SOP, developed by ABB, installed at the Exebridge 33/11 kV distribution substation interconnecting the Barnstaple and Taunton 132/33 kV primary substations has shown that it is possible to increase the capacity for distributed generation by an additional 20 MW, as well as reduce power peaks in the transmission system transformers. In addition, the expected economic benefits are in the region of £15 million as a result of avoiding reinforcements (Berry et al. 2019);

- Angle-DC (2016–2020) (SP Energy Networks 2021): the DSO, SP Energy Networks, is experimenting with the connection of the island of Anglesey with North Wales via an MVDC line in order to avoid reinforcing the system in the face of increased renewable energy production and consumption. Indeed, two transformers on the island's distribution system (Amlwch and Caergeilliog) are operating close to their thermal limit in N-1 situation, and overvoltage is expected due to high renewable production. The replacement of the AC line with the SOP allows for a 43% increase in transfer capacity due to a lower cable operating temperature.

– In China (Ji et al. 2019): the pilot project located in Tianjin is testing multi-terminal SOPs of 1 MVA, each connecting four feeders from 10/110 kV primary substations. The feedback is: reduced joule losses, better power distribution across feeders, reduced voltage deviations and improved reliability. Ji et al. (2019) also show that optimized SOP control strategies can increase the performance of distribution systems, in particular, for situations of distributed generation.

From the smart transformer to the DC hub

The smart transformer is challenging conventional transformers through advances in power electronics as it could enable the provision of additional services and the creation of AC and DC inputs and outputs. SP Energy Networks is currently coordinating the LV ENGINE project, launched in January 2018, which aims to replace six 11 kV/400V 500 kVA transformers with smart transformers in order to avoid reinforcements related to the development of low-carbon technologies. The expected benefits would be in the order of £528 million by 2050. The specifications of the services to be provided by these smart transformers include:

– the control of the LV feeder voltage;

– control of MV reactive power to provide complementary control to the voltage control already present on MV systems;

– control of power flows by better balancing the load with neighboring transformers;

– the creation of DC outputs for the connection of renewable energy production, storage systems, or electric vehicles.

The smart transformer consists of a hardware part, a solid-state transformer and an intelligent controller which collects data from the LV system and determines the settings of the solid-state transformer. The latter is composed of:

– on the MV side: an AC/DC converter to control the reactive power and to obtain an MVDC output if needed;

– on the LV side: a DC/DC converter including a high frequency step-down transformer for power flow control and an LVDC output. A DC/AC converter makes it possible to obtain an LVAC output and also to control the voltage on the LV outputs.

In areas where there is strong development of renewable energy production, connected via converters and fast charging stations for electric vehicles, the creation of a DC hub, a device based on power electronics allowing the various DC components to be interfaced with other AC and DC systems, is an emerging solution for medium- and low-voltage setups. It would allow, among other things, for the isolation of the different components in the event of electrical faults, and to control them independently (U.S. DOE Office of Electricity 2020).

Partial operation of the system in direct current

The implementation of SOPs and smart transformers also allows for the operation of an entire part of the distribution system in direct current, which could make sense with the increase in resources connected to the distribution system via converters.

China is leading the development of these medium- and low-voltage DC systems with applications being considered around smart cities and industrial microgrids (Fan et al. 2021). These pilot projects of fully DC systems connected to the AC distribution system, such as the City Energy Internet (Qu et al. 2019) in Zhuhai, highlight the existing difficulties with respect to the sizing and operation of this type of system. Specifically, in a multi-terminal system, the definition of the system architecture and the design and coordination of protection devices and control of converters is complex (Monadi et al. 2017). Nevertheless, the reduction of line losses and the improvements in voltage quality and system flexibility motivate the use of this DC architecture (Qu et al. 2019).

Although a DC-only distribution system may seem far from reality today, the benefits of progressive hybridization for many applications, such as data centers and areas with high levels of distributed generation, could lead to a portion of the distribution system moving to a DC topology in the future.

7.2.3.1.2. AC/DC architectures

In the context of a high penetration of photovoltaic parks, fast charging stations for electric vehicles and storage systems expected in the coming years, planning methodologies need to be developed in terms of the AC/DC architecture of systems, that is, the choice of topologies and voltage levels. How do we partition a given system into AC and DC clusters of different voltages in order to minimize costs and avoid congestion? Some possible topologies, shown in Figure 7.3 and inspired by Syfullah Fuad et al. (2020), Siemens Energy (2021), De Doncker (2014) and Jia et al. (2015), could be:

– AC to DC without modification of the system topology: the radial AC distribution system would become a mixed AC/DC distribution system with DC parts in both MV and LV, taking advantage of the reduction in losses and the improvement in the voltage profile, but still keeping its radial topology;

– AC to DC with SOPs between previously unconnected feeders: in addition to the advantages mentioned above, the surplus of generation connected to the distribution systems is then no longer injected into the transmission systems but exchanged directly between the distribution systems, thus avoiding constraints within the transmission systems from power flows that are too high;

– mainly DC ring distribution system for dense areas: sources, charging stations and storage systems are connected to a DC ring via different types of converters (DC/DC or DC/AC converters);

– mainly DC star distribution systems for isolated areas, like the connection of offshore wind farms.

In order to validate such structures, it is necessary to define the desired functions of the converters, to size them, to evaluate their present and future costs (through a perspective of industrialization) and to estimate the TOTEX of the migration of existing architectures toward these AC/DC architectures.

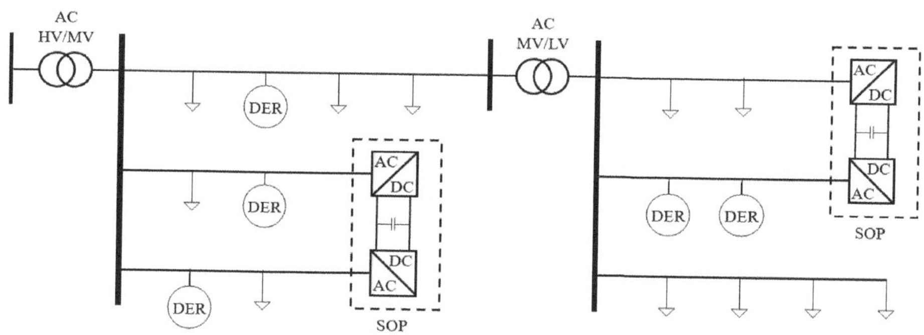

AC to DC with SOPs (Syfullah Fuad et al. 2020)

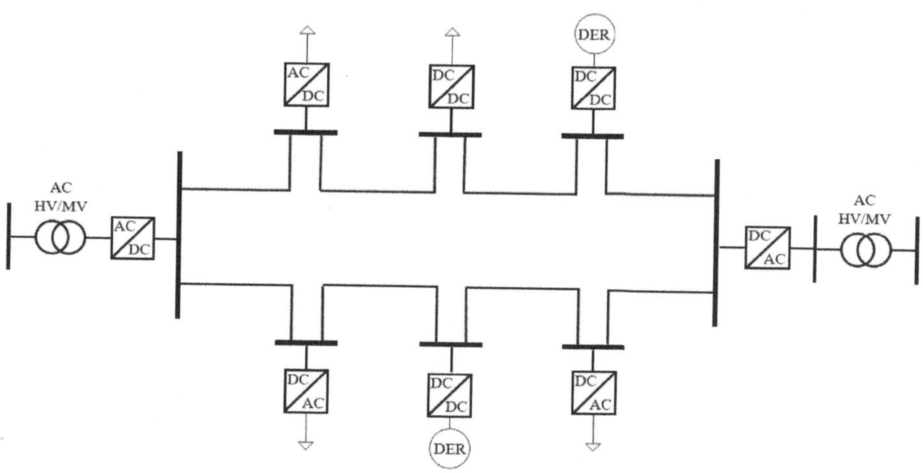

Mainly DC ring topology (Jia et al. 2015)

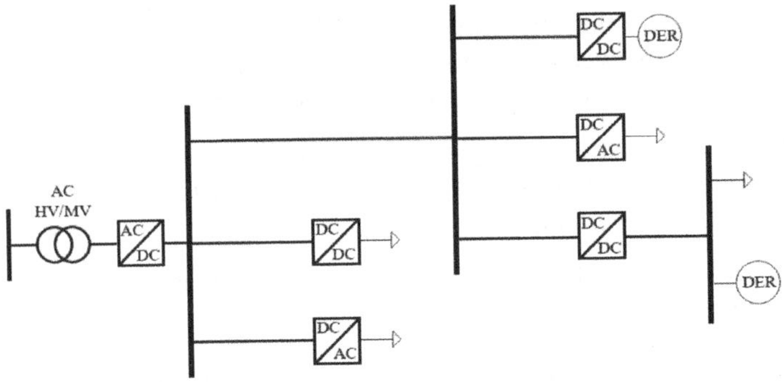

Mainly DC star topology (Jia et al. 2015)

Figure 7.3. *Possible AC/DC topologies*

7.2.3.1.3. Opportunities to evolve planning methods with DC

The various historical technical choices, such as the choice of voltage levels, architectures, operation and protection modes, for example, can be revisited since these were made on the basis of techno-economic calculations in the context of developing the AC system for the predominant supply of loads, such as lighting and rotating machines. Therefore, the benefit of DC in an existing AC system can also be:

– AC to DC retrofit: DC can be used to avoid an AC reinforcement that is needed to increase the transmitted power demand, as in the case of the Angle-DC project (SP Energy Networks 2021) mentioned above, whose objective is to reuse AC structures to create an MVDC line between the distribution system of the island of Anglesey and that of North Wales in the UK;

– AC to AC limited: smart transformers could make it possible to limit the sizing of structures, that is, to size at the theoretical nominal current of DC instead of sizing according to the short-circuit current. We could also consider a maintained fault operation but the economic significance in this kind of solution remains to be evaluated.

7.2.3.2. *Native DC for electrification*

Power systems such as those that currently exist in developed countries have been built gradually for more than a century, and it is not possible to replicate them in non-electrified and generally inaccessible areas, mainly due to prohibitive investment costs, political instability and low energy demand (Algorus Consulting, Nanoé, Tuck Foundation 2018). If we add to this the desire to use renewable resources as much as possible, then DC electrification appears to be a promising solution. Since 2010, it has already been deployed by some manufacturers via the concept of nanogrids, which are systems with a maximum power consumption of less than 5 kW (between five and 30 households) and whose main uses are lighting, mobile phone charging, television, radio and fans (HYSTRA 2017). The actors involved in their development, between 2010 and 2014, were mainly Indian (Mera Gao, Husk DC, Nature TechInfra, Mrida), Tanzanian (Devergy), Rwandan (MeshPower) and Kenyan (Steamaco) (HYSTRA 2017). After 2014, many projects are developing, including that of the French start-up Nanoé, with its concept of "scalable decentralized lateral" electrification. The principle is to deploy basic self-sustaining LV setups (nanogrids) that are then either interconnected to create larger grids (off-grid microgrids) or interconnected to a main MV system (grid-connected microgrids), if existing, with the objective of mutualizing resources to improve the reliability of the overall system, as shown in Figure 7.4. The technical characteristics of a nanogrid are detailed in Figure 7.5 and Table 7.2.

Feedback from the first installed nanogrids highlights the following benefits:

– 20–40% cost reduction compared to individual systems;

– rapid installation (between 1 and 2 days);

– scalability of nanogrids to meet uncertainties in demand;

– initial CAPEX less than €1,000 (i.e. between €167 and €250 per household) with an internal rate of return (IRR) of 3 years. HYSTRA (2017) presents similar orders of magnitude (between US$40 and US$200 per household) but without specifying the performance of nanogrids, which thus makes comparison impossible. They also indicate that the IRR can be increased (between 3 and 5 years) if the system is poorly sized or in the case of fraud.

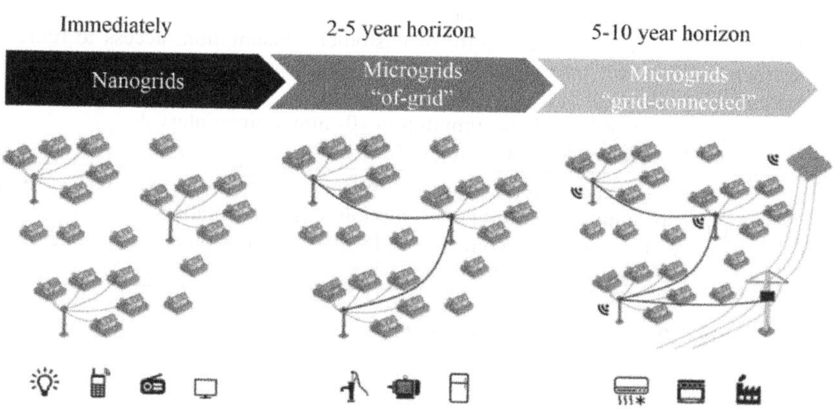

Figure 7.4. *Lateral electrification scheme (source: Nanoé (2021))*

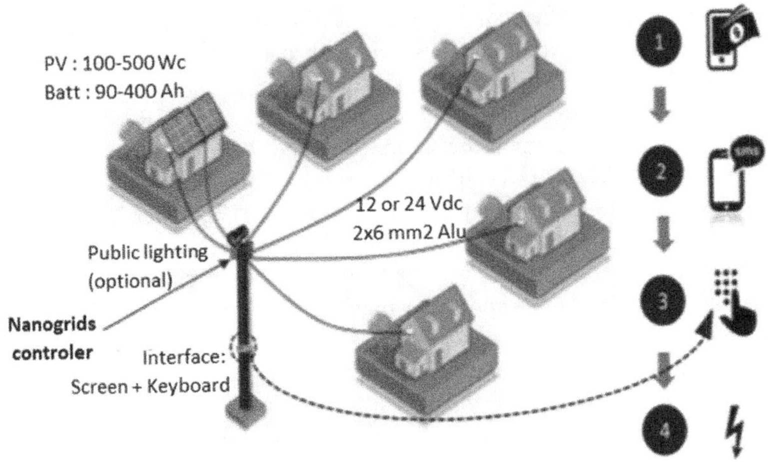

Figure 7.5. *Components of a nanogrid (source: Nanoé (2021))*

Functionality	Equipment
Consumption	– Four to six consumption areas (total peak consumption less than 1 kW) – Load type: LED, USB, radio, TV, rechargeable flashlight
Production	– 1 PV installation (between 100 and 300 W)
Storage	– One lead gel battery (between 90 and 260 Ah) – One PWM[1] regulator
Distribution	– One electrical pole – 6 mm² aluminum cables
Electricity management and metering	– One DC circuit breaker for short circuit protection of the nanogrid – One nanogrid controller – Metering and control of customer consumption: access to remaining balance, interruption in case of overflow or short circuit on customer side – Recording of consumption in 10-minute time intervals – One keyboard/screen interface
Business model	– Prepaid purchase by cell phone (choice of maximum power and consumption, several consumption options available) – Receipt of a code by SMS that the customer must enter in the keyboard/screen interface

Table 7.2. *Technical characteristics of a nanogrid (Nanoé 2021)*

1 Pulse Width Modulation (PWM): this type of regulator makes it possible to directly couple the solar panel to the battery and to control its charge and discharge. Although it is economical, the voltage is limited to 12 or 24 V (Solaris Store 2021).

7.2.3.3. *The need for systemic optimization tools*

Over the past few decades, the development of efficient power electronics technologies coupled with the massive increase in DC usage raises the question of the choice between AC, DC or mixed AC/DC distribution. The investment in converters and the protection of a DC system can therefore be offset by the long-term savings due to the reduction of line losses and the increase in system reliability in certain topologies. In addition, new performance indicators and requirements are emerging, such as improved availability and reduced environmental impact, which are challenging conventional planning choices. It is therefore necessary to develop systemic optimization tools that take into account the particularities of a mixed system, while keeping in mind a set of technical, economic and environmental objectives and constraints in order to determine, on a case-by-case basis, the optimal solution.

The presence of DC feeders significantly complicates planning optimization, increasing not only the number of variables, but also the nonlinearity and non-convexity of the problem (Liu et al. 2019). In addition, mixed-system planning must consider the possibility of active power flow management through converters, voltage regulation provided by some converters on the AC side and the increasing participation of consumers in the energy market through energy exchanges with microgrids, from local generation and storage management and the electric vehicle. These aspects make the planning of distribution systems more complex, but will lead to less costly solutions that are better suited to the modern power system paradigm.

7.2.4. *New multi-objective systemic approaches*

In July 2020, ENTSO-E (2020) published in its roadmap a methodology to evaluate individual projects in a multi-sector approach, that is, integrating interactions with other energy sectors. The main changes compared to the traditional single-sector methodology are as follows:

– the integration of the evolution of other energy sectors in the definition of the scenarios, such as the share of the different types of energy in the final consumption, the development of technologies, the capacities of transformation from one energy to another (PowerToX), the possibilities of interconnection between the energy systems;

– definition of planning strategies for each energy system independently;

– pre-analysis of planning strategies in terms of their dependencies/impacts on other energy sectors: this analysis makes it possible to define if some strategies are interdependent and need to be jointly optimized. This would reduce complexity and computation time.

However, this approach relies on an interconnection of tools optimizing the different energy subsystems rather than an integration of the different models into a single optimization tool. Indeed, the optimization of these energy subsystems independently will not lead to the optimization of the global energy system. It is therefore necessary to model the interactions and complementarities between these subsystems in order to adopt an integrated approach. One of the 14 recommendations of the Think Smartgrids recovery plan is the development of multi-system coupling (electricity, heating, cooling, hydrogen, e-fuels) (Think Smartgrids 2020), which highlights the benefits to the power system, in particular through the additional flexibility provided by Power ToX technologies to modulate the power injected and consumed in the systems because of thermal storage. In particular, three research actions are targeted:

– electrolytic hydrogen for mobility and to provide flexibility to the power system;

– thermal storage via heat pumps and thermo-fridge pumps;

– standardization of data (energy, social, economic, geographic, etc.) to facilitate the deployment of multi-energy strategies.

In order to enable an integrated multi-energy planning approach, it is necessary to define a set of performance indicators for each energy system, as well as environmental, social and economic indicators. As these indicators are not necessarily all quantitative, it is necessary to develop representation and decision support tools. Given the complexity of this level of modeling, which is partly linked to the scale of the problem, as well as the different temporal scales depending on the energy sectors, it is necessary to develop simple but representative models for each sector in order to encourage a holistic optimization.

7.3. Integrated planning tools

7.3.1. *Why integrate?*

As seen in Chapter 2, planning is a thorough techno-economic analysis split between different time horizons linked to uncertainties. The shorter the time horizon, the fewer uncertainties there are and the more data the planner has at their disposal. These data come from the different sensors installed on the system (Internet of Things (IoT)) but also from its modeling. The amount of data coming from the operation is constantly increasing, as shown in Figure 7.6, due to:

– the duration of the measurement and storage period: typically, one or more measurement points every 10 min for a communicating meter over a lifetime of 15–20 years;

– the increase in the measurement points on the sensors or communicating meters;

– the increase in the types of measurements recorded and communicated by the sensors. For example, a Linky 60A meter can deliver about 20 pieces of information at each measurement interval (Enedis 2021c).

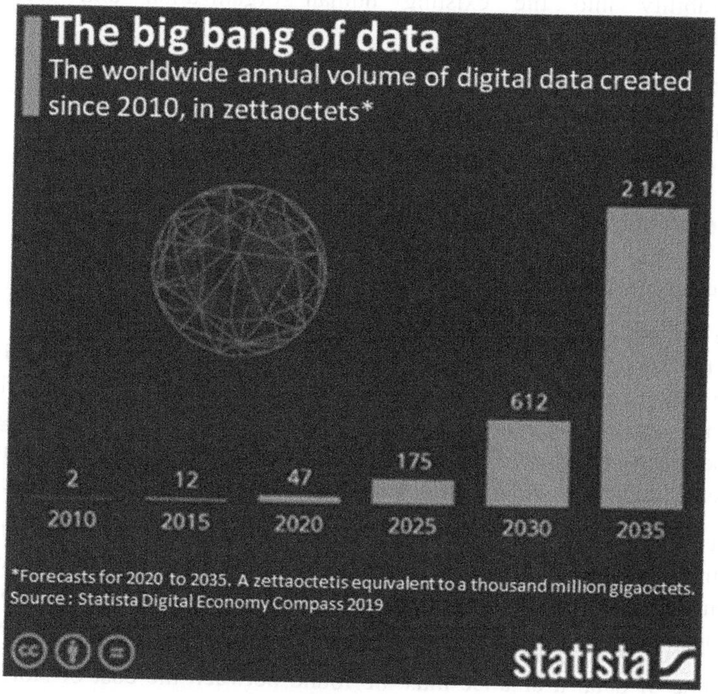

Figure 7.6. *Predicted increase in data (Statista 2021)*

A major challenge is to link system operations and condition data to planning models to make it useful data for managers so that they can make informed decisions and actions. Software solutions offer a data connection between the planning tool, the operation and maintenance tools (SCADA) and the geographic information system (GIS). As these solutions are provided by one or more software vendors, or developed by the system operator, a number of prerequisites are necessary to implement and benefit from data analytics tools. The Think Smartgrids association lists eight of these in De Buttet et al. (2019):

– accessibility of quality data through the implementation of an infrastructure;

– new professional skills such as data scientists and data architects (designers of new models based on data);

– transparency and understanding of algorithms;

– value creation across the entire infrastructure, from strategy to operations, and from the short to the long term;

– technological scalability;

– integrability into the existing human organization and technological infrastructure;

– security and cybersecurity;

– confidentiality and respect of sensitive data.

7.3.2. *The challenges of data*

The various data-related issues are as follows:

– Data collection: these have variable production frequencies, such as those linked to the 10-year planning model, to be compared with the data emitted by a voltage or power sensor every 10 min. They can have different structures, like those obtained from a communicating energy meter, or those of a medium voltage sensor, or even be unstructured, like the scanned plans of an MV substation that is more than 30 years old. The raw data collected is stored in areas called "data lakes".

– Data cleaning: the quality of data is often variable due to measurement errors or losses in transmission. They require pre-processing, called data cleansing, which can range from the simple removal of erroneous data, to the reconstruction of missing data. Without this pre-processing, the algorithms cannot deliver accurate results.

– Volume of data: a balance must be found between the cost of storing a very large amount of data and the benefits it brings. Processes are often applied to keep only the data considered to be of interest. These processes call upon data sciences, such as data analytics, or machine learning based on statistical mathematical methods. The data are then often stored in structured databases.

– Access to data: data are rarely free of rights and its use is often regulated. For example, the European Community (2016) rolled out a general regulation on the protection of personal data, known by the acronym GDPR (General Data Protection Regulation). It also launched a study in December 2018 on Big Data and digital platforms to facilitate data exchange between companies (business-to-business (B2B)) to accelerate data-driven economic growth (European Commission 2016) through two pilot projects in the healthcare and automotive sectors. The results of the study, published in March 2021 (Wavestone et al. 2021), propose an architecture for sharing health data from different sources and regions and a shared server for developing cooperative mobility. Among the barriers identified in this report, we identify in particular the issues facing interoperability between stakeholders. More and more countries or regions are regulating data, perceived to be a raw material source, such as the California Consumer Privacy Act 2018 (OAG 2021) and the Personal Information Security Specification (State Administration for Market Supervision of the People's Republic of China 2020) in China. DSOs dispose of consumer meter data only after user agreements are in place that comply with these laws, which also govern data access rights. For example, DSOs must generally provide users access to their own data without receiving any revenue, except infrastructure costs. They must seek prior consent from owners before using their data for purposes other than metering and billing, taking care to dissociate any personal data that cannot be broadcasted. In order to allow for the implementation of data analysis tools to enable the development of innovations, the professional association of EDSO (European Distribution System Operators) advocates for a regulation allowing for the following data to be made available as an open license, commonly called open source or open data (E.DSO 2018):

- meter data collected at customer sites;
- system data including information about the system, its configuration, measurements that can be taken in real time, scheduled or historical;
- market data.

We can already find data available as open license, as on the Enedis website (Enedis 2021b), EDP (2021), Fluvius (2021), ODRÉ (2021), or after the execution of an access contract and adequate infrastructure (Energinet 2021). The data are anonymized so as not to associate the name and location of the meter, pre-processed by theme (extraction, production, types of contracts, etc.), and is accessible for raw download in CSV format or via application programming interfaces (API) that can be integrated into software.

Cloud[2] solution providers offer environments and software shops that facilitate data cleansing, refinement and processing. The best known are IBM's Watson, Amazon SageMaker by AWS, Google's Compute Engine, and Microsoft's Azure. As electricity distribution becomes part of a country's critical infrastructure, additional access, security and physical storage rules are imposed to protect it from malicious acts (the concept of cybersecurity).

2 Online storage systems.

7.3.3. *Including control in planning models*

In recent years, a lot of investments have been made to increase the observability and control of transmission and distribution systems. Control systems have evolved to provide more and more functionalities based on technological platforms. Several trends have emerged, of which the following is a non-exhaustive list:

– The creation of platforms from generic tools of the IoT, operational technologies (OT) and Information Technologies (IT) from commercial or open-source solutions. This approach offers maximum flexibility to create a specific solution that meets the needs of the DSO and its technical specificities. On the other hand, this approach requires mastering all the technological aspects of infrastructure, development, operation and maintenance. It requires the implementation of intermediate layers that can be delivered through specialized technology integration providers or vendors (Wikipedia 2021):

- infrastructure as a service (IaaS): the provider hosts all of the infrastructure elements;

- platform as a service (PaaS): the provider hosts the infrastructure and the operating software for the platform;

- software as a service (SaaS): the provider hosts the infrastructure, the platform and the applications.

– The technology solutions implemented are usually complimentary products from the same vendors as the data analytics solutions, for example Microsoft's Azure and Azure IoT (Azure IoT 2021; Azure 2021) or Amazon's AWS IOT (AWS IoT 2021), or from vendors specializing in the industrial IoT such as OSISoft's PI system (OSISoft 2021).

– Fully integrated solutions developed by planning and control specialists such as DSOs. These allow for the implementation of proven solutions, respecting the standards and regulations of the energy codes, but offer less freedom in terms of the ability to hyper-customize. Development costs and the acquisition of new skills are often significant at the start of the transformation project. Typical solutions called advanced distribution management systems (ADMS) are, for example, Schneider EcoStruxure ADMS (Schneider Electric 2021a), Siemens Questa ADMS (Siemens 2021) or GE ADMS (GE Digital 2021).

– Hybrid solutions based on a mix of the two approaches require data exchange and strong interoperability. Data connections are made through connectors using standardized exchange protocols such as the Common Information Model (CIM) defined by the IEC 61970-301:2020 Standard (Statista 2021).

Because of data analytics and machine learning, these solutions and platforms offer increased functionality for system operations: detection and reduction of technical and non-technical losses, optimization of voltage plans and voltage stability, optimization of maintenance plans, parts inventory and manpower records and optimization of component lifetime.

The analysis functionalities are enriched with simulation zones (*sandbox*) to test operating scenarios using a system representation model that is consistent with the one used in operation. The connection with cost databases, the DSO or data brokers, makes it possible to perform a technical and economic analysis of maintenance and investment operations. Modeling and planning tools such as Etap (2021) are evolving to link digital models, commonly called digital twins, between planning, design and operations by interfacing with OT&IT and ADMS platforms.

Many innovations are to come because of the development of digital technology. Data analytics and machine learning approaches will offer more and more advanced analysis and prediction services. Many examples of demonstrations or implementations are given by the professional association (Utility Analytics Institute 2021a). The evolution of planning will follow the evolution of IT across all fields of data science: artificial intelligence including semantic approaches that make texts interpretable by machines, transformation of unstructured data into digitally exploitable data through graphical recognition systems, democratization of machine learning with approaches like Google Cloud AutoML (Google 2021a) and model engineering.

7.3.4. *The challenge of skills*

This evolution of planning functionalities is accompanied by the need for digital resources, which are in great demand in all industrial sectors: video games, social systems, the digitalization of customer and user experience across all industries. If the needs were historically focused on IoT and IT (Gartner 2021b) (sensors, operator mobility, infrastructure, cybersecurity, etc.), these are now becoming necessary throughout the entire DSO organization: UX (user experience) designers, data scientists, data analysts, data architects, web designers, community managers, etc.

The first challenges are to bring together experts from the power system and distribution business with data scientists in multidisciplinary teams to support cultural change centered around data and to attract digital talent (France Num 2021; Gartner 2021a). IT players are working on tools and methodologies to make certain data science technologies accessible to as many people as possible (Schatsky et al. 2021), such as Google's Cloud AutoML (Google 2021b).

New approaches based on open innovation are emerging to foster the emergence of new skills and to attract new talent, such as idea and algorithm competitions for start-ups, data scientists, students and anyone who wants to contribute. For example, the Agorize platform offers this type of challenge for all types of economic actors (Agorize 2021). The rewards for the finalists are very diverse, ranging from job interviews to remuneration or support in the development of their company.

7.4. New economic actors and new business models

7.4.1. *Diversity of actors*

Today, we find a wide variety of actors (as shown in Figure 7.7 for France), including consulting firms operating in various fields. The consulting firms Enea Consulting and E-CUBE, for example, assist various actors (manufacturers, energy companies, institutions, etc.) in developing energy transition strategies. ADEME has commissioned several publications from them, which have become important working documents for the entire electricity sector. These high-level publications, consisting of recommendations, feedback and systematic reviews, are complemented by those of the system operators Enedis and RTE, which publish their own assessments and reports on various themes and topics (RE, electric mobility, etc.).

The importance of *software tool developers* has increased in recent years. Since 2013, Enedis has replaced its internal software PRAO with the new tool ERABLE, which it developed on the basis of the Power Factory (PwF) software, with the addition of further functionalities. PwF is sold worldwide by the German company DIgSILENT GmbH. In France, other DSOs have also acquired PwF to remain consistent with the main DSO Enedis, such as GEREDIS for the Deux-Sèvres department or GreenAlp for Grenoble. There are other software solutions for system operators, such as NEPLAN, developed by the Swiss company NEPLAN AG, the software suite (transmission, distribution, production) developed by the Californian company ETAP, CYME solutions developed by EATON and completed by modules from the research institute EPRI, and FeederAll developed by ABB. Alternative solutions are also proposed by academic spin-offs, such as TOPPLAN in France, supported by the Linksium incubator. Open-source solutions also exist, such as MATPOWER, OpenDSS, GridCal or PandaPower, and which are mainly promoted by the scientific community.

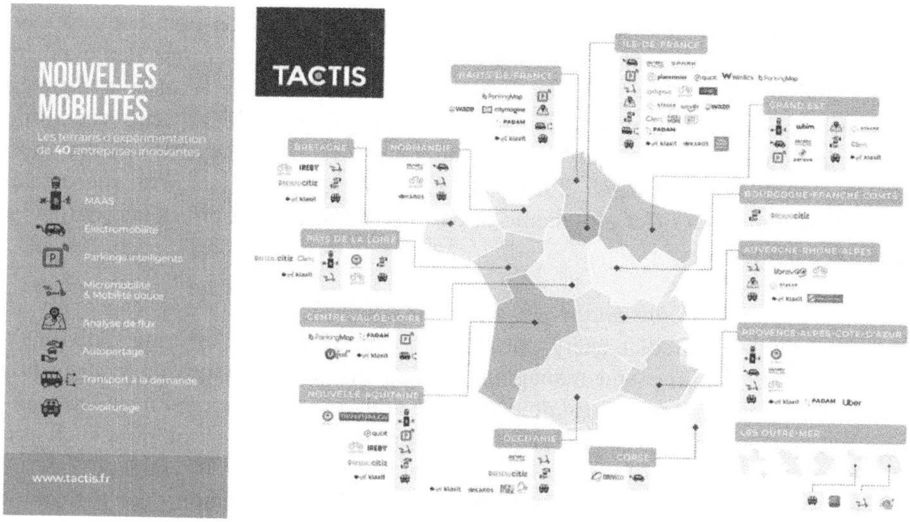

Figure 7.7. *Actors of the ecological transition in France, Tactis Innovation database (source: TACTIS (2021))*

7.4.2. Diversity of topics

Data, as presented in section 7.3.2, is a challenge at all levels for power system actors. The first step is data collection: Landis+Gyr enables automated asset management by linking them with automated digital sensors. Energisme provides a software platform in SaaS[3] mode to facilitate the acquisition of massive and heterogeneous data and its smart processing. OSISoft offers its customers solutions to collect (in real-time if necessary), clean and contextualize, share and visualize this data. Collecting and accumulating these data requires exploiting it: for example, C3.ai helps utilities to use relevant data with the help of machine learning models. In addition to data from system operator assets and smart metering, a lot of additional data are generated and added to the existing panel. For example, we can mention consumption or production forecasting data, which allows for a better integration of RE: forecasting models are proposed by the start-up meteo*swift or the spin-off EDF Store & Forecast.

The end users of the electrical system, who were once regarded as "passive loads" of the system, are now actors in the system: *prosumers*. There are many solutions to help customers better manage their consumption, ranging from the simple choice of electricity supplier to the installation of *energy boxes* to manage

3 *Software as a service*, that is, software installed on remote servers and not on the user's premises.

various equipment in the home, as well as better access to their own consumption data via the deployment of Linky smart meters and online platforms deployed by Enedis and various other suppliers (these data are therefore accessible through different channels). Dynamic pricing also makes it possible to send economic signals to consumers who can then decide how to act. In the CRE deliberation published in May 2021 (CRE 2021), the CRE obliges suppliers with more than 200,000 customers to offer a dynamic tariff indexed to market prices and defines a regulatory framework.

The topic of *self-consumption* is very important for consumers, especially in connection with local solar energy production. Companies are offering individuals solutions for controlling and managing existing electrical equipment (*energy box* coupled with sensors and control devices to be added to the electrical installation, as proposed by Comwatt for example) or the integration of new technologies, such as a "smart" radiator, which integrates a storage battery created by Lancey Energy Storage, a Grenoble-based start-up founded in 2016. By combining the themes of renewable energies, the digitization of electrical systems, the decentralization of production and the emergence of prosumers have led to the term: the *Internet of Energy* (IoE).

Flexibility is now a resource used by several manufacturers, for example, to contribute to the real-time balance of the system while receiving compensation for reducing their bills. In the residential sector, the Voltalis solution controls the radiators and electric hot water tanks of thousands of customers to contribute to this balance, without impacting the comfort of the occupants. In the industrial sector, Energy Pool analyzes the available flexibilities to optimize the consumption profile. The flexibility services provided to the power system also include compensation for the intermittency of renewable energies, alternatives to fossil fuels to guarantee the security of the system, management of congestion and incidents on the system.

7.4.3. *Diversity of business models*

One of the emerging business models is a model mixing services and software creation, often in SaaS mode. Companies such as Roseau Technologies, Odit-e and Blacklight Analytics combine electrical engineering, computer science and applied mathematics skills to create custom solutions for customers, either as a service using in-house tools or via the delivery of software that can be easily integrated at the customer's site or deployed on a server.

New services, sometimes free of charge, are also emerging from utilities, such as the Ter@ platform from Enedis, which allows users of the electricity system to easily simulate the connection of a consumption or production facility free of charge. Public actors (notably energy unions and cities) and associations are also increasingly taking up the issue of energy transition and offering online platforms for local authorities and individuals. For example, the TerriSTORY consortium provides tools for monitoring and sharing indicators useful for the development of territories.

7.5. Conclusion

This chapter has presented some new trends that could become mainstream in the planning of electrical distribution systems (definition of objectives and methods used), in the context of ever-increasing electronics-based components in electrical power systems, and the growth of heterogeneous data and associated processing and analysis tools (data analytics). First, the traditional system objectives of cost and reliability are evolving to include additional indicators related to recyclability and risk management. Similarly, the historical challenges of system codes and system resiliency may become more flexible and decentralized. The renewed interest in direct current, whether for the creation of new systems in a logic of electrification, or for the optimization of existing systems (AC/DC hybridization), could make sense because of the development of high-performance power electronics structures. Technical and economical validations are still needed, even if some industrial utilities are already implementing such solutions. Second, planning could evolve toward a more integrated model allowing for optimization at different spatial (transmission, distribution) and temporal horizons (medium- and long-term control and planning), in particular, because of the models built from the data available from communication meters and other sensors. Moreover, the development of a more systemic tool integrating other sectors (energy but also socioeconomic) would allow for the optimization of a global system. However, many challenges remain: data processing, storage, accessibility and the interoperability of the different tools.

7.6. References

ADEME (2019). Équipements électriques et électroniques. Report.

Agorize (2021). Rejoignez la plus grande communauté d'innovateurs ! [Online]. Available at: https://www.agorize.com/fr [Accessed 2021].

Algorus Consulting, Nanoé, Fondation Tuck (2018). Électrification latérale : vers un nouveau modèle d'électrification pour l'Afrique [Online]. Available at: https://www.academia.edu/36892599/Electrification_lat%C3%A9rale_VERS_UN_NOUVEAU_MODELE_DELECTRIFICATION_POUR_LAFRIQUE.

AWS IoT (2021). AWS IoT, Libérez vos données IoT et accélérez la croissance de votre entreprise. [Online]. Available at: https://aws.amazon.com/fr/iot/ [Accessed 2021].

Azure (2021). [Online]. Available at: https://azure.microsoft.com/en-us/ [Accessed 2021].

Azure IoT (2021). [Online]. Available at: https://azure.microsoft.com/en-us/overview/iot/ [Accessed 2021].

Banjar, K. (2019). Micro-réseau résilient à haute stabilité dynamique en présence d'une intégration massive des énergies renouvelables variables. PhD Thesis, Université Grenoble Alpes, Grenoble.

Berry, J., Mavrocostanti, Y., Murdoch, N., Hardman, D. (2019). System equilibrium – Closed down report [Online]. Available at: https://www.google.com/url?sa=t&rct=j&q=&esrc=s&source=web&cd=&ved=2ahUKEwiDnaeV9OPtAhWjA2MBHYu6AtoQFjAAegQIAxAC&url=https%3A%2F%2Fwww.westernpower.co.uk%2Fdownloads-view-reciteme%2F51982&usg=AOvVaw0AR7kYZJt8XJQJVlrfV_8V [Accessed 2021].

Cao, W., Wua, J., Jenkins, N., Wang, C., Green, T. (2016). Benefits analysis of Soft Open Points for electrical distribution system operation. *Appl. Energy*, 165, 36–47.

CIRED (2021). Lifetime extension – WG 2021-1 [Online]. Available at: http://www.cired.net/cired-working-groups/lifetime-extension-wg-2020-1 [Accessed 2021].

CRE (2021). Délibération 2021-135 [Online]. Available at: https://www.cre.fr/Documents/Deliberations/Decision/modalites-selon-lesquelles-l-offre-a-tarification-dynamique-prevue-au-ii-de-l-article-l332-7-du-code-de-l-energie-prend-en-compte-les-variations-de [Accessed 2021].

De Buttet, B., Lacroix, B., Strang, K., Szniten, D., Penserini, A. (2019). DATA analytics in utilities. The rise of new technologies [Online]. Available at: https://www.thinksmartgrids.fr/wp-content/uploads/2019/11/ThinkSmartgrids_livret-data_11.05-vdef-web.pdf.

De Doncker, R. (2014). Power electronic technologies for flexible DC distribution grids. In *The 2014 International Power Electronics Conference,* Hiroshima.

E.DSO (2018). E.DSO policy brief on open data [Online]. Available at: https://www.edsoforsmartgrids.eu/wp-content/uploads/EDSO-Open-Data-Policy-Brief_1812_final-1.pdf [Accessed 2021].

EDP (2021). [Online]. Available at: https://opendata.edp.com/explore/?refine.keyword=visible&sort=modified [Accessed 2021].

Enedis (2021a). [Online]. Available at: https://www.enedis.fr/greenlys [Accessed 2021].

Enedis (2021b). [Online]. Available at: https://data.enedis.fr/explore/?sort=modified [Accessed 2021].

Enedis (2021c). Sorties de télé-information client des appareils de comptage Linky utilisés en généralisation par Enedis [Online]. Available at: https://www.enedis.fr/sites/default/files/Enedis-NOI-CPT_54E.pdf [Accessed 2021].

Energinet (2021). What is the purpose of Datahub? [Online]. Available at: https://en.energinet.dk/Electricity/DataHub#Documents [Accessed 2021].

ENTSO-E. (2020). ENTSO-E Roadmap for a multi-sectorial planning support [Online]. Available at: https://www.entsoe.eu/2020/07/16/entso-e-roadmap-for-a-multi-sectorial-planning-support/.

etap (2021). Digital twin journey [Online]. Available at: https://etap.com/products/etap-digital-twin [Accessed 2021].

EUR-Lex (2021). Directive 2009/125/EC of the European Parliament and of the Council of 21 October 2009 establishing a framework for the setting of ecodesign requirements for energy-related products [Online]. Available at: https://eur-lex.europa.eu/legal-content/EN/TXT/?uri=CELEX:32009L0125 [Accessed 2021].

European Commission (2016). Règlement (UE) 2016/679 du parlement européen et du conseil du 27 avril 2016 relatif à la protection des personnes physiques à l'égard du traitement des données à caractère personnel et à la libre circulation de ces données [Online]. Available at: https://eur-lex.europa.eu/legal-content/FR/TXT/?uri=CELEX%3A32016R0679.

European Commission (2021). Big data and B2B digital platforms [Online]. Available at: https://ec.europa.eu/growth/industry/policy/advanced-technologies/industrial-applications-artificial-intelligence-and-big-data/big-data-b2b-digital-platforms_en [Accessed 2021].

Fan, Y., Chi, Y., Li, Y., Wang, Z., Liu, H., Liu, W., Li, X. (2021). Key technologies for medium and low voltage DC distribution system. *Global Energy Interconnection*, 4(1), 91–103.

Fluvius (2021). Open data [Online]. Available at: https://lokaal-bestuur.fluvius.be/nl/thema/nutsvoorzieningen/open-data#open-data-per-type [Accessed 2021].

France Num (2021). Transformation digitale et transformation des compétences : bonnes pratiques et statistiques [Online]. Available at: https://www.francenum.gouv.fr/comprendre-le-numerique/transformation-digitale-et-transformation-des-competences-bonnes-pratiques [Accessed 2021].

Fuad, K.S., Hafezi, H., Kauhaniemi, K., Laaksonen, H. (2020). Soft open point in distribution networks. *IEEE Access*, 8, 210550–210565.

Gartner (2021a). Avoid 5 pitfalls when building data and analytics teams [Online]. Available at: https://www.gartner.com/smarterwithgartner/avoid-5-pitfalls-when-building-data-and-analytics-teams/ [Accessed 2021].

Gartner (2021b). Lack of skills threatens digital transformation [Online]. Available at: https://www.gartner.com/smarterwithgartner/lack-of-skills-threatens-digital-transformation/ [Accessed 2021].

GE Digital (2021). [Online]. Available at: https://www.ge.com/digital/applications/advanced-distribution-management-solutions-adms?utm_medium=Paid-Search&utm_source=Google&utm_campaign=Grid-ADMS-ALL-MOF-GLOB-Search&utm_content=%2Badms%20%2Bsystem&gclid=EAIaIQobChMI-ZOHpaTp8AIVWunmCh1KwgOpEAAYASAAE [Accessed 2021].

Google (2021a). [Online]. Available at: https://cloud.google.com/automl [Accessed 2021].

Google (2021b). Cloud AutoML [Online]. Available at: https://cloud.google.com/automl?hl=fr [Accessed 2021].

HYSTRA (2017). Reaching scale in access to energy. Report.

IEC (n.d.). Guide 109 – aspects liés à l'environnement – prise en compte dans les normes. Report, IEC.

Ji, H., Wang, C., Li, S., Song, G., Yu, H. (2019). Operation and optimization technologies of active distribution system with multi-terminal soft open points. In *Proc. 25th Int. Conf. Electr. Distrib.* CIRED, Madrid.

Jia, L., Zhu, Y., Wang, Y. (2015). Architecture design for new AC-DC hybrid micro-grid. *Proceedings of IEEE First International Conference on DC Microgrids (ICDCM)*, Atlanta.

Klingel, J. (2020). Resiliency: An earning opportunity for the greater good of customers and utilities [Online]. Available at: https://www.utilitydive.com/news/resiliency-an-earning-opportunity-for-the-greater-good-of-customers-and-ut/576127/ [Accessed 2020].

LCIE (2021). EIME – LCA and eco-design software [Online]. Available at: https://codde.fr/en/our-software/eime-presentation [Accessed 2021].

Li, P., Ji, H., Wang, C., Zhao, J., Song, G., Ding, F., Wu, J. (2017). Coordinated control method of voltage and reactive power for active distribution networks based on soft open point. *IEEE Transactions on Sustainable Energy*, 8(4), 1430–1442.

Liu, X., Liu, Y., Liu, J., Xiang, Y., Yuan, X. (2019). Optimal planning of AC-DC hybrid transmission and distributed energy resource system: Review and prospects. *CSEE Journal of Power and Energy Systems*, 3(5), 409–422.

Long, C., Wu, J., Thomas, L. (2016). Optimal operation of soft open points in medium voltage electrical distribution systems with distributed generation. *Appl. Energy*, 184, 427–437.

MIGRATE (2020). AN EU funded project under the framework of European union's horizon 2020 [Online]. Available at: www.h2020-migrate.eu/.

Monadi, M., Rouzbehi, K., Candela, J., Rodriguez, P. (2017). DC distribution systems : A solution for integration of distributed generation systems. In *Distributed Generation Systems – Design, Opertion and Grid Integration*, Ghaarehpetien, G., Mousavi, M. (eds). Butterworth-Heineman, Oxford.

Myllymäki, J. (2020). CEER webinar on DSO development plans and system planning – Current practice of D NDPs Elenia Case Study. Report, CEER.

Nanoé (2021). [Online]. Available at: https://www.nanoe.net/ [Accessed 2021].

OAG (2021). California Consumer Privacy Act (CCPA) [Online]. Available at: https://oag.ca.gov/privacy/ccpa [Accessed 2021].

ODRE (2021). Explorer les données Open Data Réseaux Énergies (ODRÉ) [Online]. Available at: https://opendata.reseaux-energies.fr/ [Consulté en 2021].

OSISoft (2021). [Online]. Available at: https://www.osisoft.com/pi-system [Accessed 2021].

PEP Association (2021). PEP ecopassport® program [Online]. Available at: http://www.pep-ecopassport.org/ [Accessed 2021].

Qu, L., Yu, Z., Song, Q., Yuan, Z., Zhao, B., Yao, D., Chen, J., Liu, Y., Zeng, R. (2019). Planning and analysis of the demonstration project of the MVDC distribution network in Zhuhai. *Frontiers in Energy*, 13(1), 120–130.

RTE (2020a). Le projet Ringo de RTE accueille ses premiers équipements en Côte-d'Or [Online]. Available at: https://www.rte-france.com/actualites/le-projet-ringo-de-rte-accueille-ses-premiers-equipements-en-cote-dor.

RTE (2020b). *Intégration des ENR : RTE facilitateur de la transition énergétique*. Film.

Schatsky, D., Muraskin, C., Chauhan, R. (2021). Democratizing data science to bridge the talent gap (Deloitte Insights) [Online]. Available at: https://www2.deloitte.com/us/en/insights/focus/signals-for-strategists/democratization-of-data-science-talent-gap.html [Accessed 2021].

Schneider Electric (2021a). EcoStruxure™ ADMS [Online]. Available at: https://www.se.com/fr/fr/product-range-presentation/61751-ecostruxure%E2%84%A2-adms/ [Accessed 2021].

Schneider Electric (2021b). EcoStruxure Asset Advisor from data to actionable insights [Online]. Available at: https://www.se.com/us/en/work/services/field-services/ecostruxure-asset-advisor.jsp [Accessed 2021].

Siemens (2021). Fact Sheet, Questa ADMS [Online]. Available at: https://resources.sw.siemens.com/en-US/fact-sheet-questa-adms [Accessed 2021].

Siemens Energy (2021). Managing the future grid: How to improve power transfer in the future grid?. Report, Siemens Energy.

Solaris Store (2021). Principe de fonctionnement d'un régulateur solaire [Online]. Available at: https://www.solaris-store.com/content/44-principe-de-fonctionnement-d-un-regulateur-solaire [Accessed 2021].

SP Energy Networks (2021). Angle-DC [Online]. Available at: https://www.spenergynetworks.co.uk/pages/angle_dc.aspx [Accessed 2021].

Spaes, J. (2020). Projet Ringo : La première batterie du plus gros stockage d'électricité en faveur du réseau est en place [Online]. Available at: https://www.pv-magazine.fr/2020/10/06/projet-ringo-la-premiere-batterie-du-plus-gros-stockage-delectricite-en-faveur-du-reseau-est-en-place/.

State Administration for Market Supervision of the people's Republic of China (2020). Information security technology – Personal information (PI) security specification, GB/T 35273-2020 [Online]. Available at: https://www.dataguidance.com/legal-research/standard-gbt-35273-2020-information-security.

Statista (2021). Le Big Bang du Big Data; Author: Tristan Gaudiaut, 2021 [Online]. Available at: https://fr.statista.com/infographie/17800/big-data-evolution-quantite-donnees-numeriques-creees-dans-le-monde/ [Accessed 2021].

TACTIS (2021). La base de données de référence de l'innovation territoriale en France [Online]. Available at: https://www.tactis.fr/base-innovation-territoriale/ [Accessed 2021].

TenneT TSO GmbH (2021). The massive integration of power electronic devices – Enabling the energy transition by providing solutions for the technological challenges [Online]. Available at: https://www.h2020-migrate.eu/_Resources/Persistent/b955edde3162c8c5bf66 96a9a936ad06e3b485db/19109_MIGRATE-Broschuere_DIN-A4_Doppelseiten_V8_online. pdf [Accessed 2021].

Think Smartgrids (2020). Les smart grids, colonne vertébrale de la relance durable [Online]. Available at: https://www.thinksmartgrids.fr/publications.

U.S. DOE Office of Electricity (2020). Transformer resilience and advanced components (TRAC). Solid State Power Substation Technology Roadmap.

Utility Analytics Institute (2021a). [Online]. Available at: https://utilityanalytics.com/ [Accessed 2021].

Utility Analytics Institute (2021b). Building the foundation for utility asset analytics [Online]. Available at: https://utilityanalytics.com/2021/01/building-the-foundation-for-utility-asset-analytics-part-2/ [Accessed 2021].

Wavestone, ICCS, TNO, CEPS, Grimaldi Studio Legale (2021). Big data and B2B platforms: The next big opportunity for Europe [Online]. Available at: https://www.ceps.eu/ceps-publications/big-data-and-b2b-platforms-the-next-big-opportunity-for-europe/.

Western Power Distribution (2014). System equilibrium [Online]. Available at: https://www.ofgem.gov.uk/publications-and-updates/low-carbon-systems-fund-submission-western-power-distribution-%E2%80%93-system-equilibrium.

Wikipedia (2021). Cloud computing [Online]. Available at: https://en.wikipedia.org/wiki/Cloud_computing.

Conclusion

In this book, we have defined distribution systems in detail: their structure, operating modes and planning methods, taking into account their history and regulatory frameworks (Chapters 1 and 2). We have described the current and future changes impacting these systems and their development going forward through concrete case studies (Chapters 3–7). The main recommendations for the evolution of planning methods can be summarized into six categories as outlined below:

– Update the input assumptions for the modeling of the planning problem: one of the main assumptions made in planning studies is the load growth rate. Not only does this rate only concern conventional loads (electric heating, electric water heating, lighting, computer equipment, etc.), but its evolution is often assumed to be constant each year throughout the study period. It is necessary to integrate *assumptions for DER growth* (electric vehicles, production, storage, etc.) as part of the input for the planning problem. This deterministic approach does not take into account the *uncertainties* of the data and the models, which increase as the years go by. It is necessary to use *advanced stochastic optimization methods*, as detailed in Chapter 5, which require the definition of future evolution scenarios.

– Integrate DERs into planning objectives: planning methods must develop solutions that *maximize the integration of DERs at low cost* while ensuring good quality indicators (SAIDI, SAIDI and ENS).

– Diversify the planner's options and calculation methods: typical planning issues are system extension, reinforcement to overcome the manifestation of constraints or non-compliance with a technical criterion, a request for connection from a producer or a consumer and finally the establishment of a master plan. All of these solutions are based on investments whose characteristics are determined by studying the "worst-case scenario", also known as fit and forget, which consists of studying the most unfavorable case for the system over the study period, in terms of current and/or voltage constraints (historically for consumption, but more recently for generation). While this has the advantage of minimizing the risk of constraints appearing in the future close to zero, it does lead to significant investments that can be challenged by other alternatives such as *flexibility schemes*, also known as *non-wire alternatives*. These alternative solutions require the *use of time series power curves* in order to evaluate the performance of these solutions, but also to *quantify the risk* of constraints appearing when compared to a classic investment solution. Planning must therefore integrate these new options, but also allow for *breakthrough solutions* across long-term horizons, as well as a load-curve-based approach rather than reducing the number of sizing cases.

– Update master plans more frequently given the rapid dynamics of the appearance of technologies and the variability of the evolution scenarios of the system actors.

– Develop public mapping in order to freely monitor the system status: although a lot of *open data* is available from certain sources, there is a lack of visibility on the flexibility needs of the system. *Mapping* systems down to the LV feeder level would allow for different service providers and local energy communities to know the needs of the system in order to better *propose adapted flexibility schemes* that would lead to a more global optimum for the community. In this context, it is essential to *provide a regulatory framework* to define the rules of operation. This must be done through *exemptions* in order to validate the relevance of these operations at scale 1.

– Develop multi-scale and multi-temporal planning: the various system operators, whether they belong to the same energy sector (DSO and TSO) or not, work in a rather compartmentalized manner with little exchange of information. Planning must evolve toward a *more coordinated approach* through data exchanges, or even co-modulations. Moreover, at the low voltage level, planning methods must be developed since most DERs will appear at this level and these systems will therefore have an important role to play throughout the power system (especially if we consider the deployment of microgrids not connected to the main grid, as is the case in Australia). Similarly, operational and long-term planning teams must work together to *integrate short time scales* into their *long-term modeling*, especially in the case of DER flexibility integration. Frequent back and forth between these teams is necessary.

– Integrate multidisciplinary approaches: current planning is mainly based on electrotechnical rules with classical economic approaches of cost–benefit analysis and net present value calculations. However, *economic models* that include *uncertainty* and *risk* must be integrated. Moreover, as soon as the actions of end-users are considered, it becomes necessary to *integrate* their *behavioral modeling* into the planning problem. Finally, access to data, with particular thanks to the advent of smart meters, makes it possible to refine planning hypotheses (e.g. modeling of load curves), the use of flexibilities, whether in real time or for a posteriori analysis (e.g. estimation of curtailed energy) or the reconstruction of missing information (e.g. system topology). To this end, methods from *data analytics* and *artificial intelligence* should also be integrated into planning studies.

Power systems and planning are often perceived as old-fashioned with fixed technological choices, however, being at the heart of every energy, technological and societal transition, their nature and evolutionary trajectory are multiple and specific to each country, leaving room for the development of numerous innovations.

Index

A, C

D

E, F

G, H, I

L, M, N

O, P

R, S

T, U

Other titles from

iSTE

in

Energy

2022

ABOUELATTA Mohamed, SHAKER Ahmed, GONTRAND Christian
Smart Power Integration

BOUTARD Benoit
Energy Autonomy: From the Notion to the Concepts

2021

ROBYNS Benoît, DAVIGNY Arnaud, FRANÇOIS Bruno, HENNETON Antoine, SPROOTEN Jonathan
Electricity Production from Renewable Energies (2nd edition)

2020

BOISGIBAULT Louis, AL KABBANI Fahad
Energy Transition in Metropolises, Rural Areas and Deserts

SOUALHI Abdenour, RAZIK Hubert
Electrical Systems 1: From Diagnosis to Prognosis
Electrical Systems 2: From Diagnosis to Prognosis

2019

BENALLOU Abdelhanine
Energy Transfers by Convection
(Energy Engineering Set – Volume 3)
Energy Transfers by Radiation
(Energy Engineering Set – Volume 4)
Mass Transfers and Physical Data Estimation
(Energy Engineering Set – Volume 5)

LACHAL Bernard
Energy Transition

ROBYNS Benoît, DAVIGNY Arnaud, BARRY Hervé, KAZMIERCZAK Sabine, SAUDEMONT Christophe, ABBES Dhaker, FRANÇOIS Bruno
Electrical Energy Storage for Buildings in Smart Grids

2018

BENALLOU Abdelhanine
Energy and Mass Transfers: Balance Sheet Approach and Basic Concepts
(Energy Engineering Set – Volume 1)
Energy Transfers by Conduction
(Energy Engineering Set – Volume 2)

JEMEÏ Samir
Hybridization, Diagnostic and Prognostic of Proton Exchange Membrane Fuel Cells: Durability and Reliability

RUFFINE Livio, BROSETA Daniel, DESMEDT Arnaud
Gas Hydrates 2: Geoscience Issues and Potential Industrial Applications

VIALLET Virginie, FLEUTOT Benoit
Inorganic Massive Batteries
(Energy Storage – Batteries, Supercapacitors Set – Volume 4)

2017

BROSETA Daniel, RUFFINE Livio, DESMEDT Arnaud
Gas Hydrates 1: Fundamentals, Characterization and Modeling

LA SCALA Massimo
From Smart Grids to Smart Cities: New Challenges in Optimizing Energy Grids

MOLINA Géraldine, MUSY Marjorie, LEFRANC Margot
Building Professionals Facing the Energy Efficiency Challenge

SIMON Patrice, BROUSSE Thierry, FAVIER Frédéric
Supercapacitors Based on Carbon or Pseudocapacitive Materials
(Energy Storage – Batteries, Supercapacitors Set – Volume 3)

2016

ALLARD Bruno
Power Systems-on-Chip: Practical Aspects of Design

ANDRE Michel, SAMARAS Zissis
Energy and Environment

DUFOUR Anthony
Thermochemical Conversion of Biomass for the Production of Energy and Chemicals

2015

CROGUENNEC Laurence, MONCONDUIT Laure, DEDRYVÈRE Rémi
Electrodes for Li-ion Batteries
(Energy Storage – Batteries, Supercapacitors Set – Volume 2)

LEPRINCE-WANG Yamin
Piezoelectric ZnO Nanostructure for Energy Harvesting
(Nanotechnologies for Energy Recovery Set – Volume 1)

ROBYNS Benoît, FRANÇOIS Bruno, DELILLE Gauthier, SAUDEMONT Christophe
Energy Storage in Electric Power Grids

ROSSI Carole
Al-based Energetic Nanomaterials
(Nanotechnologies for Energy Recovery Set – Volume 2)

TARASCON Jean-Marie, SIMON Patrice
Electrochemical Energy Storage
(Energy Storage – Batteries, Supercapacitors Set – Volume 1)

2013

LALOUI Lyesse, DI DONNA Alice
Energy Geostructures: Innovation in Underground Engineering

2012

BECKERS Benoit
Solar Energy at Urban Scale

ROBYNS Benoît, DAVIGNY Arnaud, FRANÇOIS Bruno, HENNETON Antoine, SPROOTEN Jonathan
Electricity Production from Renewable Energies

2011

GAO Fei, BLUNIER Benjamin, MIRAOUI Abdellatif
Proton Exchange Membrane Fuel Cell Modeling

MULTON Bernard
Marine Renewable Energy Handbook

2010

BRUNET Yves
Energy Storage

2009

SABONNADIERE Jean-Claude
Low Emission Power Generation Technologies and Energy Management

SABONNADIÈRE Jean-Claude
Renewable Energy Technologies

Printed and bound by CPI Group (UK) Ltd, Croydon, CR0 4YY
16/09/2026
14963183-0001